Radiochromic Film

Role and Applications in Radiation Dosimetry

IMAGING IN MEDICAL DIAGNOSIS AND THERAPY

Series Editors: Andrew Karellas and Bruce R. Thomadsen

Published titles

Quality and Safety in Radiotherapy
Todd Pawlicki, Peter B. Dunscombe,
Arno J. Mundt, and Pierre Scalliet, Editors
ISBN: 978-1-4398-0436-0

Adaptive Radiation Therapy
X. Allen Li, Editor
ISBN: 978-1-4398-1634-9

Quantitative MRI in Cancer
Thomas E. Yankeelov, David R. Pickens,
and Ronald R. Price, Editors
ISBN: 978-1-4398-2057-5

Informatics in Medical Imaging
George C. Kagadis and Steve G. Langer, Editors
ISBN: 978-1-4398-3124-3

Adaptive Motion Compensation in
Radiotherapy
Martin J. Murphy, Editor
ISBN: 978-1-4398-2193-0

Image-Guided Radiation Therapy
Daniel J. Bourland, Editor
ISBN: 978-1-4398-0273-1

Targeted Molecular Imaging
Michael J. Welch and William C. Eckelman,
Editors
ISBN: 978-1-4398-4195-0

Proton and Carbon Ion Therapy
C.-M. Charlie Ma and Tony Lomax, Editors
ISBN: 978-1-4398-1607-3

Physics of Mammographic Imaging
Mia K. Markey, Editor
ISBN: 978-1-4398-7544-5

Physics of Thermal Therapy:
Fundamentals and Clinical Applications
Eduardo Moros, Editor
ISBN: 978-1-4398-4890-6

Emerging Imaging Technologies in
Medicine
Mark A. Anastasio and Patrick La Riviere, Editors
ISBN: 978-1-4398-8041-8

Cancer Nanotechnology: Principles and
Applications in Radiation Oncology
Sang Hyun Cho and Sunil Krishnan, Editors
ISBN: 978-1-4398-7875-0

Image Processing in Radiation Therapy
Kristy Kay Brock, Editor
ISBN: 978-1-4398-3017-8

Informatics in Radiation Oncology
George Starkschall and R. Alfredo C. Siochi,
Editors
ISBN: 978-1-4398-2582-2

Cone Beam Computed Tomography
Chris C. Shaw, Editor
ISBN: 978-1-4398-4626-1

Computer-Aided Detection and
Diagnosis in Medical Imaging
Qiang Li and Robert M. Nishikawa, Editors
ISBN: 978-1-4398-7176-8

Cardiovascular and Neurovascular
Imaging: Physics and Technology
Carlo Cavedon and Stephen Rudin, Editors
ISBN: 978-1-4398-9056-1

Scintillation Dosimetry
Sam Beddar and Luc Beaulieu, Editors
ISBN: 978-1-4822-0899-3

Handbook of Small Animal Imaging:
Preclinical Imaging, Therapy,
and Applications
George Kagadis, Nancy L. Ford, Dimitrios N.
Karnabatidis, and George K. Loudos Editors
ISBN: 978-1-4665-5568-6

IMAGING IN MEDICAL DIAGNOSIS AND THERAPY

Series Editors: Andrew Karellas and Bruce R. Thomadsen

Published titles

**Comprehensive Brachytherapy:
Physical and Clinical Aspects**
*Jack Venselaar, Dimos Baltas, Peter J. Hoskin,
and Ali Soleimani-Meigooni, Editors*
ISBN: 978-1-4398-4498-4

**Handbook of Radioembolization:
Physics, Biology, Nuclear Medicine,
and Imaging**
*Alexander S. Pasciak, PhD., Yong Bradley, MD.,
J. Mark McKinney, MD., Editors*
ISBN: 978-1-4987-4201-6

**Monte Carlo Techniques in Radiation
Therapy**
Joao Seco and Frank Verhaegen, Editors
ISBN: 978-1-4665-0792-0

**Stereotactic Radiosurgery and
Stereotactic Body Radiation Therapy**
*Stanley H. Benedict, David J. Schlesinger,
Steven J. Goetsch, and Brian D. Kavanagh,
Editors*
ISBN: 978-1-4398-4197-6

Physics of PET and SPECT Imaging
Magnus Dahlbom, Editor
ISBN: 978-1-4665-6013-0

Tomosynthesis Imaging
Ingrid Reiser and Stephen Glick, Editors
ISBN: 978-1-138-19965-1

Ultrasound Imaging and Therapy
Aaron Fenster and James C. Lacefield, Editors
ISBN: 978-1-4398-6628-3

**Beam's Eye View Imaging in
Radiation Oncology**
Ross I. Berbeco, Ph.D., Editor
ISBN: 978-1-4987-3634-3

**Principles and Practice of
Image-Guided Radiation Therapy
of Lung Cancer**
*Jing Cai, Joe Y. Chang, and Fang-Fang Yin,
Editors*
ISBN: 978-1-4987-3673-2

**Radiochromic Film: Role and
Applications in Radiation Dosimetry**
Indra J. Das, Editor
ISBN: 978-1-4987-7647-9

Radiochromic Film

Role and Applications in Radiation Dosimetry

Edited by
Indra J. Das, PhD, FIPEM, FAAPM, FACMP, FACR, FASTRO
Department of Radiation Oncology
New York University Langone Medical Center

CRC Press
Taylor & Francis Group
Boca Raton London New York

CRC Press is an imprint of the
Taylor & Francis Group, an **informa** business

CRC Press
Taylor & Francis Group
6000 Broken Sound Parkway NW, Suite 300
Boca Raton, FL 33487-2742

First issued in paperback 2020

ISBN-13: 978-1-4987-7647-9 (hbk)
ISBN-13: 978-0-367-78175-0 (pbk)

Library of Congress Cataloging-in-Publication Data

Names: Das, Indra Jeet, editor.
Title: Radiochromic film: role and applications in radiation
dosimetry / edited by, Indra J. Das.
Other titles: Imaging in medical diagnosis and therapy.
Description: Boca Raton, FL : CRC Press, Taylor & Francis Group, [2017] |
Series: Imaging in medical diagnosis and therapy
Identifiers: LCCN 2017019838| ISBN 9781498776479 (hardback ; alk. paper) |
ISBN 1498776477 (hardback ; alk. paper)
Subjects: LCSH: Radiography--Films. | Radiation dosimetry.
Classification: LCC RC78 .R228 2017 | DDC 612/.01448--dc23
LC record available at https://lccn.loc.gov/2017019838

Visit the Taylor & Francis Web site at
http://www.taylorandfrancis.com

and the CRC Press Web site at
http://www.crcpress.com

To my late grandfather, Jag Dev Das, who provided me with the dream of education to achieve my goal

and

With special gratitude and love to my wife, Sununta C. Das, my son, Avanindra C. Das, and my daughter, Anita C. Das, from whom I stole time for research during my career building.

Contents

Series preface

Since their inception over a century ago, advances in the science and technology of medical imaging and radiation therapy are more profound and rapid than ever before. Further, the disciplines are increasingly cross-linked as imaging methods become more widely used to plan, guide, monitor, and assess treatments in radiation therapy. Today, the technologies of medical imaging and radiation therapy are so complex and computer driven that it is difficult for the people (physicians and technologists) responsible for their clinical use to know exactly what is happening at the point of care, when a patient is being examined or treated. The people best equipped to understand the technologies and their applications are medical physicists, and these individuals are assuming greater responsibilities in the clinical arena to ensure that what is intended for the patient is actually delivered in a safe and effective manner.

The growing responsibilities of medical physicists in the clinical arenas of medical imaging and radiation therapy are not without their challenges, however. Most medical physicists are knowledgeable in either radiation therapy or medical imaging and expert in one or a small number of areas within their disciplines. They sustain their expertise in these areas by reading scientific articles and attending scientific talks at meetings. In contrast, their responsibilities increasingly extend beyond their specific areas of expertise. To meet these responsibilities, medical physicists periodically must refresh their knowledge of advances in medical imaging or radiation therapy, and they must be prepared to function at the intersection of these two fields. How to accomplish these objectives is a challenge.

At the 2007 annual meeting of the American Association of Physicists in Medicine in Minneapolis, Minnesota, this challenge was the topic of conversation during a lunch hosted by Taylor & Francis Group and involving a group of senior medical physicists (Arthur L. Boyer, Joseph O. Deasy, C.-M. Charlie Ma, Todd A. Pawlicki, Ervin B. Podgorsak, Elke Reitzel, Anthony B. Wolbarst, and Ellen D. Yorke). The conclusion of this discussion was that a book series should be launched under the Taylor & Francis banner, with each volume in the series addressing a rapidly advancing area of medical imaging or radiation therapy of importance to medical physicists. The aim would be for each volume to provide medical physicists with the information needed to understand technologies driving a rapid advance and their applications to safe and effective delivery of patient care.

Each volume in the series is edited by one or more individuals with recognized expertise in the technological area encompassed by the book. The editors are responsible for selecting the authors of individual chapters and ensuring that the chapters are comprehensive and intelligible to someone without such expertise. The enthusiasm of volume editors and chapter authors has been gratifying and reinforces the conclusion of the Minneapolis luncheon that this series of books addresses a major need of medical physicists.

The series *Imaging in Medical Diagnosis and Therapy* would not have been possible without the encouragement and support of the series manager, Lu Han, of Taylor & Francis Group. The editors and authors, and most of all I, are indebted to his steady guidance of the entire project.

William R. Hendee
Founding Series Editor
Rochester, MN

Preface

The intent of this book is to serve as an authoritative text on radiochromic film, covering the basic principles, advances in technology, practical methods, and applications in many scientific disciplines, but mainly on complex issues in radiation dosimetry. Radiographic films have served over a century for this purpose, but digital imaging has completely replaced radiographic film technology. Radiochromic (or GAFchromic™) films have replaced radiographic films and are now widely used in radiation dosimetry. Radiochromic film changes its color due to polymerization of chemicals in the film when exposed to ionizing radiation. Thus, it does not require any chemical or physical processing. The fact that radiochromic films are near tissue equivalent and have high spatial resolution with dose response and nearly energy independent makes them particularly suitable for a wide range of applications over conventional approaches. This book will encompass development in this technology, uses in radiation dosimetry from diagnostic X-rays, brachytherapy, and radiosurgery to external beam therapies (photons, electrons, and protons), stereotactic body radiotherapy, intensity-modulated radiotherapy, and other emerging radiation technologies.

This is the first focused book on radiochromic film dosimetry. Other books addressing the broader subject of radiation dosimetry might mention radiochromic film and its uses, but not at the level of detail given here. Here we provide a framework for understanding the basic concepts, advantages, and main applications of radiochromic films in radiation dosimetry.

There are two main sections covering the basic, background information, and radiation dosimetry applications. Part I (Basics) addresses the characteristics and fundamental physics of radiochromic film, including type classification and safety issues. It also covers scanner technology and functionality as well as correction techniques. Part II (Applications) gives a detailed discussion of all the main applications, including kilovoltage, brachytherapy, megavoltage, electron beam, proton beam, skin dose, *in vivo* dosimetry, and postal and clinical trial dosimetry. It also discusses the state of the art in microbeam, synchrotron radiation, and ultraviolet radiation dosimetry. We hope this book fills the niche in two-dimensional imaging and dosimetry that had been missing for a long time.

About the editor

Indra J. Das, PhD, is currently the vice-chair, professor, and director of medical physics in the Department of Radiation Oncology at New York University Langone Medical Center, New York. Dr. Das is an internationally acclaimed medical physicist with expertise in radiation dosimetry, dose calculation, dose specification, small-field dosimetry, treatment planning, nanoparticles, and proton beam therapy. He received his BSc and MSc degrees from Gorakhpur University, Gorakhpur, Uttar Pradesh, India, followed by his diploma in radiological physics (DRP) from Bhabha Atomic Research Center, Mumbai, India. He earned his MS in medical physics from the University of Wisconsin–Madison, and PhD from the University of Minnesota, Minneapolis.

He has previously held positions at the University of Massachusetts Medical Center, Worcester, Massachusetts; Fox Chase Cancer Center, Philadelphia, Pennsylvania; University of Pennsylvania, Philadelphia, Pennsylvania; and Indiana University School of Medicine, Indianapolis, Indiana. He was the director of proton beam therapy at Indiana University, and an adjunct professor at Purdue University, West Lafayette, Indiana; Osaka University, Suita, Japan; and Amrita Institute of Medical Science, Kochi, India. He is chair of the working group for proton beam therapy at the American Association of Physicists in Medicine (AAPM) and wrote standard of care guidelines for the American College of Radiology (ACR) proton beam working group. He is an elected fellow of the Institute of Physics and Engineering in Medicine (IPEM), the AAPM, the American College of Medical Physics (ACMP), the ACR, and the American Society for Radiation Oncology (ASTRO). He is also the recipient of numerous accolades including the AAPM's Farrington Daniels Award (1988) and the Association of Medical Physicists of India's (AMPI's) Dr. Ramaiah Naidu Award for lifetime achievement (2015).

He has published more than 400 abstracts, 21 books/chapters, and more than 200 peer-reviewed scientific papers. He serves on the editorial board of many radiation journals, including *Medical Physics, Journal of Medical Physics, British Journal of Radiology,* and *International Journal of Radiation Oncology Biology and Physics* and has delivered invited lectures throughout the world.

Contributors

Paola Alvarez
Department of Radiation Oncology
University of Texas MD Anderson Cancer Center
and
IROC Houston Quality Assurance Center
Houston, Texas

Samara Alzaidi
Department of Radiation Oncology
Chris O'Brien Lifehouse
Sydney, Australia

David Barbee
Department of Radiation Oncology
New York University Langone Medical Center
New York, New York

Martin Butson
Department of Radiation Oncology
Chris O'Brien Lifehouse
Sydney, Australia

Maria Chan
Department of Medical Physics
Memorial Sloan Kettering Cancer Center
Basking Ridge, New Jersey

Gwi Cho
Department of Radiation Oncology
Chris O'Brien Lifehouse
Sydney, Australia

Jeffrey C. Crosbie
School of Science
RMIT University
and
William Buckland Radiotherapy Centre
Alfred Hospital
Melbourne, Australia

Scott Crowe
Cancer Care Services
Royal Brisbane and Women's Hospital
and
Queensland University of Technology
Brisbane, Australia

Annemieke De Puysseleyr
Department of Radiation Oncology and Experimental Cancer Research
Ghent University
Ghent, Belgium

Carlos De Wagter
Department of Radiation Oncology and Experimental Cancer Research
Ghent University
Ghent, Belgium

Sharifeh A. Dini
Educational Consultant
Las Vegas, Nevada

Damien Dumont
Center for Molecular Imaging, Radiotherapy and Oncology
Institut de Recherche Expérimentale et Clinique
Université Catholique de Louvain
Brussels, Belgium

David Followill
Department of Radiation Oncology
University of Texas MD Anderson Cancer Center
and
IROC Houston Quality Assurance Center
Houston, Texas

Simran Gill
Department of Radiation Oncology
Chris O'Brien Lifehouse
Sydney, Australia

Steven J. Goetsch
San Diego Medical Physics
and
San Diego Gamma Knife Center
San Diego, California

Tina Gorjiara
Department of Radiation Oncology
Chris O'Brien Lifehouse
Sydney, Australia

Mamoon Haque
Department of Radiation Oncology
Chris O'Brien Lifehouse
Sydney, Australia

Robin Hill
Department of Radiation Oncology
Chris O'Brien Lifehouse
and
The Institute of Medical Physics
School of Physics
University of Sydney
Sydney, Australia

Tanya Kairn
Genesis Cancer Care Queensland
and
Cancer Care Services
Royal Brisbane and Women's Hospital
and
Chemistry, Physics and Mechanical Engineering
Queensland University of Technology
Brisbane, Australia

Tomas Kron
Department of Physical Sciences
Peter MacCallum Cancer Centre
and
Sir Peter MacCallum Department of Oncology
University of Melbourne
and
School of Science, Engineering and Technology
RMIT University
Melbourne, Australia

Elizabeth Kyriakou
Department of Physical Sciences
Peter MacCallum Cancer Centre
Melbourne, Australia

Joerg Lehmann
Department of Radiation Oncology
Calvary Mater Newcastle
and
School of Mathematical and Physical Sciences
University of Newcastle
Newcastle, Australia

and

Institute of Medical Physics
School of Physics
University of Sydney
Sydney, Australia

and

School of Science, Engineering and Technology
RMIT University
Melbourne, Australia

H. Harold Li
Department of Radiation Oncology
Washington University School of Medicine
Saint Louis, Missouri

Jessica E. Lye
Australian Clinical Dosimetry Service (ACDS)
Australian Radiation Protection and Nuclear Safety Agency (ARPANSA)
Melbourne, Australia

Thomas R. Mazur
Department of Radiation Oncology
Washington University School of Medicine
Saint Louis, Missouri

Ali S. Meigooni
Comprehensive Cancer Centers of Nevada
and
University of Nevada
Las Vegas, Nevada

Johnny Morales
Department of Radiation Oncology
Chris O'Brien Lifehouse
Sydney, Australia

Azam Niroomand-Rad
Department of Radiation Medicine
Georgetown University School of Medicine
Washington, DC

David Odgers
Department of Radiation Oncology
Chris O'Brien Lifehouse
Sydney, Australia

Evaggelos Pantelis
Medical Physics Laboratory
Medical School, National and Kapodistrian University of Athens
and
CyberKnife and TomoTherapy Center
Iatropolis Clinic
Athens, Greece

Joel Poder
Department of Radiation Oncology
Chris O'Brien Lifehouse
Sydney, Australia

and

Centre of Medical Radiation Physics
University of Wollongong
Wollongong, Australia

Dane Pope
Department of Radiation Oncology
Chris O'Brien Lifehouse
Sydney, Australia

Benjamin S. Rosen
Department of Radiation Oncology
University of Michigan Health System
Ann Arbor, Michigan

Yulin Song
Department of Medical Physics
Memorial Sloan Kettering Cancer Center
Westchester, New York

Edmond Sterpin
Laboratory of Experimental Radiotherapy
Department of Oncology
Katholieke Universiteit Leuven
Leuven, Belgium

and

Center for Molecular Imaging, Radiotherapy and Oncology
Institut de Recherche Expérimentale et Clinique
Université Catholique de Louvain
Brussels, Belgium

Iori Sumida
Department of Radiation Oncology
Graduate School of Medicine
Osaka University
Suita, Japan

Andy (Yuanguang) Xu
Department of Radiation Oncology
New York University Langone Medical Center
New York, New York

PART 1

BASICS

Introduction

INDRA J. DAS

1.1 HISTORY AND BACKGROUND

The current book represents the state-of-the-art knowledge on the fast growing field of two-dimensional dosimeter such as film for simple and complex dosimetry. Radiochromic film (RCF) is a class of polymer-based device that changes color when exposed to radiation. A lot of development in this area was conducted by a chemical company GAF, known as General Aniline & Film located in Parsippany, NJ, that produced films known as GAFchromic™. Today, GAFchromic name has become synonymous to a general name radiochromic film. Detailed description, characteristics, and usage have been described in TG-55 [1]. A revision to this task group (TG-235) is underway that may provide additional information in near future when published.

Radiographic films based on silver halides are becoming extinct in digital age, which were also used for radiation dosimetry [2]. There were numerous problems with such films, but mainly the chemical processing and energy dependence made it harder to adapt. On the other hand, RCF does not require any processing and is tissue equivalent. The demise of radiographic films, the characteristics of which have been described in TG-69 [3] and unfavorable characteristics, made RCF a compelling choice for radiation dosimetry.

1.2 SUMMARY OF CONTENTS

The necessity of a book was realized because of vast amount of research conducted on the radiochromic class of dosimeter visible clearly from publications in every radiation-related journal. In addition, there is no other book on this subject. There is a growing need to have an authoritative book for graduate-level courses taught in universities around the world on every aspect of the radiation dosimetry. Thus, the chapters of this book are divided into application based on three-dimensional conformal radiation therapy (3DCRT), intensity-modulated radiation therapy, stereotactic radiosurgery, particle beam, and other aspects of radiation fields. This book contains 20 chapters and is truly an international collaboration, in which expert authors from all over the world, including Australia, Japan, France, Belgium, Greece and USA, are represented.

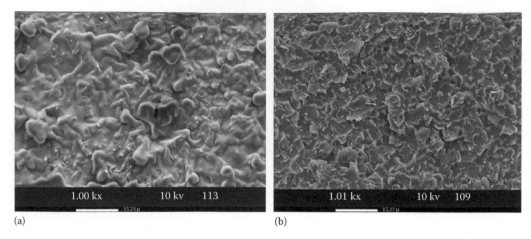

(a)   (b)

Figure 1.1 Electron micrograph of external beam therapy (EBT) film (a) unexposed and (b) exposed to 2 Gy of X-rays. (Details of techniques and magnifications are also provided in the image.)

Chapter 2 of the book provides a rare historical background and characteristics of RCF. In addition, it covers RCF action and various characteristics of films. To clearly show the action of radiation on film, Figure 1.1 shows electron micrograph of the structures in unexposed and exposed film with 2-Gy X-ray radiation. The polymerization is dose and time dependent, which is discussed in Chapter 2 of this book. Please note that polymerization reduces the structure to a flatter level, thereby changing the color from semitransparent to blue or yellow color. Additional images are also provided in other chapters whenever necessary.

The unique properties of RCF coloration are used to measure the radiation dose in every aspect of radiation dosimetry. Figure 1.2 shows the growth of research evaluated from publications searched thorough Pubmed. Until the end of 2015, there have been nearly 1000 papers published, all related to radiation-related applications of various types of RCF. These data may not reflect actual growth as Pubmed does not cover basic sciences and nonmedical journals. There are several review articles that have covered the breadth of this subject [4–6].

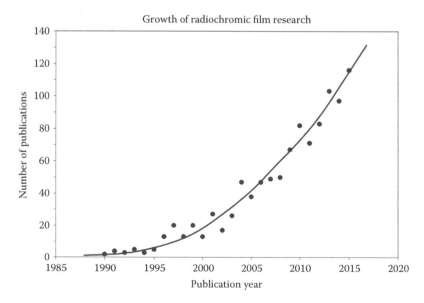

Figure 1.2 Yearly peer-reviewed publications in research journals.

The writing style is kept standard on, with introduction of modality, how RCF can be used with a summary and up-to-date references. Due to a large number of authors from various countries, uniformity of writing style was a critical factor that has been kept in mind. In addition, the presentation is made simple such that undergraduate and other students can follow on with minimum knowledge of the subject. It is hoped that this book will provide a badly needed literature for this unique topic.

REFERENCES

1. Niroomand-Rad A, Blackwell CR, Coursey BM et al. Radiochromic film dosimetry: Recommendations of AAPM radiation therapy committee task group 55. *Med Phys* 1998;25:2093–2115.
2. Das IJ. Radiographic film. In: Rogers DWO, Cyglar JE (Eds.) *Clinical Dosimetry Measurements in Radiotherapy*. Madison, WI: Medical Physics Publishing; 2009. pp. 865–890.
3. Pai S, Das IJ, Dempsey JF et al. TG-69: Radiographic film for megavoltage beam dosimetry. *Med Phys* 2007;34:2228–2258.
4. Devic S. Radiochromic film dosimetry: Past, present, and future. *Phys Med* 2011;27:122–134.
5. Devic S, Tomic N, Lewis D. Reference radiochromic film dosimetry: Review of technical aspects. *Phys Med* 2016;32:541–556.
6. Soares CG. New developments in radiochromic film dosimetry. *Radiat Prot Dosimetry* 2006;120:100–106.

Historical background, development, and construction of radiochromic films

MARTIN BUTSON AND AZAM NIROOMAND-RAD

2.1 HISTORICAL DEVELOPMENT OF RADIOCHROMIC MATERIALS

The history of development of radiochromic materials dates back to the early nineteenth century before discovery of X-ray by Roentgen (1895) and prior to the use of conventional silver halide-based radiographic films for radiographic imaging and radiation dosimetry. Radiographic films became an important tool for detecting ionizing radiation and measuring dose [1]. However, there are numerous problems associated with quantification of ionizing radiation dose using radiographic films. One problem is the energy absorption and transfer properties of radiographic films that are not similar to those of biological tissues. The radiographic films also have the disadvantages of being sensitive to room light and requiring wet chemical processing. Some of these difficulties led scientists to search and develop an alternative to radiographic films for radiation imaging and dosimetry with high spatial resolution, which did not require a special chemical processing and/or developmental procedure. The radiochromic processes always involve the direct coloration of a material by the absorption of electromagnetic radiation, without requiring latent chemical, optical, or thermal development or amplification [2–4].

In the early 1800s, a radiochromic process was first demonstrated by Joseph Niepce, a French scientist, inventor [2]. He applied a light sensitive tar-like material known as *bitumen of Judea* solution (an unsaturated hydrocarbon polymeric mixture) to a pewter plate, made of an alloy of tin, copper, and antimony. Niepce projected a view from a window onto a pewter plate and after about 8 h of exposure to sunlight he saw a permanent bitumen image on the plate. This image production was attributed to an unsaturated hydrocarbon polymeric mixture in the bitumen that underwent cross-linking upon irradiation [5].

By the mid-1800s, an important direct-imaging process was developed. This involved papers and gels impregnated with potassium dichromate, which was known to undergo photo-reduction. This development led to daguerreotype, collotype, photogravure, and photolithographic prints— a process that is an inverse of radiochromism, in that the brownish color is left in the unreduced portions of the image [6]. Since then, cross-linking organic systems have been developed for radiographic imaging. Such processes typically involve pairing of free radicals to form radiation cross-linked carbon-chain materials, resulting in covalently bonded growing chains [6]. Another type of organic free-radical imaging medium involves combination of photopolymerization with leuco dyes that produce color upon irradiation [7]. The direct-imaging effects include radiation-induced vesicular films giving light-scattering properties and the radiation-induced changes in the hardness of polymeric microcapsules containing diffusible dyes, inks, and pigments that are released mechanically [8,9]. Some other radiochromic organic image-forming systems involve *cis–trans* isomeric conversions, or dissociations resulting in enolic, ketonic, and anilic bonds and other molecular rearrangements. Such tautomerizations lead to double-bonded coloration of spiropyrans, anils, organic acids, stilbenes, and other polycyclic compounds [4,10].

About a century ago, barium platinocyanide pastille discs were developed along with color wheels to quantify absorbed dose defined as the amount of ionizing radiation required to produce visible reddening of the skin of the hand or arm [11]. Since then, triphenyl tetrazolium chloride salts were developed that could be used as direct-imaging biological stains for botanical specimens and for characterizing normal and malignant mammalian tissues [12–14]. The triphenyl-methane leucocyanides (e.g., formyl violet nitrile) salts are colorless in aqueous solutions but upon irradiation become highly colored, insoluble formazans [15]. The hydrophilic triphenyl tetrazolium chloride salts are also used for radiographic imaging in hydrocolloids and other aqueous gels, and for mapping ionizing radiation-dose distribution directly in animal tissues [16,17].

2.2 HISTORICAL DEVELOPMENT OF RADIOCHROMIC FILMS

Since the 1960s, McLaughlin et al. [18] and other investigators reported on the development of colorless solid solutions of particular materials: derivatives of the triphenyl methane molecule, which underwent radio-synthesis to produce dyes [19]. Much of these early research and development of the radiochromic materials are attributed to the U.S. National Bureau of Standards (present U.S. National Institute of Standards and Technology). The initial support for development was provided by the Division of Isotopes Development, U.S. Atomic Energy Commission, and with the assistance of the inventor of ultraviolet sensitive systems, by Chalkley [20]. The earlier forms of radiochromic media had a useful dose range of 10^3–10^6 Gy, and as such their use was limited to high-dose applications such as food irradiation, medical instrument sterilization, and other industrial applications.

The materials in the radiochromic films that are responsible for the coloration are known as crystalline polyacetylenes, in particular diacetylenes, and upon thermal annealing or radiation exposure they undergo polymerization, turning blue or red depending on their specific composition [21,22]. The hydrophobic-substituted triphenylmethane leucocyan materials, upon irradiation, undergo heterolytic bond scission of the nitrile group and form highly colored dye salts in solid polymeric solution. The basic foundation for such films is generally nylon, vinyl, or styrene-based polymer. These films have also been used for high-resolution, high-contrast radiation images and to map radiation-dose distributions across material interfaces [23]. Since the 1960s, radiochromic thin films giving permanently colored images were also developed as high-dose radiation dosimeters for 10^4–10^6 Gy dose range [6,24,25]. On the contrary, this kind of radiochromic films is not sensitive enough to be used for clinical or radiological applications.

2.3 HISTORICAL DEVELOPMENT AND CONSTRUCTION OF GAFCHROMIC™ FILMS

By the mid-1980s, a new radiochromic film medium was developed for nonindustrial applications [26]. The radiochromic film, known as GAFchromic™, was produced by the International Specialty Products (ISP) Technology, which was a division of GAF Chemical Corporation (GAF Corporation, Wayne, NJ). In 1991, the GAF Chemical Corporation was publicly listed and is now known as the International Specialty Products Inc. The particular dye in GAFchromic film was more sensitive than previous types and could be used for mapping dose distributions above 5 Gy, [27,28]. McLaughlin et al. [29] demonstrated that the mechanism of color production in GAFchromic films was a first-order solid-state polymerization of the diacetylene monomer and reported that propagation of the polymerization was complete within 2 ms of a single 20 Gy, 50 ns electron pulse. Subsequently McLaughlin et al. [30] also observed that postirradiation polymerization continued to occur, most notably within the first 24 h following exposure. The GAFchromic film was also used for dosimetry of radioactive hot particles (or spheres) of ^{60}Co and ^{90}Sr/^{90}Y with activities ranging from 1 to 300 MBq [31,32], and its usage was extended to ^{90}Sr/^{90}Y ophthalmic applicators [33]. Other usages have included (but not limited to) the dosimetry of small stereotactic radiosurgery fields and intravascular brachytherapy sources [34]. In addition, by the mid-1980s, other radiochromic films were being developed that were used for electron and proton beam dosimetry [35,36].

2.4 HISTORICAL DEVELOPMENT AND CONSTRUCTION OF RADIOCHROMIC FILMS

The ISP manufactured a wide range of GAFchromic films that were simply referred to as radiochromic films (RCFs). The word GAFchromic and radiochromic have become synonymous now.

The differences between each type of ISP RCF were initially defined by construction rather than chemicals. For example, whether the film was constructed with a single or double active layer and/or whether the film was reflective or transmissive and what was its physical dimension. Changes then occurred with the specific chemical composition of the active layer. By the mid-1980s, a special form of radiochromic film, based on polydiacetylene, was introduced for medical applications and were supplied in two types, DM-1260 (also known as HD-810 for nomenclature designation and single-layer MD-55 for the absorbed dose ranges 50–2500 and 10–100 Gy,

Figure 2.1 GAFchromic™ MD-55 film envelope with various cut and exposed films. (Courtesy of Martin Butson.)

respectively) [27,37,38]. Figure 2.1 shows GAFchromic MD-55 film envelope with various cut and exposed films. The HD-810 film, also known as DM-1260, consisted of a single 7 μm layer of the emulsion coated on 100 μm thick polyester base. This film had a dose range of 10–400 Gy.

The single-layer radiochromic films, designated as MD-55-l, was extensively investigated by McLaughlin et al. [30]. These films were studied by pulse radiolysis and flash photolysis, in terms of the kinetics of their response to ionizing radiation and ultraviolet light [37–39]. The radiochromic reaction was a solid-state polymerization, whereby the films turned deep blue color proportionately to radiation dose, due to progressive 1,4-trans additions which led to colored polyconjugated, ladder-like polymer chains. The crystal polymerization image can be seen in Chapter 1. The pulsed-electron-induced propagation of polymerization had a first-order rate constant on the order of 10^3 s^{-1}, depending on the irradiation temperature. The UV-induced polymerization was faster by about one order of magnitude ($k_{obs} = 1.5 \times 10^4$ s^{-1}). In the case of the electron beam effect, the radiation-induced absorption spectrum exhibited a much slower blue shift of the primary absorption band ($\lambda_{max} = 675 - 660$ nm) on the 10^{-3}–10^{+1} s time scale. This effect is attributed to crystalline strain rearrangements of the stacked polymer strand units. Each of these film types is colorless before irradiation, consisting of a thin, active microcrystalline monomeric dispersion coated on a flexible polyester film base. It turns progressively blue upon exposure to ionizing radiation. The radiochromic radiation chemical mechanism is a relatively slow first-order ($k \cong 10^3$ s^{-1}) solid-state topochemical polymerization reaction initiated by irradiation, resulting in homogeneous, planar polyconjugation along the carbon-chain backbone [37,38].

Later on the HD-810 and MD-55-l were replaced by double-layer MD-55, designated as MD-55-2 for medical applications (useful dose range 3–100 Gy) [30,40]. Figure 2.2 shows GAFchromic MD-55-2. The double layer MD-55-2 contained two 16-μm layers of emulsion separated by a 25-μm layer of polyester and 2×25 μm layers of adhesive, all sandwiched between two 66 μm polyester layers.

The total active layer was over double to that of the HD-810 films making it sensitive to a dose range of 2–100 Gy. These films have a single 40 μm layer of emulsion layered between two 97 μm thick layers of polyester. The outer layers provide a water barrier and again the greater thickness of active layer increases the sensitivity. The high sensitivity (HS) film also provided a dose range of 0.5–40 Gy. Niroomand-Rad et al. [41] provided a detail discussion of these films. A summary is enclosed in Table 2.1 for HD-810, MD-55-l, and MD-55-2 films.

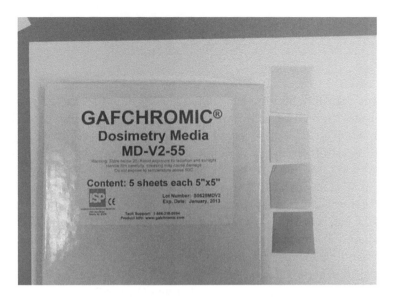

Figure 2.2 GAFchromic MD-55-2 film jacket and exposed cut films. (Courtesy of Martin Butson.)

Table 2.1 Structure, dimension, and approximate composition of radiochromic films

Film type	HD-810 (DM-1260)	MD-55-1	MD-55-2
Nuclear associates (vendor) number	_____	37-041	37-041
Standard size	$20 \times 20\ cm^2$	$12.5 \times 12.5\ cm^2$	$12.5 \times 12.5\ cm^2$
Nominal thickness (µm)	107 (Figure 2.1)	82 (Figure 2.1)	278 (Figure 2.1)
Sensitive layer(s) (µm)	7 ± 1	15 ± 1	30 ± 1
Base material (µm)	99	67	159
Sensitive layer–7 µm Adhesive layer–1.5 µm Conductive layer–0.05 am Polyester base–99 am		Sensitive layer–15 µm Polyester base–67 µm	Polyester base–67 µm Sensitive layer–15 am Pressure sensitive adhesive–44.5 am Polyester base–25 am Pressure sensitive adhesive–44.5 am Sensitive layer–15 am Polyester base–67 am
Polyester base: Carbon (45 Atom %)	Hydrogen (36 Atom %)	Oxygen (19 Atom %)	
Sensitive layer: Carbon (31 Atom %)	Hydrogen (56 Atom %)	Oxygen (8 Atom %)	Nitrogen (5 Atom %)
Adhesive layer: Carbon (33 Atom %)	Hydrogen (50 Atom %)	Oxygen (17 Atom %)	
Conductive layer: Indium Tin Oxide			

2.5 HISTORICAL DEVELOPMENT AND CONSTRUCTION OF RECENT RADIOCHROMIC FILMS

The sensitivity of HD-810 and MD-55 films decreases with decreasing energy, therefore ISP developed XR type T and R (transmissive and reflective) specifically for use in the low-energy range, 20–200 kVp. The film has the same active layer as the previous films but included a high-Z material and has a dose range of 0.1–15 Gy. By the mid-2000s, a new film was released called external beam therapy (EBT). The active layer was a variation of the monomer used in the previous films and was a hair like version of the lithium salt of pentacosa-10, 12-diyonic acid crystal [42]. The atomic composition of EBT is (42.3% C, 39.7% H, 16.2% O, 1.1% N, 0.3% Li, and 0.3% Cl). The inclusion of the moderate atomic number element chlorine ($Z = 17$) provided a Z_{eff} of 6.98 making it near tissue equivalent. The new active layer was also found to be more sensitive, providing a dose range of 0.01–8 Gy. The diacetylene monomers exposed to heat, UV, or ionizing radiation undergo progressive 1,4-polymerization leading to the production of colored polymer chains that grow in length with level of exposure as shown in Figure 2.3 [21,38,43]. The packing of the diacetylene monomers in the crystal lattice depends on the particular end groups (R_1 and R_2 in Figure 2.3), the monomers in EBT films are approximately 0.75 μm in diameter and for the polymerization to occur the triple bonds of adjacent monomers should be within 0.4 nm of each other [42]. The radiation sensitivity of the crystal is also dependent on the particular end groups, with the lithium salt of pentacosa diynoic acid used in EBT film being sensitive to doses as low as 1cGy, several orders of magnitude more sensitive than the earlier radiochromic films.

The particular diacetylene monomer used in EBT film is the lithium salt of pentacosa-10, 12-diyonic acid [42]. The new EBT film has a face lift for packaging that is shown in Figure 2.4 exposed cut film.

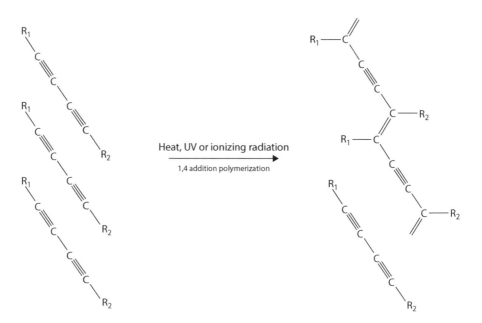

Figure 2.3 Diacetylene monomers undergo a 1,4-polymerization upon exposure to heat, UV, or ionizing radiation. (Courtesy of Azam Niromand-Rad.)

Figure 2.4 GAFchromic EBT film package along with cut exposed films showing the color changes from near transparent to dark blue. (Courtesy of Martin Butson).

As diacetylene crystals are too large to provide useful sensitivity, they are dissolved in a solvent, such as *n*-butanol solvent, which has the added advantage of improving the light resistance of the polyacetylenic crystals. The dissolved crystals are then dispersed in a binder such as an aqueous gelatin solution. After removing the alcohol solvent the binder is coated onto a substrate. Once the crystals are dried, they become fixed in orientation. The drying and aging process may take several months. Possible substrate materials include polyester, ceramics, glass, or metals, and an adhesive may also need to be used [44]. Additional coatings may be added to the active layer to reduce UV sensitivity and to act as an antioxidizing layer [22,45]. Films may be constructed as laminates with single, double, or triple emulsion layers coated onto intervening substrate layers. However, the rate of polymerization decreases with growing polymer. This results in the postexposure changes of RCFs in which the optical transmission and the rate of change of the optical transmittance of an irradiated film decrease with time as shown in Figure 2.5 for EBT2 and EBT3 films.

It should be noted that the change in optical density is proportional to log (time-after-exposure) [46]. However, as shown in Figure 2.5, after a month or more, the postirradiation changes of RCFs

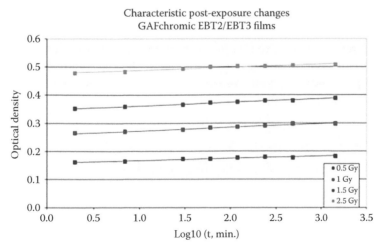

Figure 2.5 Postexposure characteristic changes for GAFchromic EBT2/EBT3 films. (Courtesy of Azam Niroomand-Rad.)

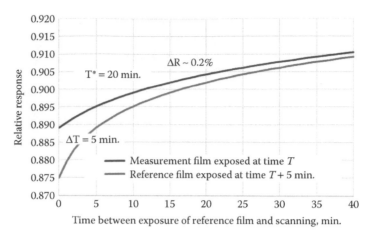

Figure 2.6 Relative response compared to measurement film exposed at time T and reference film exposed at time $T + 5$ min. (Courtesy of Azam Niroomand-Rad.)

with time are not significant. As recommended by Niroomand-Rad et al. [41], by waiting about 24 h after irradiation of calibration film and the film used for dosimetry, the polymerization is almost complete and the rate of response change is low. Therefore, so long as these postexposure changes are appreciated and the measurement films and the calibration film are treated and handled equally, the measurement uncertainties will have little effect on the measured response.

Furthermore, to eliminate the necessity of lengthy postexposure wait, Chan et al. and Lewis [47,48] recommended a *One-scan* protocol in which films can be scanned a few minutes after exposure. With this protocol, dose results can be available a few minutes after exposure as shown in Figure 2.6.

As depicted in Figure 2.6, in *One-scan* protocol, the blue line shows postexposure change of a first film, assumed to be a film from which measurements are required, exposed at time T, whereas the green line shows postexposure change of a second film (same production lot), exposed to a similar dose at time $T + \Delta T$. If ΔT is short and the two films are scanned together, immediately after the second film is exposed the response of the first film will be about 1.5% greater. However, as scanning is delayed for longer periods after the second film is exposed the response difference between the films rapidly diminishes. In practice the difference falls to about 0.2% when the delay between exposure of the second film and scanning is about $4 \times \Delta T$. By using a response measurement from a reference film exposed to a known dose a short time after an application film, adjustments can be made to the measured responses of the application film to compensate for postexposure response behavior [48]. Postexposure polymerization error can be handled by any one of the methods discussed earlier.

2.6 THE DEVELOPMENT AND CONSTRUCTION OF AVAILABLE RADIOCHROMIC FILMS

In the current section, we review available RCF models and focus on the GAFchromic RCF products by (Ashland, Bridgewater, NJ) and their use in dosimetry as they have useful dose ranges suitable for clinical radiation therapy applications. Since the 1998 publication by Niroomand-Rad et al. [41], the manufacturer of GAFchromic film has introduced additional products, which are detailed in the literature [49]. By the early 2000s several limitations of the original GAFchromic films for radiotherapy dosimetry had become apparent. These included, but not limited to, the relatively low sensitivity of the products, their lack of uniformity, the limitation on physical sizes, and their high cost [40,50–54]. Hence, the manufacturer of radiochromic films addressed most of these problems and with continual improvement, the scope of dosimetry with GAFchromic film media has been evolved substantially [49].

In 2004, a more sensitive radiochromic film, EBT film, was introduced that was several orders of magnitude more sensitive than previous RCF models. In 2009, the production of the EBT films was discontinued and replaced by the EBT2 films. The EBT2 film as shown in Figure 2.7 had the same active component as the EBT films but with a yellow dye added to the active layer and it was also constructed as a single layer instead of double as shown in Figure 2.8. The film has a slightly narrower active layer than EBT and slightly different overall atomic composition (42.37% C, 40.85% H, 16.59% O, 0.01% N, 0.10% Li, 0.04% Cl, 0.01% K, 0.01% Br). The Z_{eff} of EBT2 is 6.84 compared with 6.98 for EBT, and close to Z_{eff} of water (7.3).

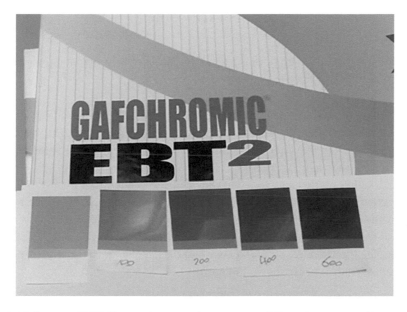

Figure 2.7 GAFchromic EBT2 film package and cut exposed films. (Courtesy of Martin Butson.)

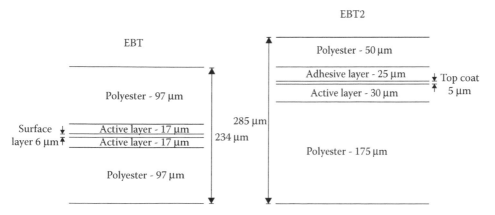

Figure 2.8 Illustration of the physical construction of the EBT and EBT2 films. Please note the substrate thickness and active layer in two films. (Courtesy of Martin Butson.)

Figure 2.9 GAFchromic EBT3 film package and cut exposed films indicating yellowish tint to the films. (Courtesy of Martin Butson.)

In 2010, it was reported that the measured response of the EBT2 film model asymmetric Type 2 configuration to be dependent upon which side of the film faced the light source in a flatbed scanner [55]. Later on, the polyester layer configuration was modified and this new RCF is now marketed as EBT3 as shown in Figure 2.9.

In addition, a yellow marker was added to the emulsion of the EBT2 to absorb light in the blue part of the spectrum (400–500 nm), well away from the dosimetric peaks, which are in the green (500–600 nm) and red (600–700 nm) parts of the spectrum. This allows the use of the blue wavelengths for a determination of the emulsion thickness for quantifying sensitivity variations due to this parameter. The spectra of yellow marker dye and the net change in the absorbance of the active component due to irradiation are shown in Figure 2.10.

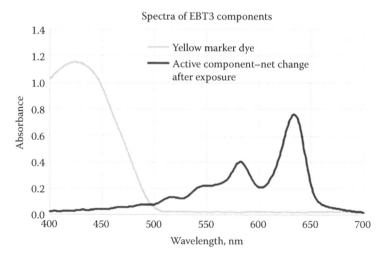

Figure 2.10 Absorbance and spectra of yellow marker dye and active component—Net change after exposure. (Courtesy of Azam Niroomand-Rad.)

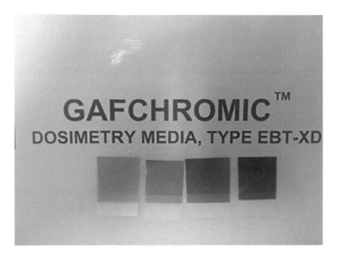

Figure 2.11 GAFchromic EBT-XD film package and cut exposed films. (Courtesy of Martin Butson.)

Moreover, minor changes in atomic composition of EBT2 and EBT3 were made to improve on the photon energy dependence of these newer RCFs [55]. Following these improvements in early 2015, a new film, EBT-XD (Figure 2.11) was developed that had similar composition and construction as EBT3. As the crystal size of the active component of EBT-XD films were small, these films were less sensitive and their increased slope of the dose–response curve at doses above 10 Gy made them more suitable for dose measurements >10 Gy [56,57].

The major difference between EBT3 and EBT-XD films is the active layer crystal structure [58]. The EBT-XD film exhibits a substantially smaller crystal structure. This characteristic determines the lower sensitivity response of the film as the degree of coloration caused by radiation is reduced per unit area within the film. This type of film construction may also lend itself to better uniformity in the lateral scanning direction of desktop scanners as it should provide lower light scattering and polarization effects compared with EBT3 film.

Presently, all available new radiochromic films have three types of configurations [49]: Type 1, Type 2, or Type 3 as depicted here in Figure 2.12.

Substrate: Depending on the intended use of film, Substrate 1, on which the active layer is coated, may be smooth transparent polyester, transparent polyester with a surface treatment containing microscopic silica particles, or an opaque white polyester substrate containing barium sulfate whitening agent. The substrate with the silica surface treatment is used in the EBT3 film family and EBT-XD film. The silica particles are <10 μm in size and prevent the formation of Newton's Rings interference patterns by maintaining an air gap, much larger than the wavelength of light, between the film and the glass window of a flatbed scanner. The silica particles constitute a tiny mass fraction of the entire film and have no discernible effect on dosimetric performance. The opaque white substrate is used with film (e.g., RTQA-2) primarily intended to aid in beam location rather than measure dose and is often viewed

Type 1 configuration	Type 2 configuration	Type 3 configuration
	Substrate #2	Substrate #2
	Adhesive layer	Active layer
Active layer	Active layer	Substrate #1
Substrate #1	Substrate #1	

Figure 2.12 Type 1, Type 2, and Type 3 configuration of radiochromic films. (Courtesy of Martin Butson.)

reflectively. These types of films are scanned in reflection. Films intended for use with kilovoltage photons (e.g., XR-R, XR-QA2) also use the opaque base as the barium sulfate-whitening agent boosts photoelectric absorption thus increasing the film's sensitivity.

The polyester overlaminate, Substrate 2, may be smooth transparent and colorless (MD-V3, EBT2), has the silica surface treatment previously described (the EBT3 family and EBT-XD), or may contain a yellow dye (RTQA, XR-R, XR-QA2). The yellow dye absorbs most of the blue and ultraviolet wavelengths and therefore reduces the light exposure to which the active component is most sensitive. The yellow dye is also complementary to the blue polymer formed by exposure of the active component to radiation and therefore enhances the visual impact of the color changes. The yellow overlaminates are used in all the film types coated on opaque white substrates. Films in which Substrate 2 contains the yellow dye can only be used for single, red-channel dosimetry. Multichannel dosimetry is not feasible.

Adhesive layer: The Type 2 configuration uses either a pressure-sensitive acrylic adhesive layer, or a water-soluble polymer to bond Substrate 2 to the active layer.

Active layer: In all RCFs, the active layers contain microcrystals of the active component dispersed in a water-soluble polymer matrix. Other components are essential to the dosimetric function. For instance EBT2, MD-V3, HD-V2, EBT-XD, and the EBT3 film family contain aluminum oxide nanoparticles in the active layer to minimize the energy dependence of film response from the MV range down to about 40 keV. These RCFs also contain nanoparticles of a special yellow dye homogeneously dispersed in the active layer. The marker dye is unchanged by exposures up to at least 500 Gy. The utility of this material, often referred to as the yellow marker dye, is to provide the means by which the film's response can be compensated for small differences in the thickness of the active layer. All film models can be used for single red-channel dosimetry, but films containing the yellow marker dye and clear transparent polyester for Substrate 2, can be used with multichannel dosimetry if scanned with a Red-Green-Blue scanner. Figure 2.10 shows the spectra of the yellow marker dye and the net change in the absorbance of the active component due to irradiation. Thus the blue part of the spectrum (400–500 nm) is dominated by information about the uniformity of the active layer whereas the green and red bands (500–600 nm and 600–700 nm, respectively) mainly contain information about the radiation dose. By providing a solution that minimizes the dose differences between the color channels, the triple-channel dosimetry method compensates for thickness nonuniformity defects in the RCF.

Active component: The active component in all RCFs has come from a class of hydrocarbons known as diacetylenes. When present in a form with molecular order, for example, crystal, micelle, absorbed monolayer, and so on, many compounds in this class undergo chain polymerization yielding an intensely colored polymer. For some diacetylenes, the polymerization can be initiated by exposure to any form of ionizing radiation. When diacetylenes are in a form with the right molecular order, the individual molecules are arranged and aligned so that chain polymerization, initiated by radiation exposure, can readily propagate down a stack of molecules [59]. This domino-type polymerization will only proceed linearly down the stack. It will not jump from stack-to-stack, or propagate in a second dimension.

The unlaminated EBT3 film products have Type 1 configuration that are especially useful for the dosimetry of radiation that are strongly attenuated by a polyester substrate, such as very low energy electrons and beta radiation as well as ions and photons [49]. The Type 2 and Type 3 films contain polyester substrates attached to both sides of the active layers to lessen effects from exposure to light and to protect the active layer from mechanical damage. In a more recent investigation,

Lewis et al. [60] showed that response values measured on EBT3 film were dependent on the distance between the film and the light source in the scanner. It was discussed that the active layer in EBT3 film, and all other RCFs, scatter some of the transmitted light; therefore, all the films exhibit so-called Callier effect [61].

The recent RCFs such as MD-V3, EBT2, RTQA-2, XR-RV3, and XRQA films have Type 2 configuration with the second polyester substrate attached by an adhesive layer. Whereas all members of the EBT3/EBT3+ products and EBT-XD films have Type 3 configuration in which both polyester substrates are directly attached to the active layer. The available RCFs currently include those that are suitable for therapy dosimetry including brachytherapy, intensity-modulated radiotherapy (IMRT), volumetric-modulated arc therapy (VMAT), stereotactic body radiation therapy (SBRT), and sterotactic radiosurgery (SRS) with photons, electrons, protons, and ions of other heavier elements and those that are suitable for diagnostic X-ray procedures and the exposures related to interventional radiology. As pointed out by Niroomand-Rad et al. [41], the dosimetric characteristics of RCFs are described by a dose–response curve. The dose–response curve may be a plot of net optical density against dose for each output channel of the readout system. For higher radiation dose, the film is darker, represented by lesser light transmission and higher optical density. Alternatively, the dose–response curve may be a plot of a scanners response versus dose. In both cases the dose–response curve is not linear. Furthermore, the dose–response characteristics of RCFs depend on other factors such as film type, radiation type, and readout system as outlined as follows:

- Film type includes film model and lot, emulsion (chemical composition), lamination configuration, absorption spectra
- Radiation type includes photon, electron, proton, and so on (including brachytherapy sources) and radiation energy (keV, MeV)
- Readout system includes wavelength range of light source (laser, LED, white light, red–green–blue colors), wavelength range that the light detector is sensitive to, film orientation on scanner bed, transmission versus reflection mode, readout system artifacts and correction, output signal resolution (e.g., 8, 12, or 16 bit depth)
- Temporal factor such as time between readout and irradiation

Some of these are discussed in respective chapters in this book.

2.7 CHARACTERISTICS OF THE CURRENTLY AVAILABLE RADIOCHROMIC FILMS

It is reported that early RCFs (HD-810, MD-55) could be damaged when exposed to temperatures >60°C before exposure and that such treatment of an exposed film results in color change from blue to red [62]. Although the newer RCFs are more robust, the manufacturer recommends storing exposed and unexposed films at room-ambient temperature (~22°C), or less, and not to exposing them to temperatures above 60°C. The best practice is to keep the film refrigerated. As shown in Figure 2.12 and Tables 2.2 and 2.3, in the majority of available RCFs the active layer is sandwiched between two polyester substrates to make the films rugged and protected from mechanical damage. This construction means that water can only diffuse into the active layer at the edges of the film. The rate of water penetration in EBT3 film is relatively small (<1 mm/h). Other film models behave similarly. This means that most films can be used in a water tank or water phantom for short periods of time. So long as the film is not damaged mechanically, the changes caused by water uptake are reversible and even the wet areas remain usable after drying [63–66].

Table 2.2 Configuration, active layer, Substrate 1, Substrate 2, adhesive layer, sizes, and dose range of the available radiochromic films for radiation therapy dosimetry

| Film model | Active layer | | | | Substrate 1 | | Substrate 2 | | Adhesive layer | Sizes | Dose range | Comments |
	Configuration	Nominal thickness, (μm)	Marker dye	Alumina	Type	Nominal thickness, (μm)	Type	Nominal thickness (μm)	Nominal thickness (μm)			
HD-V3	Type 1	12	Yes	Yes	Clear transparent polyester	97	–	–	–	8″ × 10″	10–1000 Gy	
MD-V3	Type 2	10	Yes	Yes	Clear transparent polyester	125	Clear transparent polyester	50	7	5″ × 5″	1–100 Gy	Available in special shapes/ sizes to fit CyberKnife®, Lucy, and other phantoms
EBT2	Type 2	28	Yes	Yes	Clear transparent polyester	175	Clear transparent polyester	50	20	8″ × 10″ 12.8″ × 17″	0.01–40 Gy	Available in special shapes/ sizes to fit CyberKnife, Lucy, and other phantoms
EBT3	Type 3	28	Yes	Yes	Clear transparent polyester	125	Clear transparent polyester	125		8″ × 10″ 12.8″ × 17″	0.01–40 Gy	Available in special shapes/ sizes to fit CyberKnife, Lucy, and other phantoms
EBT3, unlaminated	Type 1	14	Yes	Yes	Clear transparent polyester	125	–	–		8″ × 10″	0.02–20 Gy	

(Continued)

Table 2.2 (Continued) Configuration, active layer, Substrate 1, Substrate 2, adhesive layer, sizes, and dose range of the available radiochromic films for radiation therapy dosimetry

Film model	Configuration	Active layer			Substrate 1		Substrate 2		Adhesive layer	Sizes	Dose range	Comments
		Nominal thickness, (µm)	Marker dye	Alumina	Type	Nominal thickness, (µm)	Type	Nominal thickness (µm)	Nominal thickness (µm)			
EBT3F	Type 3	28	Yes	Yes	Clear transparent polyester	125	Clear transparent polyester	125		8″ × 10″	0.01–40 Gy	Pre-cut with fiducial marks to fit axes of Linac/treatment system
EBT3P	Type 3	28	Yes	Yes	Clear transparent polyester	125	Clear transparent polyester	125		8″ × 10″	0.01–40 Gy	Pre-cut to fit GAFchromic Quick Phantom
EBT3+	Type 3	28	Yes	Yes	Clear transparent polyester	125	Clear transparent polyester	125		8″ × 11″	0.01–40 Gy	With removable 1.5″×8″ reference strip
EBT3+P	Type 3	28	Yes	Yes	Clear transparent polyester	125	Clear transparent polyester	125		8″ × 11″	0.01–40 Gy	Pre-cut to fit GAFchromic Quick Phantom, with removable 1.5″×8″ reference strip
EBT-XD	Type 3	25	Yes	Yes	Clear transparent polyester	125	Clear transparent polyester	125		8″ × 10″	0.04–40 Gy	
RTQA-2	Type 2	17	Yes	Yes	Opaque, white polyester	97	Yellow transparent polyester	97		10″ × 10″ 12.8″ × 17″ 1.25″ × 11″	0.02–8 Gy	Available in special shapes/ sizes to fit CyberKnife, Lucy, and other phantoms

Table 2.3 Configuration, active layer, Substrate 1, Substrate 2, adhesive layer, sizes, and dose range of the available radiochromic films for diagnostic radiology applications

Film model	Configuration	Active layer			Substrate 1		Substrate 2		Adhesive layer	Sizes	Dose range	Comments
		Nominal thickness, (μm)	Marker dye	Alumina	Type	Nominal thickness, (μm)	Type	Nominal thickness, (μm)	Nominal thickness, (μm)			
XR-RV3	Type 2	17	Yes	No	Opaque, white polyester	97	Yellow transparent polyester	97	20	12.8″ × 17″	0.05–15 Gy	
XR-QA2	Type 2	25	No	Yes	Opaque, white polyester	97	Yellow transparent polyester	97	20	10″ × 12″	0.1–20 cGy	
XR-CT2	Type 2	25	No	Yes	Opaque, white polyester	97	Yellow transparent polyester	97	20	1.75 × 12cm^2	0.1–20 cGy	
XR-M2	Type 2	25	No	Yes	Opaque, white polyester	97	Yellow transparent polyester	97	20	1.06″ × 3.06″	0.1–20 cGy	

Source: Courtesy of Azam Niroomand-Rad.

In the current section, characteristics of the currently available radiochromic films are tabulated in Tables 2.2–2.5 for radiotherapy and diagnostic dosimetry as described here. Table 2.2 provides configuration, active layer, substrates, adhesive layer, sizes, and dose range of the available radiochromic films that are used for radiation therapy dosimetry. As shown, Table 2.2 [67] contains important information about the useful dose ranges indicating that the least sensitive, HD-V2, is useful for the 10–1000 Gy dose range, whereas the MD-V3 model has intermediate sensitivity, useful for the dose range of 1–100 Gy and the EBT2 and EBT3 families are useful for most clinical radiation therapy applications with dose ranges 0.01–40 Gy. It should be pointed out that the EBT-XD films with useful dose range of 0.04–40 Gy are designed for single fraction SRS and SBRT applications. However, for the dose range of interest, if there is a choice of film models, consider the RCF contrast knowing that the slope of the response curve decreases as dose increases. Therefore, for equal doses the RCF with the lesser sensitivity will have higher contrast resulting in lower dose uncertainty [49]. Thus for doses >10 Gy, EBT-XD would be preferred over EBT2 or EBT3. But when measuring nonpenetrating radiation, choose a Type 1 RCF such as HD-V2 or unlaminated EBT3 where the active layer is at the surface and expose the film from that side to avoid attenuation by the polyester substrate.

Moreover, Table 2.3 provides configuration, active layer, substrates, adhesive layer, size, and dose range of the available radiochromic films XR-RV3, XR-QA2, XR-CT2, and XR-M2 that are used for diagnostic radiology applications. As shown in Table 2.3, these films are manufactured in layer configurations, consisting of one or more polyester substrates, a layer containing the active component (active layer, also called emulsion layer) and in some film models they are called adhesive layer.

Note that for most RCFs, sheets or smaller pieces are unlikely to be perfectly flat and this can lead to substantial uncertainty in the measured response values if the film is not coplanar with the glass window in a flatbed scanner. A simple solution is to use a piece of 3–4 mm thick clear glass, the same size as the scanner's window to flatten the film. The reading issue is addressed in detail in Chapter 4.

Furthermore, Table 2.4 provides overall elemental compositions and effective atomic numbers (Z_{eff}) of HD-V2, MD-V3, EBT2, EBT3, unlaminated EBT3, EBT-XD, RTQA2, and XR-QA2 radiochromic films. Please note that the composition values are in atom (%) and not by weight (%).

Table 2.4 Overall elemental compositions, atom (%), and their effective atomic numbers (Z_{eff}) of HD-V2, MD-V3, EBT2, EBT3, EBT3 unlaminated, EBT-XD, RTQA2, and XR-QA2 radiochromic films

Film model	Composition by element and atom (%)											
	H	Li	C	N	O	Na	Al	S	Cl	Ba	Bi	Z_{eff}
HD-V2	38.8	0.1	43.5	0.0	17.5	0.1	0.0	0.0	0.1			6.73
MD-V3	38.3	0.0	43.9	0.0	17.7	0.0	0.0	0.0	0.0			6.68
EBT2	40.0	0.1	42.7	0.0	17.1	0.0	0.2	0.0	0.0			6.70
EBT3	38.4	0.1	43.7	0.0	17.7	0.0	0.2	0.0	0.0			6.71
EBT3 unlaminated	38.4	0.1	43.7	0.0	17.7	0.0	0.2	0.0	0.0			6.71
EBR-XD	38.3	0.1	44.0	0.0	17.6	0.0	0.1	0.0	0.0			6.70
RTQA2	42.1	0.0	38.2	0.0	18.5	0.0	0.1	0.5	0.0	0.5		22.71
XR-QA2	40.6	0.1	39.8	0.2	18.1	0.0	0.0	0.5	0.0	0.5	0.2	29.98

Source: Courtesy of Azam Niroomand-Rad.

Table 2.5 Specific elemental compositions, atom (%), and effective atomic numbers (Z_{eff}) of the active layer of HD-V2, MD-V3, EBT2, EBT3, EBT3 unlaminated, EBT-XD, RTQA2, and XR-QA2 radiochromic films

Film model	Composition by element and atom (%)											
	H	Li	C	N	O	Na	Al	S	Cl	Ba	Bi	Z_{eff}
HD-V2	58.2	0.6	27.7	0.4	11.7	0.5	0.3	0.1	0.6			7.63
MD-V3	58.2	0.6	27.7	0.4	11.7	0.5	0.3	0.1	0.6			7.63
EBT2	56.5	0.6	27.4	0.3	13.3	0.1	1.6	0.1	0.1			7.46
EBT3	56.5	0.6	27.4	0.3	13.3	0.1	1.6	0.1	0.1			7.46
EBT3 unlaminated	56.5	0.6	27.4	0.3	13.3	0.1	1.6	0.1	0.1			7.46
EBR-XD	57.0	0.6	28.5	0.4	11.7	1.0	1.5	0.1	0.1			7.42
RTQA2	56.5	0.6	27.4	0.3	13.3	0.1	1.6	0.1	0.1			7.46
XR-QA2	56.2	1.0	27.6	1.6	11.7	0.0			0.1		1.7	55.23

Source: Courtesy of Azam Niroomand-Rad.

Likewise, Table 2.5 enlists specific elemental compositions, atom (%), and effective atomic numbers (Z_{eff}) of the active layer of HD-V2, MD-V3, EBT2, EBT3, unlaminated EBT3, EBT-XD, RTQA2, and XR-QA2 radiochromic films that are listed in Table 2.3. Please note that in Table 2.5 the composition values are in atom (%) and not by weight (%). Also note that all RCFs produced in November 2011 or later contain alumina [68]. These films are all transparent and yellow green in color, except for RTQA-2, which is opaque white on one face and orange on the opposite face. The EBT2 film has smooth polyester substrates, whereas EBT-XD and all members of the EBT3 product family use polyester with a special surface treatment containing microscopic silica particles that prevent the formation of Newton's Rings interference patterns when films are in close proximity to another reflective surface as in a flatbed scanner. All members of the EBT3 family have the same active layer and substrates but differ in sheet size and other features. Some are precut (EBT3F, EBT3P, EBT3+P), with fiducial marks (EBT3F), or with a removable perforated reference strip (EBT3+, EBT3+P) as described in Table 2.5.

2.8 RECENT RADIOCHROMIC FILMS FOR RADIOLOGY WITH kV PHOTONS

Although dosimetry with radiochromic films plays a significant role in therapeutic applications, another important area of use includes medical diagnostics and radiology. In general, radiochromic films lack the sensitivity required for accurate dosimetry measurement in diagnostic procedures but have been found to be a valuable asset for quality assurance tasks in the kilovoltage range. These films have included the incorporation of a high atomic number material into their construction to increase the sensitivity in the kilovoltage range providing an effective dose measurement range down to 0.1 cGy. Some other specific usages of films are GAFchromic XR-CT2 film [69] designed for measuring radiation beam slice width and beam position alignment on CT scanners. On the other hand the GAFchromic XR-M2 film is specifically designed

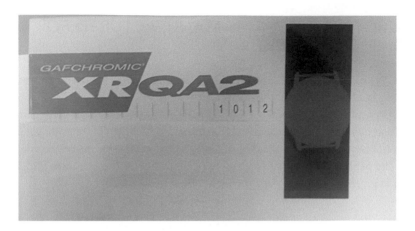

Figure 2.13 GAFchromic XRQA2 film. (Courtesy of Martin Butson.)

for mammography quality assurance testing. Using strips of XR-M2 film, the location of the light field and the radiation field, plus the position of the detector can be determined.

The GAFchromic XR-QA2 film (Figure 2.13) is designed specifically as a quality assurance tool for radiology applications in a processor-less environment. The XR-QA2 film may also be used as a dosimeter [70,71]. These are designed to be used in the energy range of 20–200 kVp and have a dynamic dose range of 0.1–20 cGy.

Finally, the GAFchromic XR-RV3 film is designed for surface-peak skin-dose measurement in interventional procedures guided by fluoroscopy [72,73]. The dose measurement range is from 5 cGy up to 15 Gy and it is designed for use in the energy range of 30 keV–30 MeV.

2.9 RADIOCHROMIC FILMS FOR NICHE APPLICATIONS

Ashland Inc. has produced many film types to cover both radiotherapy and diagnostic applications as mentioned previously. They have also developed some unique niche application products with radiochromic films. These are designed with one specific application in mind and are mentioned briefly in the following.

The Rad-Sure blood irradiation indicators provide positive, visual verification of irradiation. They can be used for blood irradiation associated with transfusion-associated graft-versus-host disease. When attached to blood products, Rad-Sure XR 15 Gy, XR 25 Gy, 15 Gy, and 25 Gy indicators show whether the blood products have been irradiated or not. Before a blood product and its attached indicator are irradiated, the indicator reads, *NOT IRRADIATED*. After the blood product and its attached indicator are irradiated, the word *NOT* in the indicator window is obscured and the indicator reads *IRRADIATED*.

The Sterin insect irradiation indicators were developed for the sterile insect technique program to provide positive, visual verification of irradiation at or above the printed minimum indicated dose of 70 Gy, 100 Gy, 125 Gy, or 145 Gy. As with the Rad-Sure product the indicator labels/radiochromic film changes from *NOT IRRADIATED* to *IRRADIATED*. However they should not be used to determine an actual applied dose level.

2.10 RADIOCHROMIC FILMS MANUFACTURED FOR COMMERCIAL USE

Besides the RCFs that are available from GAF/ISP/Ashland there have been a few other vendors of RCFs. Following is a brief outline of these vendors.

The GEX Corporation (Centennial, CO, USA) markets a series of dosimeters named the B3 film developed at Risø Laboratory in Denmark [74] and the product details are available in the reference [75]. In addition, Patel at JP Laboratories (Middlesex, NJ) was active for many years developing RCFs. This included the SIFID detector that has also been described in the literature [76]. A summary of these products is given in Sections 2.10.1 through 2.10.4.

2.10.1 RADIOCHROMIC FILMS MANUFACTURED BY FAR WEST TECHNOLOGY, INC.

One of the oldest RCF manufacturers, which are still available, is from Far West Technology (Goleta, CA). This media is based on hydrophobic substituted triphenylmethane leucocyanides and is most commonly used for high-dose applications such as radiation processing, food irradiation, and sterilization; it lacks the sensitivity for any medical applications [77]. Far West Technology produces the FWT-60 films (Figure 2.14) that are based on hydrophobic substituted triphenylmethane leucocyanides and are used for high-dose applications such as radiation processing, food irradiation, and sterilization, with a dose range of 0.5–200 kGy. These films have limited use in radiotherapy applications. The substrate that holds the dye is nylon and the film has a density of approximately 1.15 g/cm^3 and an atomic composition of (C: 63.7%, N: 12.0%, H: 9.5%, and O: 14.8%). The physical thickness of this is approximately 0.05 mm.

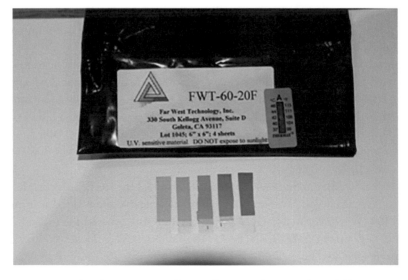

Figure 2.14 Far West Technology FWT-60 film package and cut irradiated film. (Courtesy of Martin Butson.)

Figure 2.15 B3 radiochromic film by GEX Corporation. (Courtesy of Martin Butson.)

2.10.2 RADIOCHROMIC FILMS PRODUCED BY GEX CORPORATION

GEX Corporation produces the B3 films shown in Figure 2.15 that are thin and flexible polymeric film consisting of a pararosaniline radiochromic dye embedded in a polyvinyl butyral matrix [78]. An ionizing radiation event activates the B3 dye centers, which in turn causes the B3 film to undergo a predictable color change from clear to deepening shades of pink. The amount of color change is proportional to dose and is influenced by the temperature during irradiation. The postirradiation color change can take several hours to stabilize and heat treatment of the film is recommended. The dose range of B3 is approximately 1–150 kGy, again well above the clinically useful dose range in radiation oncology.

2.10.3 RADIOCHROMIC FILMS PRODUCED BY JP LABORATORIES

JP Laboratories (Middlesex, NJ) has developed a range of radiochromic films, called SIRAD (self-indicating instant radiation alert dosimeters) as shown in Figure 2.16 and described in detail in References [79–81]. These films consist of a diacetylene monomer in a polymeric binder

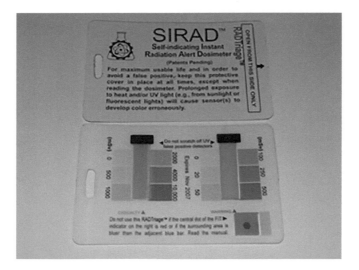

Figure 2.16 SIRAD radiochromic film/badge. (Courtesy of Martin Butson.)

coated on polyester base and laminated with an outer polyester-layer film. The SIRAD films are used for personal dose monitoring to record high-dose unintentional exposure such as in reactor or bomb.

2.10.4 RADIOCHROMIC FILMS PRODUCED FOR ULTRAVIOLET RADIATION EXPOSURE

The polysulphone films are manufactured by various institutions and have been available in limited supply commercially for many years as described in References [82,83]. Recently they have been available by request from the Queensland University of Technology (Brisbane, QLD, Australia). They are specifically designed for measurement of ultraviolet radiation, which is normally due to solar exposure. Their major use comprises human personal UV skin exposure. This is due to the films action spectrum, which relates to the sensitivity response of the film to ultraviolet radiation wavelengths, matching closely to human erythema response to UV exposure [84].

2.11 SUMMARY

Radiochromic film has become a standard dosimetry media for radiotherapy applications as well as becoming a useful tool in radiology and other niche procedures. The number of research and clinical publications associated with the examination or use of radiochromic films in radiotherapy has grown significantly over the years and is shown in Figure 2.17. These results have been compiled using SCOPUS and the search terms of radiochromic and radiotherapy. Although this is an indication of radiochromic film's use, but the increase shows a trend that radiochromic films are becoming an essential tool for medical dosimetry.

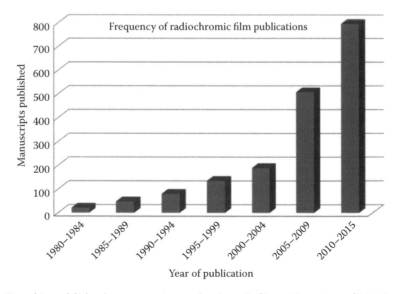

Figure 2.17 Trend in published papers using radiochromic films. (Courtesy of Martin Butson.)

REFERENCES

1. Pai S, Das IJ, Dempsey JF et al. TG-69: Radiographic film for megavoltage beam dosimetry. *Med Phys* 2007;34:2228–2258.
2. Kosar J. *Light-Sensitive Systems*. New York: Wiley; 1965.
3. Dorion GH, Wiebe AF. *Photochromism*. London, UK: The Focal Press; 1970.
4. Brown GH. *Photochromism*. New York: Wiley-Interscience; 1971.
5. Gernsheim H. The 150th anniversary of photography. *History of Photography* 1977;1:3–8.
6. McLaughlin WL. Film, dyes, and photographic systems. In: Holm NW, Berry RJ (Eds.). *Manual on Radiation Dosimetry*. New York: Dekker; 1970. Chapter 6, pp. 129–177.
7. Rust JB, Margerum JD, Miller LJ. Light-scattering imaging by photopolymerization. In: Murray RD (Ed.). *Novel Imaging Systems, Proceedings of Seminar*, Washington, DC. Boston, MA: Society for Imaging Science and Technology; 1969. Chapter 14, pp. 173–186.
8. Lawton WR. Recent advances in organic-based imaging systems. In: Murray RD (Ed.). *Novel Imaging Systems, Proceedings of Seminar*, Washington, DC. Boston, MA: Society for Imaging Science and Technology; 1969 Chapter 6, pp. 63–77.
9. Arney JS. Kinetic and mechanical descriptions of the microencapsulated acrylate imaging process. *J Imaging Sci* 1989;33:1–6.
10. Margerum JD, Miller LJ. Photochromic processes by todtomerism. In: Brown GH (Ed.). *Photochromism, Techniques of Chemistry*. New York, Vol. III, 1971. Chapter 6, pp. 557–632.
11. Friedrich W, Knipping P. Die Naturwissenschaften. *Heft* 1922;16.
12. Mattson AM, Jensen CO, Dutcher RA. Triphenyltetrazolium chloride as a dye for vital tissue. *Science* 1947;106:294–296.
13. Strodss FH, Cherionis ND, Strodss E. Demonstration of reducing enzyme systems in neoplasms and living mammalian tissues by triphenyltetrazolium chloride. *Science* 1948;108:113–115.
14. Altman FP. Review: Tetrazolium salts and formazans. *Prog Histochem Cytochem* 1976;51–88.
15. Zweig JI, Herz ML, McLaughlin WL, Bhuthimethee V. Drug activation by gamma irradiation: A new direction for molecular design. In-vitro and in-vivo studies of a substituted polyaminoaryl nitrile. *Cancer Treat Rep* 1987;61:419–423.
16. Gierlach ZS, Krebs AT. Radiation effects on 2,3,5-triphenyltetrazolium chloride solutions. *Am J Roentgenol Radium Ther* 1949;62:559–563.
17. Krebs AT. Dosimeters: Recent developments. In: Clark GL (Ed.). *The Encyclopedia of X-Rays and Gamma Rays*. New York: Reinhold; 1963. pp. 274–276.
18. McLaughlin WL, Chalkley L. Measurement of radiation dose distributions with photochromic materials. *Radiology* 1965;84(1):124–125.
19. McLaughlin WL. Novel radiation dosimetry systems. In: *High Dose Dosimetry for Radiation Processing, Proceedings of International Symposium*, Vienna, 1990, STI/PUB/846. Vienna, Austria: International Atomic Energy Agency; 1991. pp. 3–27.
20. Chalkley L. Photometric papers sensitive only to short wave ultraviolet. *J Opt Soc Am* 1952;42:387–392.
21. Tsibouklis J, Pearson C, Song YP, Warren J, Petty M, Yarwood J, Petty MC, Feast JW. *J Mater Chem* 1993;3(1):97–104.
22. Watanabe Y, Patel GN, Patel P. Evaluation of a new self-developing instant film for imaging and dosimetry. *Radiat Prot Dosim* 2006;120:121–124.
23. McLaughlin WL, Humphreys JC, Farahani M, Miller A. The measurement of high doses near metal and ceramic interfaces. In: *High Dose Dosimetry, Proceedings of International Symposium*, Vienna, 1984, STI/PUB/671. Vienna, Austria: International Atomic Energy Agency; 1985. pp. 109–133.
24. McLaughlin WL. Radiochromic dye-cyanide dosimeters. *Radiology* 1965;84:377–385.

25. McLaughlin WL, Chalkley L. Low atomic-number dye systems for ionizing radiation measurement. *Phon Sci Eng* 1965;9:159–166.
26. Lewis DF. A processless electron recording medium. In: *Electronic Imaging, Proceedings of SPSE'86 Symposium,* Arlington, VA, 1986: pp. 76–79.
27. McLaughlin WL, Yun-Dong C, Soares CG, Miller A, Van Dyk G, Lewis DF. Sensitometry of the response of a new radiochromic film dosimeter to gamma radiation and electron beams. *Nucl Instrum Methods Phys Res* 1991;A302:165–176.
28. Muench PJ, Meigooni AS, Nath R, McLaughlin WL. Photon energy dependence of the sensitivity of radiochromic film and comparison with silver halide film and LiF TLDs used for brachytherapy dosimetry. *Med Phys* 1991;18(4):769–775.
29. McLaughlin WL, Al-Sheikhly M, Lewis DF, Kovács A, Wojnárovits L. Irradiation of polymers. *Am Chem Soc* 1996;620:152–166.
30. McLaughlin WL, Puhl JM, Al-Sheikhly M, Christou CA, Miller A, Kovács A, Wojnarovits L, Lewis DF. Novel radiochromic films for clinical dosimetry. *Radiat Prot Dosim* 1996;66:263–268.
31. Soares CG, Darley PJ, Charles MW, Baum JW. Hot particle dosimetry using extrapolation chambers and radiochromic foils. *Radiat Prot Dosim* 1991;39(1–3):55–59.
32. McWilliams FF, Scannell MJ, Soares CG, Coursey BM, Chabot GE. Hot particle dosimetry using ^{60}Co spheres. *Radiat Prot Dosim* 1992;40:223–234.
33. Soares CG. Calibration of ophthalmic applicators at NIST: A revised approach. *Med Phys* 1991;18:787–793.
34. McLaughlin WL, Soares CG, Sayeg JA, McCullough EC, Kline RW, Wu A, Maitz AH. The use of a radiochromic detector for the determination of stereotactic radiosurgery dose characteristics. *Med Phys* 1994;21:379–388.
35. Uribe RM, Vargas-Aburto C, McLaughlin WL, Walker ML, Dick CE. Electron and Proton Dosimetry with Custom-Developed Radiochromic Dye Films. *Radiat Prot Dosim* 1993;47:693–696.
36. Buenfil-Burgos AE, Uribe RM, de la Piedad A, McLaughlin WL, Miller A. Thin plastic radiochromic dye films as ionizing radiation dosimeters. *Radiat Phys Chem* 1977;22:325–332.
37. McLaughlin WL, Al-Sheikhly M, Lewis DF, Kovács A, Wojnárovits L. A radiochromic solid-state polymerization reaction. In: Clough RL (Ed.). *Polymer Preprints 35, Proceedings of the Symposium on Radiation Effects on Polymers,* Washington, DC (unpublished). pp. 920–921.
38. McLaughlin WL, Al-Sheikhly M, Lewis DF, Kovács A, Wojnárovits L. Radiochromic solid-state polymerization reaction. In: Clough RL, Shalaby SW (Eds.). *Irradiation of Polymers.* Washington, DC: American Chemical Society; 1996. pp. 152–166.
39. McLaughlin WL, Puhl JM, Al-Sheikhly M, Christou CA, Miller A, Kovács A, Wojnárovits L, Lewis DF. Novel radiochromic films for clinical dosimetry. In: Peto A, Uchrin G (Eds.). *Proceedings of the 11th International Conference on Solid State Dosimetry, II.* Budapest, Hungary, July 1995.
40. Zhu Y, Kirov AS, Mishra V, Meigooni AS, Williamson JF. Quantitative evaluation of radiochromic film response for two-dimensional dosimetry. *Med Phys* 1997;24:223–231.
41. Niroomand-Rad A, Blackwell CR, Coursey BM et al. Radiochromic film dosimetry: recommendations of AAPM Radiation Therapy Committee Task Group 55. American Association of Physicists in Medicine. *Med Phys* 1998;25:2093–2115.
42. Rink A, Lewis DF, Varma S, Vitkin IA, Jaffray DA. Temperature and hydration effects on absorbance spectra and radiation sensitivity of a radiochromic medium. *Med Phys* 2008;35:4545–4555.
43. Rink A, Vitkin IA, Jaffray DA. Suitability of radiochromic medium for real-time optical measurements of ionizing radiation dose. *Med Phys* 2005;32:1140–1155.
44. Lewis DF, Moskowitz ML, Purdy SE. US Patent No. 5002852 (March 26, 1991).
45. Lewis DF, Schenfele RD, Winkler T. US Patent No. 4985290 (January 15, 1991).

46. http://www.filmqapro.com/Documents/GafChromic_EBT-2_20101007.pdf, accessed January 2016.

47. Chan MF, Chiu-Tsao S, Li J, Schupak K, Parhar P, Burman C. Confirmation of skin doses resulting from bolus effect of intervening alpha cradle and carbon fiber couch in radiotherapy. *Technol Cancer Res Treat* 2012;11:571–581.

48. Lewis DF, Micke A, Yu X, Chan MF. An efficient protocol for radiochromic film dosimetry combining calibration and measurement in a single scan. *Med Phys* 2012;39:6339–6350.

49. https://www.aapm.org/meetings/09SS/documents/23Soares-RadiochromicFilm.pdf, accessed January 2016.

50. Butson MJ, Yu PKN, Metcalfe PE. Effects of read-out light sources and ambient light on radiochromic film. *Phys Med Biol* 1998;43:2407.

51. Butson MJ, Yu PKN, Cheung T, Metcalfe P. Radiochromic film for medical radiation dosimetry. *Mater Sci Eng R* 2003;41:61–120.

52. Niroomand-Rad A, Chiu-Tsao ST, Soares CG, Meigooni AS, Kirov AS. Comparison of uniformity of dose response of double layer radiochromic films (MD-55-2) measured at 5 institutions. *Phys Med* 2005;21:15–21.

53. Soares CG. History of personal dosimetry performance testing in the United States. *Radiat Prot Dosim* 2007;125:1–4.

54. Devic S. Radiochromic film dosimetry: Past, present, and future. *Phys Med* 2011;27:122–134.

55. Desroches J, Bouchard H, Lacroix F. Potential errors in optical density measurements due to scanning side in EBT and EBT2 Gafchromic film dosimetry. *Med Phys* 2010;37:1565–1570.

56. Grams MP, Gustafson JM, Long KM, de los Santos LE. Technical note: Initial characterization of the new EBT-XD Gafchromic film. *Med Phys* 2015;42:5782–5786.

57. Palmer AL, Dimitriadis A, Nisbet A, Clark CH. Evaluation of Gafchromic EBT-XD film, with comparison to EBT3 film, and application in high dose radiotherapy verification. *Phys Med Biol* 2015;60:8741–8752.

58. http://www.ashland.com/Ashland/Static/Documents/ASI/Advanced%20Materials/EBTXD_Specifications.pdf.

59. Wegner G, Hatfield WE. *Molecular Metals*. New York: Plenum Press; 1979. pp. 209–242.

60. Lewis D, Chan MF. Correcting lateral response artifacts from flatbed scanners for radiochromic film dosimetry. *Med Phys* 2015;42:416–429.

61. Callier A. Absorption and scatter of light by photographic negatives. *J Phot* 1909;33:200.

62. Koulouklidis AD, Cohen S, Kalef-Ezra J. Thermochromic phase-transitions of GafChromic films studied by z-scan and temperature-dependent absorbance measurements. *Med Phys* 2013;40:112701.

63. Butson MJ, Cheung T, Yu PK. Radiochromic film dosimetry in water phantoms. *Phys Med Biol* 2001;46:N27–31.

64. van Battum LJ, Hoffmans D, Piersma H, Heukelom S. Accurate dosimetry with GafChromic EBT film of a 6 MV photon beam in water: What level is achievable? *Med Phys* 2008;35:704–716.

65. Aldelaijan S, Devic S, Mohammed H, Tomic N, Liang LH, DeBlois F, Seuntjens J. Evaluation of EBT-2 model GAFCHROMIC film performance in water. *Med Phys* 2010;37:3687–3693.

66. Arjomandy B, Tailor R, Zhao L, Devic S. EBT2 film as a depth-dose measurement tool for radiotherapy beams over a wide range of energies and modalities. *Med Phys* 2012;39:912–921.

67. http://www.ashland.com/Ashland/Static/Documents/ASI/Advanced%20Materials/EBTXD_Specifications.pdf, accessed November 20, 2015.

68. David Lewis—Private communication.

69. Butson MJ, Cheung T, Yu PK. Measurement of energy dependence for XRCT radiochromic film. *Med Phys* 2006;33:2923–2925.

70. Pereira J, Sousa MJ, Cunha L et al. Gafchromic XR-QA2 film as a complementary dosimeter for hand monitoring in CTF-guided biopsies. *J Appl Clin Med Phys* 2016;17:5725.

71. Di Lillo F, Mettivier G, Sarno A, Tromba G, Tomic N, Devic S, Russo P. Energy dependent calibration of XR-QA2 radiochromic film with monochromatic and polychromatic X-ray beams. *Med Phys* 2016;43:583–588.
72. Farah J, Trianni A, Ciraj-Bjelac O et al. Characterization of XR-RV3 Gafchromic(®) films in standard laboratory and in clinical conditions and means to evaluate uncertainties and reduce errors. *Med Phys* 2015;42:4211–4226.
73. Bordier C, Klausz R, Desponds L. Patient dose map indications on interventional X-ray systems and validation with Gafchromic XR-RV3 film. *Radiat Prot Dosim* 2015;163:306–318.
74. Miller A, Batsberg W, Karman W. A new radiochromic thin-film dosimeter system. *Radiat Phys Chem* 1988;31:491–496.
75. http://www.gexcorp.com/pro-b3-radiochromic.php.
76. Watanabe Y, Patel GN, Patel P. Evaluation of a new self-developing instant film for imaging and dosimetry. *Radiat Prot Dosim* 2006;120:121–124.
77. http://www.fwt.com/racm/fwt60ds.htm, accessed November 20, 2015.
78. Image from http://www.sunplume.com/GEX/.
79. Butson MJ, Cheung T, Yu PK. Visible absorption spectra of radiation exposed SIRAD dosimeters. *Phys Med Biol* 2006;51:N417–N421.
80. Riel GK, Winters P, Patel G, Patel P. Self-indicating radiation alert dosimeter (SIRAD). *Radiat Prot Dosim* 2006;120:259–262.
81. Cheung T, Butson MJ, Yu PK. X-ray energy dependence of the dose response of SIRAD radiation dosimeters. *Appl Radiat Isot* 2007;65:814–817.
82. Parisi AV, Kimlin MG. Personal solar UV exposure measurements employing modified polysulphone with an extended dynamic range. *Photochem Photobiol* 2004;79:411–415.
83. Siani AM, Casale GR, Modesti S, Parisi AV, Colosimo A. Investigation on the capability of polysulphone for measuring biologically effective solar UV exposures. *Photochem Photobiol Sci* 2014;13:521–530.
84. Olds W. Elucidating the links between UV radiation and vitamin D synthesis using an in vitro model. PhD Thesis, Queensland University of Technology, Queensland, Australia, 2010.

Physics and characteristics of radiochromic films

MARTIN BUTSON, GWI CHO, SIMRAN GILL, AND DANE POPE

3.1 INTRODUCTION

In the current chapter, we will briefly discuss some of the physical characteristics of radiochromic films (RCFs). This chapter is intended to be an overview of some of the main principles and properties of film for imaging and dosimetry purposes. Other chapters will deal with these properties in more detail as required. RCFs have a unique measurement niche in radiotherapy and diagnostic dosimetry due to their low-energy dependence, high spatial resolution, and automatic coloration properties. They can provide a detailed dose map over a two-dimensional area or if curved or curled can be used over a three-dimensional volume of analysis. Even within the RCF range, a large variety of properties of the films are displayed, and part of the medical physicists job is to decide which film will provide the optimal analysis response of the task at hand.

3.2 PRINCIPLE OF POLYMER REACTIONS

The materials in RCFs that are responsible for the coloration are known as crystalline polyacetylenes, in particular diacetylenes, and upon thermal annealing or radiation exposure, they undergo polymerization, turning blue or red depending on their specific composition [1,2]. The particular diacetylene monomer used in the external beam therapy (EBT) range of GAFchromic™ film is the lithium salt of pentacosa-10,12-diyonic acid [3]. More detailed information concerning RCF materials and polymer reactions is given in Chapter 2, with historical background, development, and construction of RCFs.

3.3 CHEMICAL FORMULATIONS/TISSUE EQUIVALENCE

The chemical formulation of RCFs has evolved over the years to match the clinical needs arising within the medical radiation environment. For dosimetric purposes, films require low, medium, and high sensitivities depending upon the need. For example, stereotactic radiosurgery techniques utilize large doses thus need less-sensitive film. The other extreme includes diagnostic procedures in which small radiation doses are delivered requiring as high a sensitivity as possible. To achieve these practical considerations for dosimetry, the chemical structure and thus tissue equivalence of various RCFs products are varied. Detailed discussion of specific RCF will be provided in respective chapters, for example, diagnostic kilovoltage—Chapter 5 and stereotactic radiosurgery—Chapter 9.

In summary, Table 3.1 shows the overall elemental composition of most currently available GAFchromic film products used in medical dosimetry applications.

Table 3.2 shows specific percent atomic compositions, and effective atomic numbers (Z_{eff}) of the active layer of the RCFs listed in Table 3.1. In this table, the composition values are in atom (%) and not by weight (%). It should also be noted that all GAFchromic film products produced after November 2011 contain aluminum [4]. More details on chemical composition and tissue equivalence are given in Chapter 2.

3.4 HANDLING TECHNIQUES AND STORAGE

This section provides a brief summary of handling and storage. Other chapters will deal with certain aspects of this in more detail.

Table 3.1 Elemental compositions, atom (%), and their effective atomic numbers (Z_{eff}) of HD-V2, MD-V3, EBT2, EBT3, EBT3 unlaminated, EBT-XD, RTQA2, and XR-QA2 radiochromic films

Film model	Composition by element and atom (%)											
	H	Li	C	N	O	Na	Al	S	Cl	Ba	Bi	Z_{eff}
HD-V2	38.8	0.1	43.5	0.0	17.5	0.1	0.0	0.0	0.1			6.73
MD-V3	38.3	0.0	43.9	0.0	17.7	0.0	0.0	0.0	0.0			6.68
EBT2	40.0	0.1	42.7	0.0	17.1	0.0	0.2	0.0	0.0			6.70
EBT3	38.4	0.1	43.7	0.0	17.7	0.0	0.2	0.0	0.0			6.71
EBT3 unlaminated	38.4	0.1	43.7	0.0	17.7	0.0	0.2	0.0	0.0			6.71
EBR-XD	38.3	0.1	44.0	0.0	17.6	0.0	0.1	0.0	0.0			6.70
RTQA2	42.1	0.0	38.2	0.0	18.5	0.0	0.1	0.5	0.0	0.5		22.71
XR-QA2	40.6	0.1	39.8	0.2	18.1	0.0	0.0	0.5	0.0	0.5	0.2	29.98

Source: Courtesy of Azam Niroomand-Rad.

Table 3.2 Specific elemental compositions, atom (%), and effective atomic numbers (Z_{eff}) of the ActiveLayer of HD-V2, MD-V3, EBT2, EBT3, EBT3 unlaminated, EBT-XD, RTQA2, and XR-QA2 radiochromic films

Film model	Composition by element and atom (%)											
	H	Li	C	N	O	Na	Al	S	Cl	Ba	Bi	Z_{eff}
HD-V2	58.2	0.6	27.7	0.4	11.7	0.5	0.3	0.1	0.6			7.63
MD-V3	58.2	0.6	27.7	0.4	11.7	0.5	0.3	0.1	0.6			7.63
EBT2	56.5	0.6	27.4	0.3	13.3	0.1	1.6	0.1	0.1			7.46
EBT3	56.5	0.6	27.4	0.3	13.3	0.1	1.6	0.1	0.1			7.46
EBT3 unlaminated	56.5	0.6	27.4	0.3	13.3	0.1	1.6	0.1	0.1			7.46
EBR-XD	57.0	0.6	28.5	0.4	11.7	1.0	1.5	0.1	0.1			7.42
RTQA2	56.5	0.6	27.4	0.3	13.3	0.1	1.6	0.1	0.1			7.46
XR-QA2	56.2	1.0	27.6	1.6	11.7	0.0			0.1		1.7	55.23

Source: Courtesy of Azam Niroomand-Rad.

3.4.1 BASIC HANDLING PROCEDURES

When dealing with any dosimetric medium that requires an optical analysis technique, certain procedures should always be followed to control the errors that can arise, caused by our handling of the dosimeter. Here, some simple yet practical details on using RCF are given. These types of rules are generally applicable to most RCF types.

First and foremost, it is a good practice not to handle films with bare hands as skin oils transferred to the films can cause uncertainties in optical-density measurements. Normal procedures in the clinic would be the use of cotton gloves, or even attaching small strips of paper to the sides of each film piece to minimize skin contact with the film. However, it should also be noted that GAFchromic film is quite robust and can be easily cleaned with the use of cleaning cloths and as such, finger prints can be removed without any scratching or damaging effects to the film before analysis. Because of this, various techniques can be adopted in a clinic as long as it suits your needs and removes any contamination from the film before optical analysis without damaging the film in the process.

3.4.2 RADIOCHROMIC FILM CUTTING

The preferred method of cutting GAFchromic film products is the use of clean sharp scissors. This is specifically to minimize the damage and breaking apart of the laminated layers of the film. Cuts should be made carefully with the scissors orthogonal to the film as this lessens damage to the films edges. Details concerning damage to film edges caused by cutting are given by Yu et al. in 2006 [5].

3.4.3 RADIOCHROMIC FILM IN WATER

A valuable aspect of most radiotherapy dosimetric RCFs (e.g., EBT3, and XD) is the ability to make measurements directly in water phantom. GAFchromic emulsions turn into milky color when wet, but when dried, the clear color is mostly restored. It should be noted that prolonged immersions will cause water to seep into the film emulsion at the cut edges to a greater degree than short (on the order of minutes) immersions. For this reason, care should be taken in making readings close to cut edges in films that have been irradiated in water. Butson et al. [6] examined these effects and found a water-penetration rate of approximately 0.5 mm per hour for GAFchromic EBT film when immersed in water.

3.4.4 TEMPERATURE DURING AND AFTER IRRADIATION

The time and temperature dependence are well documented during irradiation of GAFchromic film and during postirradiation and in 1998, Reinstein and Gluckman [7] provided a great reference document in this area. In summary, good practice dictates using controlled temperature to achieve the optimal level of accuracy. Extreme temperature makes RCF turn orange at 42°C and then red at above 50°C.

3.4.5 AMBIENT-LIGHT EXPOSURE

RCF products are relatively insensitive to ambient-light sources; however, they do exhibit some coloration caused by ultraviolet (UV) light exposure (see Chapter 20). As fluorescent lights contain a UV component to some degree as well as UV sources such as scattered solar light through windows and skylights, it is recommended to keep RCFs in a light-tight container at all times when they are not being exposed or analyzed. How long films can be left out in light conditions depends significantly on the level of UV contribution in the light sources being used. That is, some fluorescent lights have UV filters placed over them, whereas others do not. In most clinical situations, however, handling films for periods of 20 min to half an hour should cause negligible differences in measured absorbed dose. It is important however to make sure your calibration dose films receive the same amount of light exposure though, to the control film or 0 cGy film receives the same amount of coloration as all other calibration and experimental film pieces.

3.4.6 FILM STORAGE

RCF products are sensitive to UV-light exposure as well as excessive temperatures. As such, films should be stored in a dark, temperature-controlled environment to prolong their life and accuracy for radiation dosimetry. They should also be kept in an environment where they have pressure or movements applied to them. For example, if the films are kept in a cupboard drawer underneath a heavy object that is removed and replaced often, visible damage could occur to the film caused by the films rubbing together when the object is removed and replaced. Small scratches on the film can become a significant problem is film analysis is performed at a high spatial resolution or over a small area. As a common precaution, films should be kept in the original box in a temperature controlled, dust, smoke, and moisture free area, without any radiation and magnetic field environment.

3.5 ABSORPTION SPECTRA

Absorption spectrum is defined as the spectrum formed by electromagnetic radiation that has passed through a medium in which radiation of certain frequencies is absorbed [8]. Most RCFs developed and used today produce variations in their absorption spectra in the visible wavelengths when exposed to either ionizing or nonionizing radiation. As such, we can visually detect a change in color and thus a change in exposure. These films include but are not limited to RCFs produced by Ashland Inc (GAFchromic films, NJ), Far West Technology (FWT-60), GEX corporation, (B3), and SIRAD film badges. Other RCFs such as polysulphone films produce an absorption spectra change in the UV-wavelength range and thus the change is not visible but must be detected with a photo spectrometer.

The absorption spectrum of the RCF is primarily determined by the atomic and molecular composition of the film materials. Visible radiation is more likely to be absorbed at frequencies that match the energy difference between two quantum mechanical states of the molecules [9]. The absorption that occurs because of a transition between two states is referred to as an absorption line; however, a spectrum is typically composed of many lines. The frequencies in which absorption lines occur, as well as their relative intensities, primarily depend on the electronic and molecular structure of the film that depends on the composition as shown in Tables 3.1 and 3.2. The frequencies will also depend on the interactions between molecules in the sample, the crystal structure in solids, and on several environmental factors that can include but are not limited to temperature, pressure, and applied electromagnetic field. The lines will also have a width and shape that are primarily determined by the spectral density or the density of states of the system.

The readout process is dependent on the absorption of transmission of the light frequencies that have been described in Chapter 4. To analyze RCF absorption spectra, a photo spectrometer can be used [10]. Normally, these devices can be used to assess transmission-absorption spectra or reflected-absorption spectra. The type of spectra required will depend on the film type and how it is to be used. For example, EBT3 film is most commonly assessed with transmission measurement on a desktop scanner. Thus, the transmission absorption spectra will provide the most accurate representation. Whereas, RTQA film has an opaque backing material on it and thus a reflective-absorption spectra would provide the most beneficial information. As will be seen in the representative-absorption spectra given in the following, the sensitivity of the film at different wavelengths of light is significantly different. A desktop scanner provides broad spectrum-absorption spectra information that is normally divided into three main wavelength groups. Those cover the red, green, and blue components of the visible spectrum. These types of measurement will be dealt with in more detail in Chapter 4. It is suffice to say that the sensitivity or dose response of a RCF product can be modified to suit an application depending on which wavelength of light is used for analysis.

Figure 3.1 shows some representative absorption spectra of various types of RCFs, highlighting the spectral response and sensitivity response of these films to photon radiation. In each case, the peak wavelengths as well as general absorption properties are defined by the active radiation-sensitive chemicals and additives within the film that have been optimized to their specific application [11]. It shows the absorption spectra for MD-55-2 film. Absorption peaks are located at approximately 675 and 635 nm. These peaks vary in position with temperature and dose and will be dealt with in more detail in other chapters.

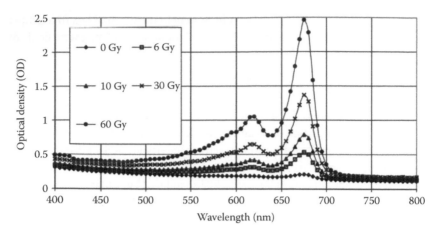

Figure 3.1 GAFchromic absorption spectra of MD-55-2 film exposed to various dosages. Note that spectra remain the same and only intensity increases with dose. (Adopted from Butson, M.J. et al., *Mater. Sci. Eng. R.*, 41, 61–120, 2003.)

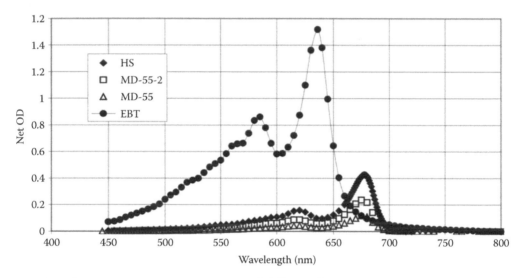

Figure 3.2 Absorption spectra comparison of different films; HS, MD-55, MD-55-2, and EBT. (Adopted from Butson, M.J. et al., *Phys. Med. Biol.*, 50, N135–N140, 2005.)

By comparison, the absorption spectra for MD-55-1, MD-55-2, HS, and EBT film are shown in Figure 3.2 [12]. Of note are the same characteristic absorption peaks for MD-55-1, MD-55-2, and HS but with different sensitivities, whereas the active chemicals in EBT film were significantly different producing absorption peaks at approximately 585 and 635 nm.

Figure 3.3 shows the absorption spectra of GAFchromic EBT film at different dose levels [12], and Figure 3.4 shows the absorption spectra for EBT2 film at similar dose levels [13]. Note the significant changes in spectra for both films.

The significant difference between EBT and EBT2 film is the incorporation of the yellow dye into the active layer which produced the large absorption tail at wavelengths below around 500 nm.

Similarly, absorption spectra of XR type T film are shown in Figure 3.5. The differences are obvious even with a similar dye in its construction, but the sensitivity to low-energy photons was

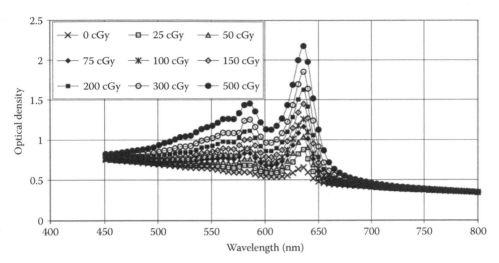

Figure 3.3 GAFchromic absorption spectra at different dose levels from EBT film. (Adopted from Butson, M.J. et al., *Phys. Med. Biol.*, 50, N135–N140, 2005.)

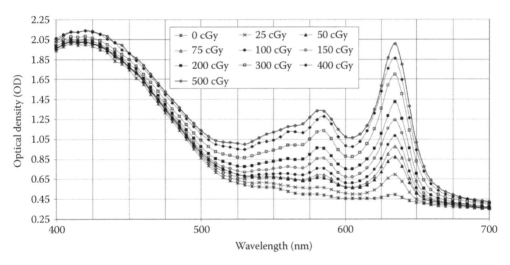

Figure 3.4 Absorption spectra at different dose levels from EBT2 film. (Adopted from Butson, M.J. et al., *Australas. Phys. Eng. Sci. Med.*, 32, 196–202, 2009.)

significantly enhanced because of other elements used in the original active chemicals used in MD-55 and HD films.

The XR T (therapy film) and XR R (diagnostic film) show distinct different spectra as shown in Figures 3.5 and 3.6. Figure 3.6 provided the spectral reflectance spectra for GAFchromic XR type R film measured by a reflectance spectrometer instead of a transmission photo spectrometer [14].

Figures 3.7 and 3.8 show the absorption spectra for XR QA film and XRCT film, respectively, adopted from Butson et al. [15,16]. Again, highlighting high-sensitivity response in the reflectance spectra for XRCT film (Figure 3.8).

Other nonspecific films to radiation oncology and radiology as discussed in Chapter 2 provide very distinct spectra. The FWT-60 film has one absorption peak located at approximately 600 nm

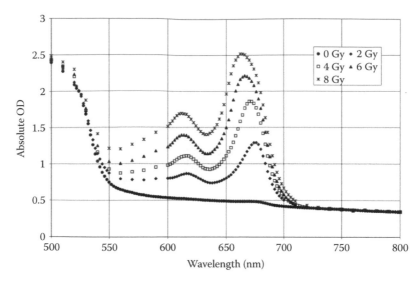

Figure 3.5 Absorption spectra from XR type T. (Adopted from Butson, M. et al., *Phys. Med. Biol.*, 49, N347–N351, 2004.)

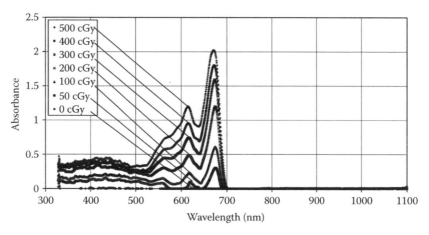

Figure 3.6 Reflectance spectra of XR type R at various dosages. (Adopted from Cheung, T. et al., *Appl. Radiat. Isot.*, 63, 127–129, 2005.)

giving it a blue color when irradiated as shown in Figure 3.9 [17], whereas SIRAD film used in personal monitoring has two absorption peaks at 565 nm and 615 nm in the blue visible region as shown in Figure 3.10 [18].

Of interest in UV-radiation dosimetry is polysulphone film. Its absorption spectra are given in Figure 3.11 adopted from Geiss et al. [19]. Significantly, the *color* change occurs at wavelengths mostly below 400 nm in the UV region. Therefore, the color change is not apparent to our eyes but should be measured using a UV spectrometer.

It is clearly obvious from Figures 3.1–3.11 that with varying absorption and transmission spectra, a reader should be sensitive enough to provide high fidelity and accurate reading needed for dose measurements. Selection of the scanner or densitometer is discussed in Chapter 4.

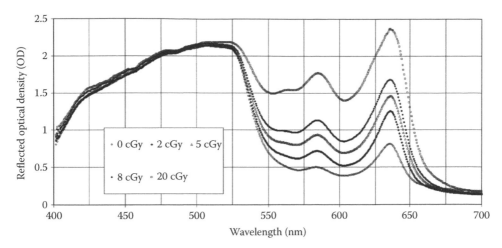

Figure 3.7 GAFchromic XR QA reflectance spectra. (Adopted from Alnawaf, H. et al., *Radiat. Measure.*, 45, 129–132, 2010.)

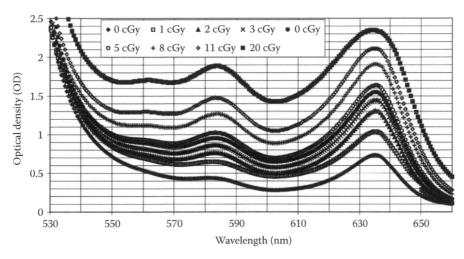

Figure 3.8 GAFchromic XR CT reflectance spectra. (Adopted from Butson, M.J. et al., *Phys. Med. Biol.*, 51, 3099–3103, 2006.)

3.6 DOSE RESPONSE

To perform accurate dosimetry with RCF, the dose response must be characterized. As film undergoes irradiation, there is a change in the absorbance of the film that occurs. The change in the films absorbance can be quantified through the measurement of the optical density (OD), which is defined as OD = $\log_{10}(I_0/I)$, where I_0 is the initial light intensity and I is the transmitted light intensity [20]. For EBT3 film, the OD increases (becomes optically less transparent) with increasing radiation dose as a result of the radiation-induced polymerization of diaceteylene monomers [21]. As a result, the dose response of film can be determined for a particular source of radiation by delivering a range of known (absolute) doses and calculating the associated OD for each dose based on measurement of initial and transmitted light intensity through the film. From this relationship, relative dose measurements can be performed within the measured dose range.

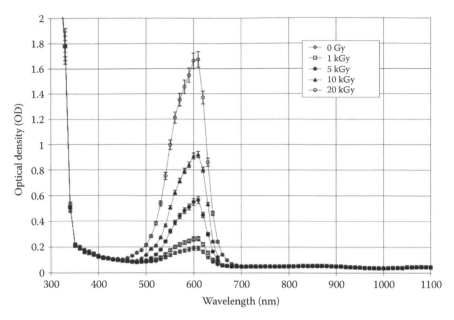

Figure 3.9 FWT-60 absorption spectra. (Adopted from Butson, M. et al., *Phys. Med. Biol.*, 49, N377–N381, 2004.)

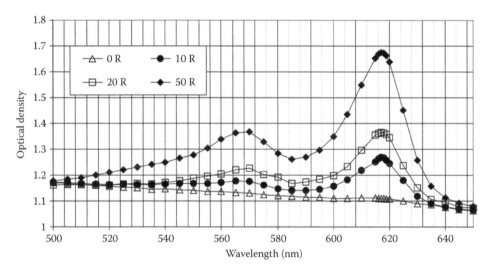

Figure 3.10 SIRAD film badge absorption spectra. (Adopted from Butson, M.J. et al., *Phys. Med. Biol.*, 51, N417–N421, 2006.)

Ideally, the OD–dose relationship should be linear; however, this is not typical and this relationship is best obtained through polynomial fits, or empirical models with variable coefficients [20]. It is also important to note that OD values are only valid for the specific wavelength or waveband of light analyzed [20]. Due to this, different scanner and densitometer systems will result in variations of measured OD, whereas it is also possible to produce a number of OD–dose curves by analyzing film with multiple wavelengths. Multichannel analysis is frequently employed for film dosimetry in which red, green, and blue channel curves are produced. This can be useful as the different channels have varying sensitivity as shown in Figure 3.12. The absolute signal from a film is dependent

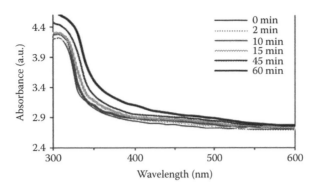

Figure 3.11 Polysulphone film absorption spectra. (Adopted from Geiss, O., Gröbner, J., Rembges, D., *Manual for Polysulphone Dosimeters: Characterisation, Handling and Application as Personal UV Exposure Devices*, Physical and Chemical Exposure (PCE) 2003, EUR 20981 EN.)

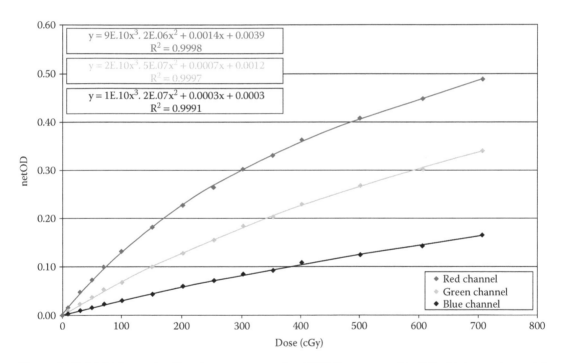

Figure 3.12 Multichannel calibration curves up of an EBT3 film exposed to 7 Gy. Note a polynomial fitting function to each channel. (Adopted from Casanova, B.V. et al., *J. Appl. Clini. Med. Phys.*, 14, 212–217, 2013.)

on many parameters including, but not limited to, the scanner characteristics, position on the scanner, and orientation of the film and air gaps. For this reason, film does not provide absolute dose assessment, rather relative dose.

The dose response curves of EBT type film have been assessed in numerous studies [22,23]. For EBT3 film, a polynomial fit for the dose response is valid up to nearly 40 Gy for IMRT dose-verification applications [22]. Please note the fitting parameters are not universal but rather vary from scanner to scanner and are affected by various parameters including but not limited to film batch, scanner, orientation, frequency, and reading condition.

Although GAFchromic film has many advantages, the nonlinear dose response is not ideal. The study of Devic et al. [24] demonstrated the use of a unique functional argument to linearize the inherently nonlinear response of EBT, EBT2, and EBT3 films. The invariance of the mathematical functional, based on one scanner model, was also tested with one film type [24]. The findings showed that the dose response could be linearized using the functional form and that this linear behavior was insensitive to both film model and flatbed scanner model used. In this way, relative dose measurements could be performed using RCF dosimetry without the need of establishing a calibration curve [24].

3.7 ENERGY DEPENDENCE

The energy dependence of RCF has been widely investigated as it can affect the dosimetric properties of the film in the presence of an unknown spectrum of beam energies ranging from kV to MV. Most commercial RCFs have lower atomic number material in the composition; thus, they are inherently energy independent compared with the older radiographic films [25–27].

The older models of the RCF such as MD55 and HS mostly contained low Z elements, thus showing a lower energy response to kV photons than higher energy MV photons. The first generation EBT film has been widely studied and was found to show very low-energy dependence over kV–MV photon energy range. The addition of minor amounts of moderate Z elements such as chlorine in the atomic composition of the EBT film helped to boost the photoelectric absorption of the kV photons, thus making it less energy dependent than its predecessor radiochromic models [28]. Similar results were observed in another study in which the GAFchromic HS film was found to show energy dependence of up to 30% as compared with 10% for the EBT film over the range of 50 kVp to 10 MV X-rays [29].

The newer GAFchromic EBT2 film was shown to have very low-energy dependence with a 6.5% variation in dose response for energy spectrum of 50 kVp to 10 MV as compared with 7.7% for the EBT film [30].

In a similar study, the energy dependence of the EBT2 film was tested over a range of energy beams comprising kilovoltage X-rays, Cs-137, Co-60 gamma rays, megavoltage X-rays (6 and 18 MV), electron beams (6 and 20 MeV), and proton beams (100 and 250 MeV). The EBT2 film was found to show negligible energy dependence with a global spread of ± 4.5% over all energies and modalities and a maximum difference of 18% between photon and proton beams [31,32].

The latest GAFchromic EBT3 film has a similar composition to the EBT2 film with the added feature of a symmetric construction. Therefore, the EBT3 film is expected to perform similarly to the EBT2 GAFchromic film. Studies performed for the EBT3 film energy dependence over a range of proton, electron, and photon beams showed a similar global spread of up to 3% with a maximum deviation of 11% as the aforementioned study performed with EBT2 [33]. Other studies for the EBT3 film have showed similar results with a maximum difference of 6% observed between photon and proton beam calibration curves [34].

Energy dependence has also been shown to be affected by absorbed dose, spatial resolution, and the color of channel used for film analysis. Dose response curves of the newer radiochromic EBT3 were found to be weakly energy dependent for the megavoltage photon beams. For the kilovoltage beams on the other hand, there were differences of up to 11% depending on the dose, resolution, and color of channel used [35].

3.8 DOSE FRACTIONATION

Dose fractionation is a common occurrence for radiotherapy treatment and sometimes dosimetry is warranted over a series of fractions to determine average doses or variations in delivery techniques. RCF has the advantage of being a self-processing film, and color changes occur without the need for chemical developing and fixing. This advantage makes the film reusable, which is especially useful for fractionated dose deliveries and therefore could be an ideal dosimetry candidate for such evaluations.

To determine the effects of fractionated versus single-exposure dose delivery, authors have investigated effects such as total change in OD [36–41] as well as dependency of transient OD at short times after irradiation takes place [42].

It is generally agreed that RCFs can be used to assess fractionated dose delivery [36–42] whereby the film can be irradiated multiple times to form a dose-accumulation measurement. However, there are variations in the reported magnitude of fractionated effects. Original studies on early versions of RCFs such as HD-810 [43] report minor differences, up to 1% in total measured dose when fractionation is delivered anywhere from 12 min to 24 h apart compared with single exposures for high-energy photon irradiations. Su et al. [36] measured similar results for electron beam (6 MeV–22 MeV) irradiations with negligible differences determined within uncertainty values for dose levels up to 2 Gy as seen in Figure 3.13. This was measured when the irradiations were performed with exposures varying from 1 single exposure of 200 cGy (MU Set = 1) to 20, 10 cGy exposures (MU Set = 6) using GAFchromic EBT film. The other values for MU Set were different fractionations between these values quoted earlier.

However, other authors have reported larger discrepancies in measured OD and thus dose when fractionated doses are delivered [42]. Ali et al. [42] performed measurements on the influence of fractionation and protracted dose delivery on OD as a function of total dose and the exposure-to-densitometry time interval for MD-55-2 film. Both their measurements and models demonstrate that fractionation induces transient OD over responses, which can be as large as 20% when small time differences between exposure and readout occurred, but rapidly dissipated within 24 h. However, as shown in Figure 3.14, their superposition model predicts 2%–5% over responses that persist as long as 700 h and was validated by their experimental procedures.

These results highlight the need to evaluate fractionated effects for RCFs within your own clinic depending on the type of film used as well as the magnitude of the fractionated measurements to be performed.

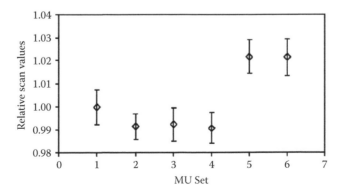

Figure 3.13 Effect of dose fractionation on radiochromic color change. (Adapted from Su, F.C. et al., *Appl. Radiat. Isot.*, 65, 1187–1192, 2007.)

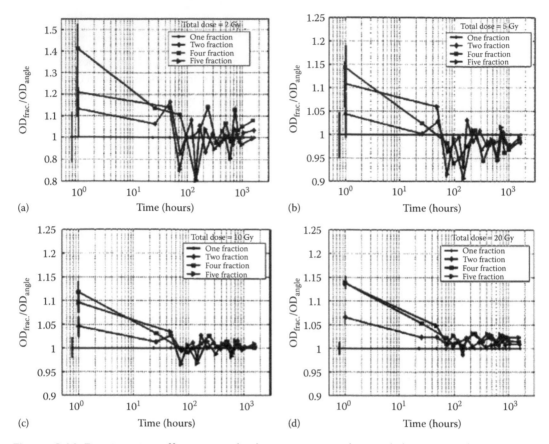

Figure 3.14 Fractionation effects on radiochromic measured optical density in relation to time after exposure of (a) 2 Gy, (b) 5 Gy, (c) 10 Gy and (d) 20 Gy. (Adapted from Ali, I. et al., *Appl. Radiat. Isot.*, 62, 609–617, 2005.)

3.9 DOSE-RATE DEPENDENCE

An ideal detector should be dose-rate independent. Thus, for RCF to be appealing, dose rate is a critical parameter. For any dosimeter, it is ideal that its response is independent of dose rate [43]. This is of particular importance for the high doses produced in short pulses by linear accelerators [43]. The dose-rate dependence of radiochromic film has been evaluated in a number of studies across various generations of the film [44–46].

In the study of Rink et al. [44], a small dose-rate dependence was measured for the EBT films when an eight-fold difference in dose rate was introduced. The study comprises delivered doses in the range of 5 –100 cGy range with dose rates varying between 16 and 520 cGy min^{-1}. Combining the OD measurements for a given dose irrespective of the dose rate used (within the studied range), the introduced dose-rate dependent uncertainty was in the order of ~1%. Across all tested dose rates and irradiated doses, the standard deviation of the OD was found to be less than 4.5%. Although this result discovered a statically significant effect in dose-rate changes, the uncertainty in measurements using this type of EBT film is satisfactory for many of its applications [44].

Twork and Sarfehnia [45] studied the effect of different dose rates on the OD for commercially available EBT3 film for a Novalis TX machine. The study compared dose rates of between 50 and 1000 Gy/min for doses of 50, 100, and 500 cGy. Findings of the study indicated a smaller dependence on dose rate; however, this was found to be within the experimental uncertainty of clinical film dosimetry [45]. As a result, the study reported that EBT3 films are dose-rate independent to within 3% for a wide dose rate irrespective of the total dose delivered. It is important to note that the study evaluated the dose rates and delivered doses that are commonly used in the clinical setting.

With the increased use of very high-dose rates offered by flattening filter-free beams, it is important that dose-rate independence over a large wider range for modern films. The dose-rate dependence of EBT films was evaluated in filter flattening free beam profiles for a Varian TrueBeam system [46]. Given the uncertainty in scanner reproducibility and film stability, the dose-rate effects were found to be negligible [46].

The dose-rate dependence for different dosimeters and detectors was compared by Karsch et al. [47] in which dose rates in the order of 15×10^9 Gy/s were used. Although this study is focused on the dose rates offered by laser-based acceleration technology, which is not clinically common at present, the results did find that for EBT film, a dose-rate independence of within 5% was still maintained.

3.10 POSTIRRADIATION COLORATION: TEMPORAL RESPONSE

The color changes in RCF are due to polymerization that is a rather slow process. The postirradiation coloration is defined as the time in which changes in the color of the film postexposure are small enough to introduce significant errors in clinical applications [48].

The RCFs may take some time to develop full color and this varies depending on the humidity, exposure time, and radiation energy. Typically at 24 h, the entire color will have developed [40]. For the older MD-55 film, the dose absorbance can increase by up to 16% during the first 24 h after irradiation, with only a slight rise (4%) thereafter, for up to about two weeks [49].

Studies performed with the EBT films have shown to produce an approximate 6%–9% increase in postirradiation OD within the first 12 h of irradiation. This has been observed within the 1–5 Gy dose range. For the older film types, this increase was noted to be approximately 13%, 15%, and 19% for MD-55-2, XR type T, and HS RCFs, respectively. The EBT film has a reduced postirradiation growth effect that stabilizes to 1% within the first 6 h. Therefore, for highest level of accuracy the film must be left at least 6 h postirradiation and the calibration films must be read out with the same postirradiation time [50].

As per the manufacturers' guidelines, for EBT2 film, it is recommended to wait at least 1–2 h postirradiation, but if feasible, up to 24 h as a delay of 2–3 h thereafter has an insignificant impact on dose accuracy [51].

Studies performed on the latest EBT3 film showed similar behavior as the EBT2 film. Between 1 and 24 h postirradiation times, the change in net OD was found to be less than 1% for a dose range of 30–400 cGy [48]. Irrespective of the temporal response, simple function can be used to correct OD postirradiation as shown in many references [23,52]. To be on a safer side, films should be read after 24 h of postirradiation in which polymerization reaches saturation level within <1% accuracy without any correction for most films.

3.11 POLARIZATION

By definition, polarized lights are light waves in which their vibrations occur in a single plane [53]. When Klassen et al. [54] reported on the effect of polarized light on the reading results of radiochromic film, they concluded that some layers of MD-55-2 film act as polarizers. As most light sources are linearly polarized to some extent, it is therefore important to maintain consistency in film orientation during readout especially relative to the direction of the incident light. Otherwise, variations in transmission OD can occur, which is unrelated to absorbed dose delivered.

As an example, when evaluating transmission OD with a nonpolarized light source compared with a fully linearly polarized light source for EBT film, the following two results (Figures 3.15 and 3.16) are achieved [55].

A significant difference in OD is seen with the polarized light source at angle highlighting the degree of polarization properties within the EBT film structure that increases with applied dose.

Desktop scanner normally uses partially polarized light source for film scanning and thus the magnitude of the polarization effect lies somewhere between these two Figures (3.15, 3.16) given

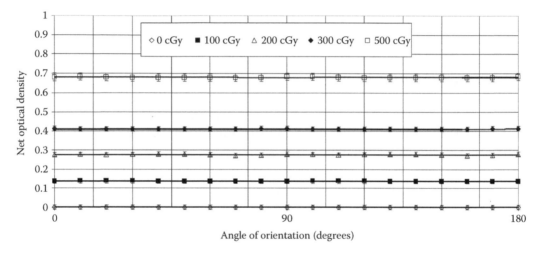

Figure 3.15 Net optical density for EBT irradiated to different dose levels and evaluated with a nonpolarized light source. (Adopted from Butson, M.J. et al., *Australas. Phys. Eng. Sci. Med.*, 32, 21, 2009.)

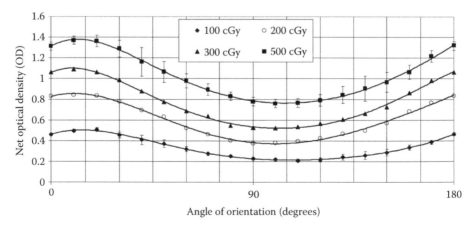

Figure 3.16 Net optical density for EBT irradiated to different dose levels and evaluated with a fully linearly polarized light source. (Adopted from Butson, M.J. et al., *Australas. Phys. Eng. Sci. Med.*, 32, 21, 2009.)

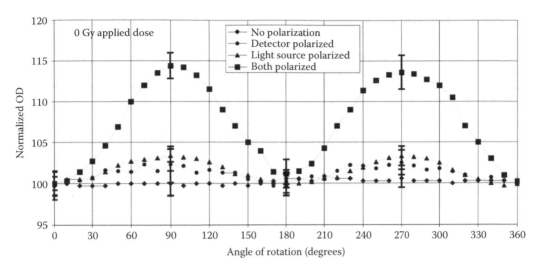

Figure 3.17 Reduced polarizing effects of transmitted optical density on the smaller crystal size active layer of GAFchromic HS film. (Adopted from Butson, M.J. et al., *Phys. Med. Biol.*, 48, N207–N211, 2003.)

earlier. This will be dealt in more detail in Chapter 4. Needless to say, the film orientation with respect to the scanner when the light source is partially or fully polarized plays a significant part in the accuracy of RCF dosimetry.

The degree of polarization effects depends on the crystal structure of the RCF evaluated. A larger or longer polymer structure within the active crystal leads to the film performing more like a polarizer and produces a larger polarizing effect. EBT films, to increase their sensitivity, where created and manufacturer to have longer crystal structures and thus polarization effects are larger. Older films like GAFchromic HS, in which the crystal structure was much smaller, produced lower polarization effects as shown in Figure 3.17 [56].

The impact of polarization still plays a significant role in RCF dosimetry today in the clinic due to the inherent polarization properties of desktop scanners that are the main instrument used for film-dose analysis. Specifically, the lateral-scatter effect [57–59] is an effect caused in part by polarization and needs to be accounted for in two-dimensional dosimetry applications. Chapter 4 deals with these effects in detail.

3.11.1 WATER IMMERSION AND PENETRATION

The ability to perform dosimetric measurements in water is advantageous; however, for many dosimeters this is limited, as their construction does not provide sufficient protection against water damage.

As RCFs are constructed in layers when the films are cut, the pressure can cause delamination to occur [20], with fractures or separation of the layers extending up to 8 mm from the edge but typically 1–2 mm [60]. With this layer design and the delamination issues, water penetration is of concern. Although the polyester sandwich-type GAFchromic films are water resistant, water penetration occurs at the sides between the layers at the site of delamination [20].

Measurements on the impact of water immersion and penetration have been well documented for a range of films. It has been shown for an older model MD-55 film that penetration from water into the outer area of the film is less than 0.5 mm/h [61].

Studies on EBT model film [62] identified considerable effects along the edges of the film after a 15-min immersion in water; however, after allowing the film to dry for an hour at room temperature, no such water trace was detectable even after scanning. As theses scans were performed for a considerable period after immersion took place, it has been suggested that additional precautions are not required for water penetration effects.

It has been reported for EBT2 film, the depth of penetration can reach as much as 9 mm around the edges of the film over a 24-h period [63]; however, this is an extreme situation. No one wants to leave film in water for irradiation more than 30 min. It was also observed that the anticipated dose error due to the change in OD resulting from water immersion appeared to be negligible for short-immersion periods of around 30 min; however, as the immersion time increases, the anticipated dose error can be around 7% [63]. It should be noted that the magnitude of the impact depends on many parameters such as film size, initial OD, and postimmersion waiting time prior to scanning and immersion time [63].

Based on the slow penetration rate, it is feasible to use RCFs to perform routine measurements in water. If the immersion times are short, the effect of water immersion should be neglected. Immersion in water overnight has also been reported as a simple means of separating the layers to gain access to the individual components [20]; however, such practice is beyond the scope of clinically accepted norm.

Although there are limited studies on the penetration of water for EBT3 films, according to producer's notes [64], the construction of EBT3 is more robust and allows for water immersion, with no limitations specified in regards to this. Based on previous results, it would be reasonable to expect similar, if not improved, performance to that of older generation films.

3.12 RESOLUTION

RCF dosimetry involves being able to quantify and evaluate accurate dose. These film dosimeters require the ability to accurately measure the absorbed dose and in doing so maintain good resolution. Theoretically, the limit of resolution of the film depends on the dimension of the activated molecule in the film composition, whereas practically the spatial resolution is determined by the sampling rate/resolution in the densitometer/scanner software [65].

There are two distinct methods for film readout. The first one being a point densitometer and the other method is the use of multiple detection systems such as scanners using charge-coupled device (CCD) technology. One of the main aspects of the use of these devices is the pixel size of a given scanner.

The spatial resolution of a scanning system is defined by several factors such as physical resolution, light source size, scattering effects from the film, and other sources of stray/ambient light within the reader. The size of the film for a particular measurement should also be taken into consideration when choosing the spatial resolution so as not to slow down the scanning process. Another important consideration is choosing the correct bit level of the data-acquisition system to achieve the required accuracy of measurement [65].

There have been various studies assessing the effect of scanner resolution to evaluate the optimal scan parameters. In a study performed with the EBT film and an Epson Expression 1680 CCD scanner, a resolution of 75 dpi or 0.3387 mm per pixel was employed [66]. A study performed with the Epson Expression 1680 professional scanner employed a scanner resolution of 127 dpi or 0.2 mm/pixel [67]. In summary, the user should match the scanning resolution to that required for the analysis process.

Using a similar study performed with the GAFchromic EBT film and the Epson Expression 10000XL scanner fully assessed the performance of the scanner to quantify corrections and parameters needed for accurate dosimetry with minimum uncertainties. Scanner resolution of 50, 75, 150, and 300 dpi was tested with all filters turned off. No significant differences in calibration curves were observed by varying the scanning parameters. To obtain a compromise between image resolution and noise, a resolution of 75 dpi was selected as default for this study [68].

3.13 FILM UNIFORMITY AND BATCH DEPENDENCE

The RCF is a 2D dosimeter, in which its exposure to uniform irradiation should give rise to a uniform dose response for accurate dosimetric applications. The film nonuniformity is characterized by local and regional fluctuations in the film response [49] under the uniform irradiation condition. The local nonuniformity is concerned with variations in film readings within the measurement region of interest and is quantified by the percent (relative) standard deviation of the film response to the region of interest mean [69]. This small scale, localized nonuniformity is influenced by the active layer of the film itself [40] and the scanner spatial and signal resolutions, pixel size, pixel depth, and electronic noise from the scanner [20,49]. The larger scale, regional variations in film response may be evaluated qualitatively via graphical display of a film image. The ratio of the maximum-minimum response within the entire region can provide a quantitative measure of the regional nonuniformity [49]. The regional nonuniformity is mainly influenced by the nonuniformity of the active layer, such as coating, particle alignment, active layer thickness across the region [70], and the readout system, for example, the inhomogeneous scanner response introduced during the film digitization process [68]. Readers are referred to general overview of this topic in various references [20,40,71] as well as sections in this book.

The uniformity of RCFs has been investigated for various RCF models and the scanning systems. The reported results are summarized in Table 3.3 for film models MD-55, MD-55-2, EBT, EBT2, and EBT3. Significant intrasheet regional nonuniformity has been reported for the film models MD-55 and HS, in the range up to 15% [72] and 6%–8% [40], respectively. Compared with EBT, earlier batches of EBT2 film model were shown to have intrasheet nonuniform dose response up to ± 8.7% at 1 Gy [73].

To mitigate the possible dosimetric errors originating from film nonuniformity, several correction approaches have been proposed. These include (1) double-exposure technique proposed by Zhu et al. [74], in which a film is exposed to uniform dose to obtain pixel-by-pixel sensitivity correction matrix prior to exposure to unknown radiation. The resulting OD from the unknown dose distribution is recovered by subtracting the uniform radiation OD from the combined OD (unknown and uniform) and division by the derived correction factor pixel by pixel; (2) pixel-wise evaluation of net OD, as the logarithm of the ratio of post- to preexposure

Table 3.3 Summary of Radiochromic Film Uniformity Testing

Film	Scanner and scanning parameters	Channel	Local	Regional intrasheet	Regional intersheet	Batches	Batch No	References
MD-55	Zeineh Soft Laser Scanner Densitometer	R		15.0% (OD, max/min)				Meigoon et al. [72]
MD-55	Personal densitometer	R	3.0%, 2.0%, 1.3% (0.05, 0.25, 0.5 mm pixel resolution)	8–15% (OD, max–min/ Mean × 100%)	4%–11% (OD)		920813	Zhu et al. [74]
MD-55-2	Lumisys Lumiscan 75 Vision Ten V Scan LKB Pharmacia Ultrascan XL Zeineh Soft Laser Scanning Densitometer Molecular Dynamic Personal Densitometer 8-12-bit	R	2.4%–5.8% (~ 9 Gy) 1.2%–4.3% (~20 Gy) or Mean ± S.D: 3.5±0.9 and 5.1±0.6 (~9 Gy) 1.6±0.4 and 3.4±0.6 (~20 Gy)		0.80–0.92(~9 Gy) 0.87–0.96 (~20 Gy) or Mean ± S.D: 0.87±0.05 and 0.81±0.02 (~9 Gy) 0.92±0.02 and 0.89±0.01(~20 Gy)		37-041	Niroomand-Rad et al. [69]
EBT	Epson Expression 1680 Pro 48-bit RGB, 150 dpi Transmission	G		<1.0%	<1.0%		35146-003AI	Fuss et al. [75]
EBT	Epson Expression 1680 Pro 48-Bit RGB, 72 dpi Transmission	R			<1% (OD, unexposed)	Up to 13% (OD, unexposed)	35322-(002-004) 36076-001 36124-(003-004) 36348-04I 37115-02I	van Battum et al. [62]

(Continued)

Table 3.3 (Continued) Summary of Radiochromic Film Uniformity Testing

Film	Scanner and scanning parameters	Channel	Local	Regional intrasheet	Regional intersheet	Batches	Batch No	References
EBT	Epson Expression 10000XL 48-Bit RGB, 72 dpi Transmission	R		1.0% (0.3 Gy) 0.5% (1 Gy)	Same order of magnitude to the intra-sheet uniformity		37122-04I	Martišíková et al. [76]
EBT2	Epson Expression 10000XL 48-bit RGB, 72 dpi Transmission	R		±6.0 to ±8.7% (1 Gy)			F03110903	Hartmann et al. [73]
EBT2	Epson Expression 10000XL 48-bit RGB, 72 dpi Reflective Transmission	R		1.8% (ADC, Reflective) 2.4% (ADC, Transmission)	1.2% (ADC, Reflective) 1.6% (ADC, Transmission)		F02060902B	Richley et al. [77]
EBT2	Epson Perfection V700 48-bit RGB, 150 dpi	R		<1.2% (2 Gy)	<0.3% (PV, unexposed)		F04090903	Huet et al. [78]
EBT2	Epson GT-X970 48-bit RGB, 72 dpi	R		0.9%–4.8% (2 Gy)	(for 2 Gy) 1.3% 1.2% 0.5% 0.4% 1.2%	0.4%–1.3% (2 Gy)	F12170902A A052810-02AA A08161005A A09171002 A05131001B	Mizuno et al. [79]
EBT3	Epson Perfection V700 48-bit RGB, 1200 dpi Transmission	R	<1.2% (PV, unexposed)	<1.5% (0.5–5.1 Gy)			A11021102	Reinhardt et al. [34]

pixel values from the red channel by Kairn et al. [80]; and (3) the recommendation by the EBT2 film manufacturer (International Specialty Products [ISP], NJ) [81] to utilize the blue channel response to correct the nonuniformity in the active layer thickness. The application of manufacturer recommended correction method to EBT2 film has been reported to increase the noise, therefore, dosimetric uncertainty [80,82].

Significant batch-to-batch variation has been reported for EBT film, in spite of small variation in intersheet uniformity of less than 1% [62]. Characterization of individual film batch, according to established departmental film dosimetry protocol, could reduce the dosimetric uncertainty.

3.14 UV/AMBIENT LIGHT EFFECTS

The active layers of most RCFs are sensitive to ultraviolet light. UV is found in high intensities in solar radiation but it is also found in fluorescent light sources as well. As such, UV can be present within the clinic from UV penetrating through windows/sky lights as well as the internal light sources and environmental lighting [83–87].

The rate of color change in RCFs associated with UV radiation exposure differs depending on film type and the amount of additives or coatings used with the film to either reduce or sometimes increase UV sensitivity. For example, Reinstein [84] measured the effects of standard fluorescent lights in their clinic and found that MD-55-2 films OD increased at a rate of 0.0007 OD units per hour. This equates to a dose effective sensitivity of 3.5 cGy/h. Butson et al. [87] reported on the successful use of MD-55-2 for performing dosimetry of UVA radiation, wavelength 320–400 nm. The manufacturer (International Specialty Products, ISP, NJ) changed its chemical formulation in the EBT2 film to include a yellow dye. This dye along with uses for dosimetric accuracy improvements also inhibited UV sensitivity and decreased the OD changes that occurred. This is shown in results by Andrés et al. [23] (Figure 3.18)

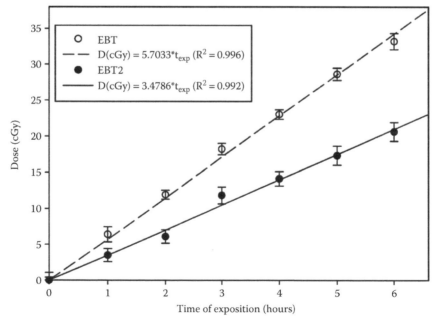

Figure 3.18 UV exposure response of EBT and EBT2 film. (Adopted from Andrés, C. et al., *Med. Phys.*, 37, 6271, 2010.)

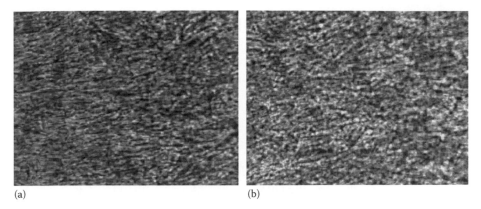

(a)   (b)

Figure 3.19 Clumping of polymers (a) without magnetic field; (b) with 0.35 Tesla magnetic field. (Adopted from Reyhan, M.L. et al., *J. App. Clin. Med. Phys.*, 16, 325–332, 2015.)

in which the UV response of EBT2 in the presence of a UV-producing lamp was approximately 50% of that seen for EBT film.

The take-home message is that room lights can affect RCF coloration and cause unwanted film darkening from UV exposure. Exposure to room and ambient light should be limited as much as possible to reduce UV-based coloration. However, small exposures to ambient room lights that would occur during normal film handling will normally produce negligible effects.

3.15 MAGNETIC FIELD EFFECTS

There is a strong interest in the radiation oncology community for acquisition of a Linac-magnetic resonance imaging (MRI)-based accelerator. Systems have been developed in Canada [88], the Netherlands [89], and in other countries such as Australia [90].

The advantages of magnetic resonance imaging (MRI) quality imaging combined within a Linac for cancer treatment should allow greater targeting and thus treatment accuracy for patients' benefits. One aspect of their design however is the constant presence of a magnetic field during the treatment process. Thus, both quality assurance dosimetry and in vivo dosimetry are performed in the presence of high magnetic fields.

Although previous experimental work using RCFs has been performed in the presence of magnetic fields [91,92], it was only recently that Reyhan [93] began initial testing on the effects of magnetic fields on the accuracy of RCF dosimetry.

In their research, they examined variations in OD changes related to similar absorbed dose levels given using EBT2 film when the film had been previously exposed to MRI magnetic fields or not. Their results observed a variation of up to 4% in final measured dose when the films were either exposure to a 1.5 T field before the irradiation or after the irradiation. With the use of microscopic imaging, Reyhan et al. [93] were able to show a discernable difference in the structure of an irradiated EBT2 film as shown in Figure 3.19. Similar findings were also shown by Reynoso et al. [94]. They demonstrated that a nonexposed film showed more clumping of the darkened polymers compared with an exposed film.

Obviously, more experimental work is required in this area of radiochromic dosimetry especially if Linac-MRI machines increase in number.

3.16 THE DOS AND DON'TS OF DOSIMETRY WITH RADIOCHROMIC FILMS

As a simple guide for RCF dosimetry, the following table provides a list of Dos and Don'ts that can be applied in radiation dosimetry:

Dos

- Know the dose levels to be measured.
- Select correct film type for job performed.
- Use same batch for a set of measurement including calibration.
- Take appropriate care when handling film.
- Calibrate the film response.
- Control all scanning parameters.
- Disable automatic image adjustments on scanner to be used.
- Scan all films in same orientation.
- Keep film out of light when not being exposed or analyzed.
- Track time period after irradiation till analysis and incorporate corrections.
- Mark films for orientation and registration purposes for dosimetry.
- Use the middle of the scanner.
- Develop a protocol and stick to it.
- Acknowledge limitations in accuracy.

Don'ts

- Treat film inappropriately—fingerprints, scratches, high temperatures, and so on.
- Mix up batches.
- Use the wrong film for the wrong job.
- Take shortcuts in the dosimetry process.
- Exposure films to high temperatures.
- Expose film to excessive room light and UV.

3.17 SUMMARY

RCFs are suitable dosimeters for two-dimensional dosimetric application. If proper precaution such as batch, storage, temperature, exposure conditions, and proper scanning methods with a reliable scanner are used, accuracy <2% can be easily achieved. Film dosimetry is more of an art than science and hence meticulous handling and patience are needed for good results. Care, consistency, and repeatability are three essential characteristics required for accurate film dosimetry, and these should be adopted in the clinic to fully utilize RCFs for dosimetry applications.

REFERENCES

1. Tsibouklis J, Pearson C, Song YP, Warren J, Petty M, Yarwood J, Petty MC and Feast WJ. Pentacosa-10,12-diynoic acid/henicosa-2,4-diynylamine alternate-layer Langmuir-Blodgett films: Synthesis, polymerisation and electrical properties. *J Mater Chem* 1993;3:97–104.
2. Watanabe Y, Patel GN, Patel P. Evaluation of a new self-developing instant film for imaging and dosimetry. *Radiat Prot Dosim* 2006;120:121–124.

3. Rink A, Lewis DF, Varma S, Vitkin IA, Jaffray DA. Temperature and hydration effects on absorbance spectra and radiation sensitivity of a radiochromic medium. *Med Phys* 2008;35:4545–4555.

4. David Lewis—Private communication.

5. Yu PK, Butson M, Cheung T. Does mechanical pressure on radiochromic film affect optical absorption and dosimetry? *Australas Phys Eng Sci Med* 2006;29:285–287.

6. Butson MJ, Cheung T, Yu PK. Radiochromic film dosimetry in water phantoms. *Phys Med Biol* 2001;46:N27–N31.

7. Reinstein LE, Gluckman GR, Meek AG. A rapid colour stabilization technique for radiochromic film dosimetry. *Phys Med Biol* 1998;43:2703–2708.

8. http://www.dictionary.com/browse/absorption-spectrum.

9. http://www.andor.com/learning-academy/absorption-transmission-reflection-spectroscopy-an-introduction-to-absorption-transmission-reflection-spectroscopy.

10. Butson M, Cheung T, Yu KN. Visible absorption properties of radiation exposed XR type T radiochromic film. *Phys Med Biol* 2004;49:N347–N351.

11. Butson MJ, Yu PKN, Cheung T, Metcalfe P. Radiochromic film for medical radiation dosimetry. *Mater Sci Engi R* 2003;41:61–120.

12. Butson MJ, Cheung T, Yu KN. Absorption spectra variations of EBT radiochromic film from radiation exposure. *Phys Med Biol* 2005;50:N135–N140.

13. Butson MJ, Cheung T, Yu PKN, Alnawaf H. Dose and absorption spectra response of EBT2 Gafchromic film to high energy X-rays. *Australas Phys Eng Sci Med* 2009;32:196–202.

14. Cheung T, Butson M, Yu KN. Reflection spectrometry analysis of irradiated GAFCHROMIC XR type R radiochromic films. *Appl Radiat Isot* 2005;63:127–129.

15. Alnawaf H, Cheung T, Butson MJ, Yu PKN. Absorption spectra response of XRQA radiochromic film to X-ray radiation. *Radiat Meas* 2010;45:129–132.

16. Butson MJ, Cheung T, Yu KN. Absorption spectra of irradiated XRCT Radiochromic film. *Phys Med Biol* 2006;51:3099–3103.

17. Butson M, Cheung T, Yu KN. Absorption spectra analysis of exposed FWT-60 radiochromic film. *Phys Med Biol* 2004;49:N377–N381.

18. Butson MJ, Cheung T, Yu PK. Visible absorption spectra of radiation exposed SIRAD dosimeters. *Phys Med Biol* 2006;51:N417–N421.

19. Geiss O, Gröbner J, Rembges D. *Manual for Polysulphone Dosimeters: Characterisation, Handling and Application as Personal UV Exposure Devices.* Physical and Chemical Exposure (PCE) 2003, EUR 20981 EN.

20. Williams M, Metcalfe P. Radiochromic film dosimetry and its applications in radiotherapy. *Concepts Trends Med Radiat Dosim AIP Conf Proc* 2001;1345:75–99.

21. Callens M, Crijins W, Simons V et al. A spectroscopic study of the chromatic properties of GafChromic™EBT3 films. *Med Phys* 2016;43:1156.

22. Casanova BV, Pasquino M, Russo G, Grosso P, Cante D, Sciacero P, Girelli G, Rosa La PM, Tofani S. Dosimetric characterization and use of GAFCHROMIC EBT3 film for IMRT dose verification. *J Appl Clin Med Phys* 2013;14:212–217.

23. Andres C, Del Castillo A, Tortosa R, Alonso D, Barquero R. A comprehensive study of the Gafchromic EBT2 radiochromic film. A comparison with EBT. *Med Phys* 2010;37:6271–6278.

24. Devic S, Tomic N, Aldelaijan S, DeBlois F, Seuntjens J, Chan MF, Lewis DF. Linearization of dose-response curve of the radiochromic film dosimetry system. *Med Phys* 2012;39:4850–4857.

25. Das IJ. Radiographic film. In: Rogers, DW, Cyglar, J (Eds.). *Clinical Dosimetry Measurements in Radiotherapy.* Madison, WI: Medical Physics Publishing; 2009. pp. 865–890.

27. Pai S, Das IJ, Dempsey JF, Lam KL, Losasso TJ, Olch AJ, Palta JR, Reinstein LE, Ritt D, Wilcox EE. TG-69: Radiographic film for megavoltage beam dosimetry. *Med Phys* 2007;34:2228–2258.

28. ISP Gafchromic EBT film, Technical Reference Guide, 2005.

29. Butson MJ, Cheung T, Yu PKN. Weak energy dependence of EBT gafchromic film dose response in the 50 kVp-10 MVp X-ray range. *Appl Radiat Isot* 2006;64:60–62.

30. Butson MJ, Yu KN, Cheung T, Alnawaf H. Energy response of the new EBT2 radiochromic film to X-ray radiation. *Radiat Meas* 2010;45:836–839.

31. Arjomandy B, Tailor R, Anand A, Sahoo N, Gillin M, Prado K, Vicic M. Energy dependence and dose response of Gafchromic EBT2 film over a wide range of photon, electron, and proton beam energies. *Med Phys* 2010;37:1942–1947.

32. Zhao L, Das IJ. Gafchromic EBT film dosimetry in proton beams. *Phys Med Biol* 2010;55:N291–N301.

33. Sorriaux J, Kacperek A, Rossomme S, Lee JA, Bertrand D, Vynckier S, Sterpin E. Evaluation of Gafchromic EBT3 films characteristics in therapy photon, electron and proton beams. *Phys Med* 2013;1:82–87.

34. Reinhardt S. Comparison of Gafchromic EBT2 and EBT3 films for clinical photon and proton beams. *Med Phys* 2012;39:5257–5262.

35. Massillon-JL G, Chiu-Tsao S, Domingo-Murioz I, Chan MF. Energy dependence of the new Gafchromic EBT3 film – dose response curves for 50 kV, 6 and 15 MV x-ray beams. *Int J Med Phys Clin Eng Rad Oncol* 2012;1:60–66.

36. Su FC, Liu Y, Stathakis S, Shi C, Esquivel C, Papanikolaou N. Dosimetry characteristics of GAFCHROMIC EBT film responding to therapeutic electron beams. *Appl Radiat Isot* 2007;65:1187–1192.

37. McLaughlin WL, Puhl JM, Al-Sheikhly M, Christou CA, Miller A, Kovács A, Wojnarovits L, Lewis DF. Novel radiochromic films for clinical dosimetry. *Radiat Prot Dosim* 1996;66:263–268.

38. Rink A, Vitkin IA, Jaffray DA. Suitability of radiochromic medium for real-time optical measurements of ionizing radiation dose. *Med Phys* 2005;32:1140–1155.

39. Saylor MC, Tamargo TT, McLaughlin WL, Khan HM, Lewis DF, Schenfele RD. *Int J Radiat Appl Instrum C Radiat Phys Chem* 1988;31:529–536.

40. Butson MJ, Peter KN, Cheung T, Metcalfe P. Radiochromic film for medical radiation dosimetry. *Mater Sci Eng Rep* 2003;41:61–120.

41. Williams M, Metcalfe P. Radiochromic film dosimetry and its applications in radiotherapy concepts and trends in medical radiation dosimetry. *Proc SSD Summer School AIP Conf Proc* 2011;1345:75–99.

42. Ali I, Williamson JF, Costescu C, Dempsey JF. Dependence of radiochromic film response kinetics on fractionated doses. *Appl Radiat Isot* 2005;62:609–617.

43. Metcalfe P, Kron T, Hoban P. *The Physics of Radiotherapy X-Rays and Electrons*. Madison, WI: Medical Physics Publishing; 2007.

44. Rink A, Vitkin IA, Jaffray DA. Intra-irradiation changes in the signal of polymer-based dosimeter (GAFCHROMIC EBT) due to dose rate variations. *Phys Med Biol* 2007;52:N523–N529.

45. Twork G, Sarfehnia A. Evaluation of the dose-rate dependency of GAFCHROMIC EBT3. *Med Phys* 2013;40:223.

46. Oyewale S, Ahmad S, Ali I. Dose rate and energy dependence of EBT, EBT2, EDR2 films, and Mapcheck2 diode arrays in beam profiles from a varian truebeam system. *Med Phys* 2012;39:3722.

47. Karsch L, Beyreuther E, Burris-Mog T, Kraft S, Richter C, Zeil K, Pawelke J. Dose rate dependence for different dosimeters and detectors: TLD, OSL, EBT films, and diamond detectors. *Med Phys* 2012;39:2447–2455.

48. Borca VC, Pasquino M, Russo G, Grosso P, Cante D, Sciacero P, Girelli G, Porta MRL, Tofani S. Dosimetric characterization and use of Gafchromic EBT3 film for IMRT dose verification. *J Appl Clin Med Phys* 2013;14:287.

49. Niroomand-Rad A, Blackwell CR, Coursey BM, Gall KP, McLaughlin WL, Meigooni AS, Nath R, Rodgers JE, Soares CG. Radiochromic dosimetry: Recommendations of the AAPM radiation therapy committee task group 55. *Med Phys* 1998;25:2093–2115.

50. Cheung T, Butson M, Yu PKN. Post irradiation coloration of Gafchromic EBT radiochromic film. *Phys Med Biol* 2005;50:N281–N285.

51. ISP Gafchromic EBT2 film, Technical Reference Guide, 2009.

52. Zhao L. Temporal response of Gafchromic EBT2 radiochromic film in proton beam irradiation. *IFMBE Proc* 2012;39:1164–1167.

53. http://www.physicsclassroom.com/class/light/Lesson-1/Polarization.

54. Klassen NV, van der Zwan L, Cygler J. GafChromic MD-55: Investigated as a precision dosimeter. *Med Phys* 1997;24:1924–1934.

55. Butson MJ, Cheung T, Yu PKN. Evaluation of the magnitude of EBT Gafchromic film polarization effects. *Australas Phys Engi Sci Med* 2009;32:21–25.

56. Butson MJ, Yu PKN, Cheung T, Inwood D. Polarization effects on a high-sensitivity radiochromic film. *Phys Med Biol* 2003;48:N207–N211.

57. Lewis DF, Chan MF. Technical note: On GAFChromic EBT-XD film and the lateral response artifact. *Med Phys* 2016;43:643.

58. Lewis D, Chan MF. Correcting lateral response artifacts from flatbed scanners for radiochromic film dosimetry. *Med Phys* 2015;42:416–429.

59. Poppinga D, Schoenfeld AA, Doerner KJ, Blanck O, Harder D, Poppe B. A new correction method serving to eliminate the parabola effect of flatbed scanners used in radiochromic film dosimetry. *Med Phys* 2014;41:21707.

60. Yu PK, Butson M, Cheung T. Does mechanical pressure on radiochromic film affect optical absorption and dosimetry? *Australas Phys Eng Sci Med* 2006;29:285–287.

61. Butson MJ, Cheung T, Yu PK. Radiochromic film dosimetry in water phantoms. *Phys Med Biol* 2001;46:N27–N31.

62. van Battum LJ, Hoffmans D, Piersma H, Heukelom S. Accurate dosimetry with GafChromic™ EBT film of a 6 MV photon beam in water: What level is achievable? *Med Phys* 2008;35:704–716.

63. Aldelaijan S, Devic S, Mohammed H, Tomic N, Liang LH, DeBlois F, Seuntjens J. Evaluation of EBT-2 model GAFCHROMIC film performance in water. *Med Phys* 2010;37:3687–3693.

64. Ashland, KY. GafChromic radiotherapy films. Available from: http://www.ashland.com/products/gafchromic-radiotherapy-films.

65. Butson MJ, Yu PKN, Cheung T, Metcalfe P. Radiochromic film for medical radiation dosimetry. *Mater Sci Eng* 2003;R41:61–120.

66. Lynch BD, Kozelka J, Ranade MK, Li JG, Simon WE, Dempsey JF. Important considerations for radiochromic film dosimetry with flatbed CCD scanners and EBT GAFCHROMIC film. *Med Phys* 2006;33:4551–4556.

67. Saur S, Frengen J. GafChromic EBT film dosimetry with flatbed CCD scanner: A novel background correction method and full dose uncertainty analysis. *Med Phys* 2008;35:3094–3101.

68. Ferreira BC, Lopes MC, Capela M. Evaluation of an Epson flatbed scanner to read Gafchromic EBT films for radiation dosimetry. *Phys Med Biol* 2009;54:1073–1085.

69. Niroomand-Rad A, Chiu-Tsao S, Soares CG, Meigooni AS, Kirov AS. Comparison of uniformity of dose response of double layer radiochromic films (MD-55-2) measured at 5 institutions. *Phys Med* 2005;21:15–21.

70. White Paper, Radiochromic Film. Available from: http://www.filmqapro.com/Documents/Lewis_Radiochromic_Film_20101020.pdf.

71. Devic S, Tomic N, Lewis D. Reference radiochromic film dosimetry: Review of technical aspects. *Phys Med* 2016;32:541–556.

72. Meigooni AS, Sanders MF, Ibbott GS, Szeglin SR. Dosimetric characteristics of an improved radiochromic film. *Med Phys* 1996;23:1883–1888.

73. Hartmann B, Martišíková M, Jäkel O. Technical note: Homogeneity of GafchromicVR EBT2 film. *Med Phys* 2010;37:1753–1756.

74. Zhu, S. Quantitative evaluation of radiochromic film response for two-dimensional dosimetry. *Med Phys* 1997;24:223–231.

75. Fuss M, Sturtewagen E, De Wagter C, Georg G. Dosimetric characterization of GafChromic EBT film and its implication on film dosimetry quality assurance. *Phys Med Biol* 2007;52:4211–4225.
76. Martišíková M, Ackermann B, Jäkel O. Analysis of uncertainties in Gafchromic® EBT film dosimetry of photon beams. *Phys Med Biol* 2008;53:7013–7027.
77. Richley L, John AC, Coomber H, Fletcher S. Evaluation and optimization of the new EBT2 radiochromic film dosimetry system for patient dose verification in radiotherapy. *Phys Med Biol* 2010;55:2601–2617.
78. Huet C, Dagois D, Derreumaux S, Trompier F, Chenaf C, Robbes I. Characterization and optimization of EBT2 radiochromic films dosimetry system for precise measurements of output factors in small fields used in radiotherapy. *Radiat Meas* 2012;47:40–49.
79. Mizuno H, Takahashi Y, Tanaka A, Hirayama T, Yamaguchi T, Katou H, Takahara K, Okamoto Y, Teshima T. Homogeneity of GAFCHROMIC EBT2 film among different lot numbers. *J Appl Clin Med Phys* 2012;13:198–205.
80. Kairn T, Aland T, Kenny J. Local heterogeneities in early batches of EBT2 film: A suggested solution. *Phys Med Biol* 2010;55:L37–L42.
81. ISP 2010 Correction Protocol for Gafchromic EBT2 Dosimetry Film. Available from: http://online1.ispcorp.com/layouts/Gafchromic/content/products/ebt2/pdfs/GAFCHROMICEBT2CorrectionProtocol-Rev1.pdf (accessible from http://gafchromic.com).
82. McCaw TJ, Micka JA, DeWerd LA. Characterizing the marker-dye correction for Gafchromic® EBT2 film: A comparison of three analysis methods. *Med Phys* 2011;38:5771–5777.
83. McLaughlin WL, Puhl JM, Al-Sheikhly M, Christou CA, Miller A, Kovács A, Wojnarovits L, Lewis DF. Novel radiochromic films for clinical dosimetry. *Radiat Prot Dosim* 1996; 66:263–268.
84. Reinstein LE, Gluckman GR, Amols HI. Predicting optical densitometer response as a function of light source characteristics for radiochromic film dosimetry. *Med Phys* 1997;24:1935–1942.
85. Butson MJ, Yu PK, Metcalfe PE. Measurement of off-axis and peripheral skin dose using radiochromic film. *Phys Med Biol* 1998;43:2407–2412.
86. Butson MJ, Cheung T, Yu PK, Abbati D, Greenoak GE. Ultraviolet radiation dosimetry with radiochromic film. *Phys Med Biol* 2000;45:1863–1868.
87. Butson MJ, Cheung T, Yu PK. Fluorescent light effects on FWT-60 radiochromic film. *Phys Med Biol* 2005;50:N209–N213.
88. http://www.mp.med.ualberta.ca/linac-mr/.
89. http://medicalphysicsweb.org/cws/article/research/56853.
90. http://sydney.edu.au/medicine/radiation-physics/research-projects/MRI-linac-program.php.
91. Raaijmakers AJ, Raaymakers BW, Lagendijk JJ. Experimental verification of magnetic field dose effects for the MRI-accelerator. *Phys Med Biol* 2007;52:4283–4291.
92. Damrongkijudom N, Oborn B, Butson M, Rosenfeld A. Measurement of magnetic fields produced by a "magnetic deflector" for the removal of electron contamination in radiotherapy. *Australas Phys Eng Sci Med* 2006;29:321–327.
93. Reyhan ML, Chen T, Zhang M. Characterization of the effect of MRI on Gafchromic film dosimetry. *J App Clin Med Phys* 2015;16:325–332.
94. Reynoso FJ, Curcuru A, Green O, Mutic S, Das IJ, et al. Technical Note: Magnetic field effects on Gafchromic-film response in MR-IGRT. *Med Phys* 2016;43:6552–6559.

Radiochromic film digitizers

BENJAMIN S. ROSEN

4.1 INTRODUCTION

The current chapter discusses the various types of radiochromic film digitizers and their advantages and limitations. For qualitative radiochromic film use, the human eye alone is able to determine the spatial distribution of color change about the active film layer. When using radiochromic film to confirm linear accelerator radiation and light-field coincidence, for example, the human eye is typically adequate. In other cases, such as personnel dosimetry, individual point density measurements may be required. In such instances, a manual optical densitometer capable of providing readouts proportional to the quantity of transmitted or reflected light at a single point at a fixed area is used. The wide spread use of radiochromic film is associated with high-resolution two-dimensional (2D) dosimetry that requires a 2D optical measurement and recording system. In the context of quantitative dosimetry using radiochromic film, precise 2D digitization is essential. This chapter focuses on the technology used to perform an accurate radiochromic film readout and on the advantages and limitations of various scanning techniques.

For the radiochromic film user, the current chapter also provides an overview of the current film digitizer technologies so that appropriate equipment decisions can be made, on the basis of individual cost, speed, and precision requirements. A reliable digitization platform affords radiochromic film users a dosimetry system capable of a vast array of applications in both clinical and research settings.

4.2 IMAGE ACQUISITION

4.2.1 MEASURED QUANTITIES

Optical light intensity measurements are used to quantify radiochromic film response. Depending on scanner type and functionality, transmission, reflection, or a combination of such measurements may be used to quantify film response. As radiochromic film generally has a broad spectrum of responsive wavelengths, it is typically necessary to specify the frequency or bandwidth of the light source used. Radiochromic film measurement accuracy is dependent on measurements reproducibility, so that the difference between intensity measurements before and after irradiation may be reliably related to radiation absorbed dose. Light *transmission* (T) is expressed as the fraction of light transmitted by a sample, such that

$$T = \frac{I}{I_0} \tag{4.1}$$

where I and I_0 are the *measured transmitted* and *incident light intensities*, respectively. *Optical density* (OD) is historically the fundamental and more useful measurement quantity, as photographic film exhibits a clear relationship between OD and exposure to light in a fixed exposure range. For convenience, radiographic and radiochromic film dosimetry measurements are often based on OD, defined as

$$OD = \log_{10}\left(\frac{1}{T}\right) \tag{4.2}$$

It should be highlighted that the desired physical quantity is *absorbance*, as changes in absorbed dose response are accompanied by changes in film absorbance. Yet, OD is simply a measurement of the logarithmic ratio of the attenuation of transmitted radiant power, which includes absorption, but also scattering and other physical processes such as reflection, diffraction, polarization, and so on. Thus, it is important to perform a characterization of the film-dosimetry system, including all of the processes that may contribute to changes in OD measurements. Although the relationship between OD and true absorbance is complex, OD is typically reported in absorbance units (AU). An optical density of 1 AU corresponds to a drop in signal intensity of one order of magnitude. Thus, typical optical instruments are capable of performing accurate measurements up to a range of approximately 4 AU (corresponding to 0.01% transmission of incident light intensity). The limiting factor is often the magnitude of the signal relative to the measurement noise, which is difficult to suppress to lower than about 0.1%.

Analog instruments such as manual optical densitometers (a device that measures the optical fluence) provide raw values that are proportional to measured light intensity. When using such instrumentation, obtaining the sample's AU using Equation 4.1 is straightforward. Digital instruments, on the other hand, often employ algorithms that bin intensity values into conventional computer memory units. Known as pixel depth, the number of bins corresponds to the number of possible pixel values, or shades of gray, that can be assigned to a given intensity measurement. To conform to computer architecture convention, pixel depths used for radiochromic film dosimetry are usually between 8 and 16 bits, corresponding to 256 and 65536 intensity values, respectively. Pixel value is, thus, used as a surrogate for light intensity, and digitization (rounding) error should be taken into account when choosing optimal pixel depth. It is recommended that at least 12 bits (4096 shades) be used for image analysis [1]. If multiple light frequencies or bandwidths are used, each is accompanied by a different set of pixel values. Thus, for a three bandwidth RGB (red, green, and blue) color scanner, using images with 48-bit pixel depth is warranted. The particular measured quantity and signal resolution are largely dependent on the light source and detector characteristics.

4.2.2 LIGHT SOURCES

The lighting source is essential to optical densitometry and has a considerable effect on the usefulness of film measurements. Spectral matching between the light source emission and the radiochromic film absorption, by definition, provides the most sensitive measurement, but it is not always practical to find a perfect match. Physical size, frequency, uniformity, intensity, stability, and polarization are all important aspects that must be considered for any light used for quantitative radiochromic film measurements [1]. In addition, extraneous film response induced by the light source itself, such as infrared and ultraviolet light, should be limited or avoided completely. Permanent darkening due to such phenomena has been reported [2]. OD measurements comprise the weighted average of a number of factors, including the intensity across the area of measurement and the spectral response of the source, detector, and sample. It is important to understand the principles of operation and emission characteristics of the various light sources used in radiochromic film optical densitometry.

4.2.2.1 FLUORESCENT LAMPS

The phenomenon of electrically stimulated luminescence has been studied since the 1840s. Present-day fluorescent lamps operate under the principle of noble gas discharge within an otherwise completely evacuated glass tube, usually 0.3% atmospheric pressure [3]. Visible light is produced through a two-step process. First, thermionic emission of electrons causes ionization of the gas vapor (usually mercury or xenon), which produces a cascade effect that emits photons in the ultraviolet (UV) energy range. Second, these UV photons collide with phosphor powder, often comprising various combinations of rare earth compounds, coated on the walls of the glass tube. As the UV photons are incident with an average energy of approximately 4.9 eV, the full visible spectrum (1.8–3.1 eV) may be produced as the electron kinetic energy is converted to visible light through inelastic collisions with phosphor electrons. The particular phosphor composition determines the fluorescent light emission spectrum.

In the context of radiochromic film dosimetry, fluorescent lamps are used in most color charge-coupled device (CCD) flatbed scanners. As discussed, UV radiation can produce a small but measurable effect on radiochromic film absorbance [1]. Users should store films in the dark whenever possible to reduce contamination from sunlight or stray room light as a possible source of UV radiation. Advantages of fluorescent lamps over other light sources are that they are cost effective and can provide spectral information at a variety of light frequencies. Thus, the dynamic range of radiochromic film response may be extended, as high-intensity measurements of wavelengths with lower absorption can be used. The reduced sensitivity, however, is often accompanied by an increase in the relative measurement noise. The popular Epson Expression line of desktop scanners (Seiko Corporation, Japan) employs a cold cathode fluorescent lamp, and a plot of the spectral emission of the Epson Expression 10000XL is shown in Figure 4.1. Further discussion on the particular characteristics of the Epson Expression 10000XL scanner will be discussed, as it is currently a scanner recommended for use by the manufacturer of radiochromic film used for radiation dosimetry of therapeutic applications. A vast array of literature on radiochromic film use with flatbed scanners has been published in the past decade [4–9]. Provided that appropriate handling, scanning, and analysis procedures are followed, excellent experience with such scanners can be expected.

4.2.2.2 LASERS

Lasers are unique light sources because they produce spatially and temporally coherent light through the process of stimulated emission. A number of different types of lasers have been used for radiochromic film densitometry, but all lasers have particular defining characteristics such as incident laser power and wavelength. The optimum laser power is found by minimizing the probability of damaging the film and inducing unintended polymerization catalysis and maximizing

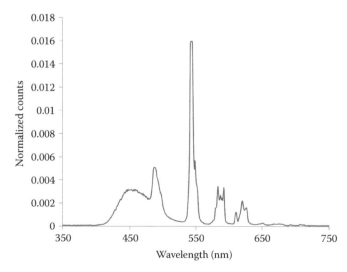

Figure 4.1 Measured emission spectrum of the xenon cold cathode fluorescent lamp of an Epson Expression 10000XL (measured by author).

the incident power to provide sufficient illumination of the sample (and, hence, lower measurement noise). Red lasers (635–670 nm) in the range of 1–10 mW are commonly employed for radiochromic film densitometry, as the GAFchromic™ (Ashland Advanced Materials, Bridgewater, NJ) line of radiochromic films exhibits maximum absorption in this range. The primary advantage of laser light is its spatial coherence, which allows for very small film regions of film to be illuminated. In this way, 2D film measurements are taken as the concatenation of many ideally independent point measurements, with minimal scatter or reflections from adjacent film areas. This is not the case for fluorescent light sources, in which the diffuse nature of the light causes significant positional dependence, especially at the furthest edges of the source. Laser light also generally provides the optimal measurement sensitivity, as a very narrow bandwidth tuned to the maximum film absorption peaks can be used. Figure 4.2 shows a representative spectral measurement of a 635-nm He–Ne laser. It should be mentioned that dosimetric accuracy can be compromised if the absolute position of the film absorption peak is susceptible to changes in absorbed dose, temperature, or other factors such as scattering, reflection, Newton's ring artifacts, polarization, and so on. For very narrow light emission bandwidths, small shifts in the spectral absorbance of the film can cause large variations in measurements. In addition, as the polydiacetylene mechanism of GAFchromic radiochromic films involves planar polarization of polymeric crystals, measurements using coherent laser light suffer from significant polarization artifacts. Scanning densitometers utilizing laser light sources with computerized translation of various optical components are discussed in Section 4.3. Due to the relatively high cost and limited availability of such scanners, use of scanning laser densitometers is not common in clinical settings. However, a number of prototype scanning systems do employ narrow bandwidth measurement light to maximize the absolute film response. Such systems will be discussed in Section 4.3.

4.2.2.3 LIGHT-EMITTING DIODES

Light-emitting diodes (LEDs) are semiconductor light sources capable of producing light with spectral bandwidth wider than lasers but generally narrower than fluorescent light. LEDs provide a practical compromise between diffuse white light sources and highly coherent laser light. As their production has increased, LEDs also offer a cheaper alternative to high-powered lasers.

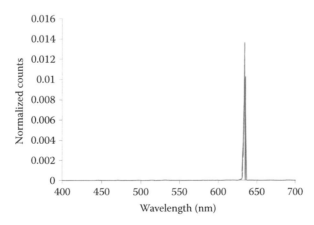

Figure 4.2 Measured emission spectrum from a 635 He–Ne laser. (Adapted from Wikipedia, https://en.wikipedia.org/wiki/Helium%E2%80%93neon_laser#/media/File:Helium_neon_laser_spectrum.png [used under the Creative Commons Attribution-Share Alike 3.0 Unported license].)

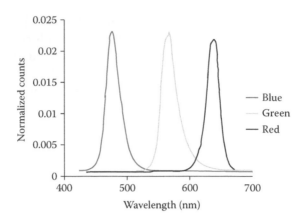

Figure 4.3 Representative measured spectrum of three single-color LEDs (blue, green, and red). (Adapted from Wikimedia Commons, https://commons.wikimedia.org/wiki/File:Red-Yellow-Green-Blue_LED_spectra.png [used under the Creative Commons Attribution-Share Alike 3.0 Unported license].)

These are available in a wide range of colors and brightness and can be made very compact, which has led to their increased use in some consumer flatbed scanners. LED scanners are less common than laser or fluorescent densitometers but may play a significant role in the future of radiochromic film densitometry. Some film scanners use bands of three or more LED colors so that a complete color spectrum can be measured. Figure 4.3 shows a representative spectral measurement of three single-color LEDs (RGB). Comparisons between the spectral characteristics of fluorescent lamps, lasers, and LEDs demonstrate the wide range of options for digitizing radiochromic film [10].

4.2.3 Light detectors

4.2.3.1 PHOTODIODE

A photodiode converts light energy to electrical current. Photodiodes have the advantage of very fast readout, as a bias voltage quickly removes any free electrons liberated by energetic light photons or holes residing in the semiconducting bandgap region. Current measurements are proportional to the light intensity incident on the active area. Photodiodes may be physically scanned opposite a light source and film sample to provide intensity measurements about the sample. The use of focusing optics is critical if the area of the desired point measurement (spatial resolution) is smaller than the active area of the photodiode. Photodiodes are typically employed in simple manual densitometers, as spatial resolution is not typically important for point measurements.

4.2.3.2 PHOTOMULTIPLIER TUBES

Photomultiplier tubes (PMTs) operate using the principle of the photoelectric effect and secondary emission from the collision of electrons with electrodes, called dynodes. PMTs are typically constructed in evacuated tubes, containing a photocathode, dynodes, and a collecting anode. The incoming photon interacts with the photocathode, and through the photoelectric effect, electrons are liberated. The electrons are directed toward an array of dynodes, and an avalanche occurs, which greatly increases the voltage at the collecting anode. PMTs are useful in a variety of applications, in which very high sensitivity is required. High-end drum scanners and scanning laser densitometers typically use PMTs, as they have very fast response with high gain and low noise compared with other image sensors.

4.2.3.3 CHARGE-COUPLED DEVICE

The CCD was invented in 1969 by researchers at Bell Labs (AT&T) [11,12]. CCDs are found in most camera systems and use a series of capacitors to store charge collected in a given sequence. The charge is passed to its neighbor using shift registers, and a sequence of voltages may be readout for image digitization, storing, or other processing. CCDs are very useful for providing high-resolution measurements, as capacitive materials can be manufactured in very fine crystalline layers. Spectral information can be obtained using a diffraction grating or other refraction medium, and using the resulting separation into color components in physically distinct CCD regions. As different sizes and coatings can be used, a wide variety of one- and two-dimensional CCDs exist. Scanning CCD desktop scanners employ a linear array CCD array that is translated in the direction of the motion of an extended line light source, so that each consecutive line of data is taken and reconstructed to form a 2D image. As an entire line of the image is taken at a time, there is the possibility of pixel crosstalk and other optical artifacts. Such artifacts will be discussed in Section 4.4.

4.3 IMAGE PROCESSING

4.3.1 LIGHT-INTENSITY VERIFICATION

It is important to verify the reproducibility and output of the digitizing instrumentation at regular intervals. Historically, known optical density tablets were used daily to verify that the light source and detector were stable within a given tolerance. This ensured that scanner measurements taken that day could be directly compared with those taken at other points of time. Whenever a batch calibration is applied to a given scanner/film combination, it is recommended that such *sanity-checks* be performed. The Epson Expression 10000XL, for example, performs a similar test during its warm-up routine. Intensity measurements of completely open and completely masked scanner regions are used to normalize the light-intensity measurements for that session. However, it is important to ensure that the intended areas are indeed free of any obstructions so that this test can be performed properly. As it is feasible that a miscalibration of the scanner using the self-test method may occur, it is still advisable to scan known density tablets periodically to ensure an overall stability of the imaging system.

4.3.2 SMOOTHING/FILTERING

The purpose of performing pixel value and dose calibration is to obtain the average response for a given combination of film batch and scanner. Small imperfections in the film and/or scanner manufacturing in addition to Gaussian noise limit the useful resolution of 2D measurements. It is only through the aggregate response of neighboring active elements that useful dosimetric information may be extracted. For this reason, it is important to choose imaging resolutions that properly balance the need for many spatially distinct measurement points with the decrease in light-intensity measurements, and most likely increased measurement noise. This, by definition, requires customized scanning protocols based on the particular measurement in question. For example, in the case of stereotactic radiosurgery quality assurance, the importance of high spatial resolution typically outweighs that of dosimetric accuracy. In routine quality assurance applications, minimizing the cost of high spatial resolution and superior dosimetric accuracy is warranted. By intentionally oversampling, digital smoothing, and filtering may be applied without significantly depreciating the robustness of intensity measurements. A typical radiochromic film imaging scheme for intensity-modulated radiation therapy quality assurance uses spatial resolutions on the order of 72 dpi (~55 microns), with a 5 × 5 median or Weiner filter [13].

4.3.3 RELATIONSHIP TO DOSE

For light-intensity measurements to provide useful information, the relationship with absorbed dose must be established. The process of determining the relationship between film darkening and dose is often termed *film calibration*. For a particular application, the film and digitizer may be calibrated in conjunction, as opposed to separately calibrated to a specific standard. Whether or not the system is calibrated together, care must be taken to ensure that the environmental variables remain constant between the calibration and measurement. Alterations in the environmental conditions between calibration and measurement may lead to inaccurate results and should be carefully studied. Uncertainty analysis, including such environmental factors, is considered in the next section.

The film calibration process, used to establish the relationship between dose and measured light intensity, consists primarily of the following three steps: irradiating film pieces to known dose levels, measuring the change in light intensity between irradiated films and un-irradiated controls (or, alternatively, the difference between the film intensity before and after irradiation), and applying a film calibration curve, used to interpolate dose from subsequent intensity measurements that fall between calibration points. As radiochromic film response is inherently nonlinear across a wide dose range [7], it is imperative that the predetermined calibration dose levels properly bound in magnitude all desired dose measurements. For instance, if quality assurance of an intensity-modulated radiation therapy (IMRT) plan is to be performed, the user must decide *a priori* what minimum and maximum dose levels are to be used for analysis. It is a good practice to bind the measurement dose levels by at least 15% to account for any changes in transient beam quality between calibration and measurement irradiations [13]. For a direct measurement in terms of absolute dose, the calibration film irradiations should be verified using some other dosimeter, traceable to national dosimetric standards, such as an ionization chamber [13].

The user must decide what functional form should be used to interpolate between calibration dose levels. The best fitting function (i.e., that which provides the smallest residual error between measured and true delivered dose) is somewhat dependent on scanner characteristics [14] and application [7]. For instance, some calibration curves may be more applicable in a specific dose region, whereas others are better for wider dose range. A variety of fitting functions has been used in literature [6,14–16], and each has inherent advantages and limitations. Two popular choices for fitting functions are power fits, in which exponential constants describe empirical scanner and film nonlinearities handled strictly as fitting parameters, and rational functions, which attempt to describe the film/scanner response in a semianalytical fashion [16,17]. Though the derivation of such functional forms may differ, fitting parameters are most often determined empirically through the use of calibration film irradiations and measurements [7]. It is recommended that each film lot receive a separate calibration [13], but it is up to the user to establish the dosimetric uncertainty associated with applying calibration fits across film pieces of the same lot. To perform such a calibration, unirradiated films are scanned, and the OD is calculated using Equation 4.2, to provide $OD_{unirrad}$. The calibration films may be the same pieces used for $OD_{unirrad}$, but a single, large unirradiated piece is usually used. The number of calibration points required to establish a useful calibration curve varies on the basis of the dose range and desired accuracy. The number of calibration points typically ranges from just a few points to over a dozen [6,7,18]. Films are typically irradiated under well-established reference condition (i.e., in 10-cm depth of water or water-mimicking material at the machine calibration distance from target). After irradiation and a sufficient wait time (typically at least 12 h), the films are scanned, and the net change in OD, ΔOD, is calculated according to

$$\Delta OD = OD_{irrad} - OD_{unirrad} \tag{4.3}$$

Once ΔOD is calculated for each irradiated dose, D, fitting parameters are calculated by optimizing a calibration function. Some common functional forms given in literature are provided in the following equations, where A, B, C, and n are fitting constants:

$$\Delta OD = \left(\frac{A+B}{D-C} \right) \tag{4.4}$$

$$D = A + B\,\Delta OD + C\,\Delta OD^{n} \tag{4.5}$$

Other functions may be used, but it recommended that the user fully characterize the estimated uncertainties for a given film model and application. A number of investigators have studied the effect of various empirical and semianalytical models to describe film response, and the reader is encouraged to refer to such works when deciding which calibration curve to use for a particular application [15–21]. Once the conversion between film darkening and dose is established, measurement films may be used for absolute dosimetry applications. The following sections describe what factors may affect the dosimetric and spatial accuracy of such measurements.

4.3.4 UNCERTAINTY ANALYSIS

Van Battum et al. [18] reported on the overall accuracy of EBT® film exposed in water using a flatbed scanner for readout. They observed an overall random uncertainty on the order of 1.3% (1 sigma). This was in agreement with Saur and Frengen [22], who reported a 2-sigma uncertainty of approximately 4%, with the maximum accuracy at 2 Gy for absolute dosimetry in water. These studies scanned a limited area of film and provide uncertainty estimated under the best possible conditions for absolute dosimetry. Systematic errors and random imaging artifacts were largely mitigated to obtain such favorable uncertainties. Thus, it is important to benchmark a scanning system prior to use so that a clear understanding of the dosimetric results in addition to realistic uncertainty estimates may be attained.

4.4 DIGITIZER TYPES AND FUNCTIONALITY OF READOUT

4.4.1 EVOLUTION AND HISTORICAL REVIEW

In the early 1990s, the original devices used for radiochromic film scanning were scanning densitometers, most commonly used for electrophoresis gel and liquid sample analysis. Scanning densitometers, such as those formerly manufactured by Molecular Dynamics (Sunnyvale, CA) and Pharmacia (Stockholm, Sweden), are essentially motorized versions of manual densitometers, in which the light source, detector, or sample is translated, and in which a 2D matrix of measurement values is formed. Mirrors, lenses, and other optics are often used to direct transmitted and incident light, so that source and detector components remain stationary during the scan. These instruments commonly utilize laser light sources in conjunction with a PMT detector. Significant limitations of scanning densitometers are cost, usually in the range of $20K–$40K, and availability. Classical laser scanning densitometers also exhibited significant artifacts, as described by Dempsey et al. [10] owing primarily to their hardware characteristics and method of readout. The issues with cost and laser-based artifacts warranted a search for new technologies to digitize radiochromic films. The natural progression was directed toward scanners already used to digitize conventional X-ray radiographs, such as those manufactured by Howtek (Nashua, NH) and Vidar (Herndon, VA). X-ray film digitizers typically translate the film using a document feed, and the digitization takes place a single line at a time. With the advent and prompt adoption of digital radiography in the late 1990s, the availability and cost of radiographic film digitizers became prohibitive, and radiochromic film users were again in search of a suitable device. This led to the widely popular use of low-cost flatbed photograph and document scanners. Flatbed document and photograph scanners range between approximately $100 and $3000, with the higher end scanners having increased functionality and features. Thus, the two main classes of scanners used for radiochromic film are flatbed imagers and scanning densitometers. The two classes share similar characteristics, as they both aim to create 2D digitized images of the radiochromic response. However, they differ primarily in the choice of light source, detector, and translation mechanisms. The following sections discuss some specific scanners, their functionality, primary advantages, and reported limitations. This should not be taken as a comprehensive list, but instead a compilation of relevant literature and technology integral to the accurate use of radiochromic film.

4.4.2 SCANNING DENSITOMETERS

4.4.2.1 COMMERCIAL DEVICES

The Pharmacia LKB (Stockholm, Sweden) and Molecular Dynamics Personal Densitometer (Sunnyvale, CA) are shown in Figures 4.4 and 4.5, respectively. Though both units are obsolete, some are still in use and a number of similar devices are available through third-party retailers.

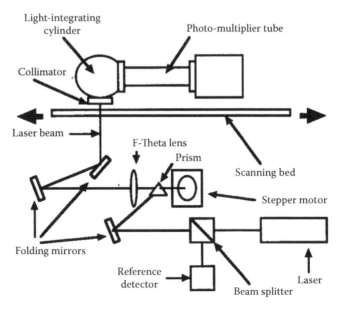

Figure 4.4 Optical system of molecular dynamics personal scanning laser densitometer. (From Dempsey, J.F. et al., *Med. Phys.*, 26, 1721–1731, 1999. With permission.)

Figure 4.5 Pharmacia LKB densitometer (used at NIST).

Due to their relatively simple design, it is usually straightforward to maintain, repair, and alter certain components of scanning densitometers, providing a degree of flexibility that current scanners lack. Dempsey et al. [10] performed a comprehensive characterization of various scanning artifacts attributed to the Molecular Dynamics scanner. The light-spread artifact is characterized by the spreading of transmission light in high OD gradient regions and can produce errors as large as 50% [10]. Interference patterns stemming from the multiple reflections between the various film layers and glass resting plate create nonreproducible artifacts, known as Newton's rings, creates readout error. In addition, the previously discussed optical-polarization artifact results in reproducible changes in measured absorbance with changes in the relative angle between the emulsion and coherent light polarization. Dempsey et al. [23] used image-correction techniques and light-diffusing material to mitigate some of the artifacts and successfully validated low- and high-energy acute radiation exposures with submillimeter spatial resolution and accuracy within a few percent. The Molecular Dynamics scanner employs an He–Ne red laser coupled to a PMT. An optical component system is used to increase readout speed (Figure 4.6). Gluckman and Reinstein [24] evaluated three high-resolution scanning densitometers: a Lumiscan 75 (Heidelberg, Germany), a Vidar VXR-16 (Herndon, VA), and a Howtek MultiRAD 460 (Nashua, NH). The digitizers were compared and characterized by the measured response per unit dose, signal-to-noise ratio, presence

Figure 4.6 Molecular dynamics personal densitometer (used at the University of Wisconsin ADCL).

of artifacts, and dosimetric accuracy. Similar to the Molecular Dynamics scanner, the Lumiscan employs an He–Ne laser light source. In contrast, the Vidar VXR and Howtek MultiRAD use a fluorescent and a high-intensity red LED light source, respectively. It was found that the Howtek MultiRAD provided the best signal-to-noise ratio and sensitivity and also eliminated the afore-mentioned Moire interference artifact, as no glass bed was present. In a more comprehensive study, Devic et al. [14] compared the Vidar VXR-16, Molecular Dynamics Personal Densitometer, and the following five other digitizers: the LKB Pharmacia UltroScan XL, Photoelectron Corporation CMR-604 (North Billerica, MA), Nuclear Associates Radiochromic Densitometer Model 37-443 (Cleveland, OH), Laser Pro 16 (Radlink, El Segundo, CA), and AGFA Arcus II (Mortsel, Belgium) document scanner. All scanners exhibited similar precision estimates, and experimental measurement uncertainty ranged from 1% to 5% with an additional parametric fit uncertainty of approximately 2%–7%, depending on film type and digitizer combinations. With the introduction of increasingly sensitive radiochromic film models, most commercially available radiochromic film digitizers are capable of providing measurements with acceptable limiting resolution and dosimetric uncertainty.

4.4.2.2 PROTOTYPES

A number of prototype radiochromic film digitizers have been reported in the literature [25–27]. The primary motivation for constructing a prototype device is to have a densitometry system that is perfectly tuned to the desired characteristics of measurement. In addition, the ability to interchange various components to match any film modifications that may affect measurements is crucial to accurate dosimetry. Rosen et al. [26] described a translating densitometer that eliminates artifacts caused by the glass bed and is designed for reference dosimetry using OD values, traceable to national standards. The primary motivation for such a device is to standardize film absorbance measurements so that absolute OD values accurately describe changes in film response across various users. The device successfully mitigated the glass interference and positional dependence artifacts. Ranade et al. [25] developed a scanning densitometer using a stationary red LED, stationary dual silicon photodiode detectors, and translating film plate. The scanner successfully eliminated the film orientation dependence found in other scanners, due to the point measurements using incoherent light. Though commercial and prototype densitometers were the original radiochromic film digitizers, their limited availability paved the way for the widespread use of CCD flatbed scanners.

4.4.3 CHARGE-COUPLED DEVICE FLATBED DOCUMENT SCANNERS

Significant improvements in quality and reproducibility of images from flatbed scanners, in addition to substantial advances in analysis and scanning techniques, has allowed the use of these scanners to surpass other forms of radiochromic film digitization for most general applications. The convenience and low cost of flatbed document scanners have also contributed to their status as the current workhorse for radiochromic film digitization. Typical CCD imaging devices such as photocopiers and desktop scanners use reflection scanning to digitize media. In this mode, the light source and detector are located on the same side of the scanned media. Higher end and specialized film scanners use transparency adapters, which allow direct transmission measurements. Reflective scans provide pixel values that are inversely proportional to the film transmittance. Figure 4.7 shows a schematic of a typical reflection flatbed scanner. High-end CCD document scanners have the ability to perform transmission measurements by providing a transparency adapter, which includes a light source incident on the film side opposite to the detector. Traditionally, transmission measurements

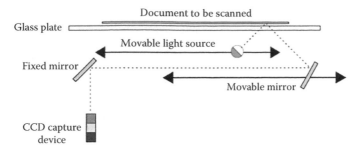

Figure 4.7 Schematic of a reflection flatbed scanner. (From Wikimedia Commons, https://commons.wikimedia.org/wiki/File:CPT_Hardware-Input-scanner-flatbed.svg [Used under Free Art License].)

are used for quantitative dosimetry, as this mode imitates densitometer measurements. However, current studies have shown that reflection measurements may be less susceptible to artifacts and are more accurate in low-dose regions by reducing the relative noise associated with low film response [28]. Transmission measurements, on the other hand, generally have better uncertainty estimates over a wide dynamic range. Most of the subsequent discussion on scanner artifacts and response characteristics are in relation to CCD flatbed scanners operating in transmission mode, as this is currently the most commonly employed radiochromic film digitization scheme.

4.5 IMAGING ARTIFACTS

The presence and magnitude of various radiochromic film digitization artifacts are functions of the scanning device and methodology. This section discusses some common scanning artifacts; however, the reader should be cautioned that this is not an exhaustive list, and full consideration of scanner limitations and inadequacies should performed prior to any clinical or research application.

4.5.1 LATERAL RESPONSE ARTIFACT

The lateral response artifact (LRA) was first reported by researchers in 2008 [22,29,30], which refers to the phenomenon of reduced light collected from areas away from the center of the extended line light source. Typical CCD flatbed scanners with transmission adapter have a fluorescent line light source, slit opening, and light recording unit underneath moving in tandem, as in the schematic in Figure 4.8. The image is formed by lines of image data perpendicular to the direction of scanning motion. As shown in Figure 4.9, the magnitude of the artifact is dependent on the dose in the surrounding region and somewhat dependent on the particular scanner model [2]. Due to the complex nature of the artifact, it is insufficient to calculate a simple 2D correction matrix from calibration film images to be applied to measurement film images. Instead, a model of the effect must be created. Schoenfeld [31] performed a comprehensive characterization of the physics behind this effect and presented detailed evidence indicating that anisotropic light scattering of the stacked needle-like polymer units is the underlying cause of the LRA. In summary, the parabolic nature of the artifact is in fact because of a limitation in scanner design, as the anisotropic light scatter subtends angles larger than the maximum acceptance angle of the optics used to focus line images on the CCD array. In principle, novel scanning technologies may have the ability to physically eliminate or greatly reduce the artifact.

Figure 4.8 Directionality of the moving parts of an Epson 10000XL flatbed document scanner. The recording mechanism (not shown) is under the slit opening. (Adapted from Schoenfeld, A.A. et al., *Phys. Med. Biol.*, 59, 3575, 2014. With permission.)

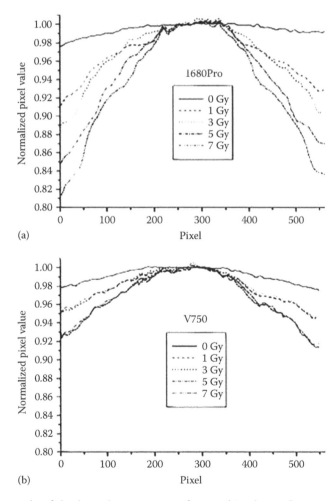

Figure 4.9 An example of the lateral response artifact and its dependence on dose and scanner models: Epson (a) 1680Pro and (b) V750. (Adapted from Menegotti, L. et al., *Med. Phys.*, 35, 3078–3085, 2008. With permission.)

Various investigators have proposed techniques to mitigate or physically suppress the LRA [30,32]. Poppinga et al. [32] proposed and validated a correction model that is obtained by scanning films exposed to known doses at various locations on the scanner bed. The algorithm essentially transforms measurements at any point within the scanning area to provide the intensity value that would be measured if the point were located in the center of the scanning area. Lewis and Chan [33] described a similar technique to model the LRA and mitigate the effect but also generalized the methods for application to triple-channel dosimetry and established the validity of using a single film lot to characterize the LRA model for a given scanner. It is important to note that even among the same scanner model, the magnitude of the fit parameters varied in excess of measurement uncertainties. From these works, it is suggested that a single characterization of a CCD flatbed scanner is sufficient to significantly mitigate the LRA effect. If left uncorrected and single-channel dosimetry is used; however, dose error in excess of 18% at the extreme lateral edges of the scan window may be observed [30]. Such dosimetric errors have the ability to propagate throughout a quality assurance process and result in failures for even relatively dose insensitive quality assurance metrics, such as gamma analysis. Lewis and Chan [33] demonstrated a 2%/2-mm gamma passing rate increasing from 76% to 92% when the method described by Poppinga et al. [32] was applied.

4.5.2 ORIENTATION DEPENDENCE

Another well-known artifact is the dependence of CCD flatbed scanner response on the orientation of the film on the scanner bed. As explained by Schoenfeld [34], the physical basis of the orientation artifact is equivalent to the LRA. The anisotripic light scatter caused by the parallel aligned polymer stacks is preferentially oriented in the directions at right angles to the coating direction that is constantly changing with every evolution of RCF product (see Chapter 2). Thus, whole film sheets and pieces cut at right angles from the sheet ends will exhibit the differences changes in pixel measurements when scanned in orthogonal orientations on the scanner bed depending on the type of film, for example, HD, HS, EBT, EBT2, EBT3, and EBT-XD. The most straightforward method of mitigating such an artifact is by paying close attention to the relative orientation of calibration and measurement films on the scan bed. It is recommended to use square or rectangular films sections with clearly marked orientation fiducials, so that measurement and calibration films are as closely aligned as possible, relative to the scanning bed and film coating direction.

4.5.3 INTERFERENCE PATTERNS

When measuring optical density of a particular sample, a portion of the measurement light undergoes multiple reflections between the surrounding media and the sample itself before reaching the detector. For radiochromic film, this interference artifact is due to the varying size of the air gap between the glass bed of most flatbed scanners and the film medium, for example, type of film. Using antireflective glass or light diffusers is possible options to remedy the artifact. The newest model of GAFchromic EBT film (EBT3) has an antireflective coating consisting of matte silica to essentially lift the films, so it is not in direct contact with the glass bed. This has been shown to successfully eliminate interference artifacts but likely reduces the limiting resolution of the scanner film system as the film is removed from the optimal focus position of the optical recording system.

4.5.4 ENVIRONMENTAL FACTORS

Various extraneous factors during the process of film digitization may contribute to increased uncertainty, such as temperature, humidity, and repeat scan dependence. Though Paelinck et al. [35] did not measure any appreciable changes in pixel value in scans subsequent to a single warm-up scan of an Epson Pro 1680 scanner, McCaw et al. [36] did find a detectable increase in blue-channel pixel value (about 1% per scan) with increasing numbers of scans on an Epson 10000XL scanner. This phenomenon may result from a complex interplay of factors, such as fluorescent tube output, film spectral sensitivity with temperature, and the particular frequency binning algorithms of the CCD array and associated electronics. For dosimetry applications requiring the use of multiple channels, a comprehensive characterization of the phenomenon is recommended. Such characteristics should be assessed during commissioning of any new film/scanner dosimetry system.

4.6 CONCLUSIONS

Radiochromic films are a useful tool for verifying 2D radiation dose distributions, due to its high spatial resolution, near water equivalence, and minimal fluence perturbation. However, the 2D distribution of film response must be faithfully digitized to be used for quantitative analysis. Digitization systems may vary widely in their particular characteristics, but their goal is to provide a high fidelity representation of the dose distribution. Pixel depth, dynamic range, cost, and availability are all important factors to be considered for the digitization system. Scanning laser and other monochromatic densitometers were used in the early days of radiochromic film digitization, but the availability and quality of flatbed document scanners led to their current status as the workhorse of radiochromic film dosimetry. Though a number of possible artifacts must be considered, current advances in film analysis and optical scanning techniques have afforded radiochromic film/flatbed scanner dosimetry systems a level of accuracy unattainable in previous film dosimetry systems. Future advancements of radiochromic film digitization techniques and analysis are destined to further improve the utility and functionality of radiochromic film in a variety of settings.

REFERENCES

1. Soares CG. Radiochromic film dosimetry. *Radiat Meas* 2006; 41:S100–116.
2. Butson MJ, Cheung T, Peter KN et al. Ultraviolet radiation dosimetry with radiochromic film. *Phys Med Biol* 2000; 45:1863.
3. Bloomfield LA. *How Things Work*. New York: Wiley; 2006.
4. Borca VC, Pasquino M, Russo G et al. Dosimetric characterization and use of GAFCHROMIC EBT3 film for IMRT dose verification. *J Appl Clin Med Phys* 2013; 14:158–171.
5. McCabe BP, Speidel MA, Pike TL, Lysel VAMS. Calibration of GafChromic XR-RV3 radiochromic film for skin dose measurement using standardized X-ray spectra and a commercial flatbed scanner. *Med Phys* 2011; 38:1919–1930.
6. Micke A, Lewis DF, Yu X. Multichannel film dosimetry with nonuniformity correction. *Med Phys* 2011; 38:2523–2534.
7. Devic S. Radiochromic film dosimetry: Past, present, and future. *Med Phys* 2011; 27:122–134.
8. Richley L, John AC, Coomber H, Fletcher S. Evaluation and optimization of the new EBT2 radiochromic film dosimetry system for patient dose verification in radiotherapy. *Phys in Med Biol* 2010; 55:2601–2617.

9. Matney JE, Parker BC, Neck DW et al. Evaluation of a commercial flatbed document scanner and radiographic film scanner for radiochromic EBT film dosimetry. *J Appl Clin Med Phys* 2010; 11:198–208.

10. Dempsey JF, Low DA, Kirov AS, Williamson JF. Quantitative optical densitometry with scanning-laser film digitizers. *Med Phys* 1999; 26:1721–1731.

11. Boyle W, Smith G, US Patent 3,792,322: Buried channel charge coupled devices, Google Patents, 1974.

12. Boyle W, Smith G, US Patent 3,796,927: Three dimensional charge coupled devices, Google Patents, 1974.

13. Niroomand-Rad A, Blackwell CR, Coursey BM et al. Radiochromic film dosimetry: Recommendations of AAPM radiation therapy committee task group 55. *Med Phys* 1998; 25:2093–2115.

14. Devic S, Seuntjens J, Hegyi G et al. Dosimetric properties of improved GafChromic films for seven different digitizers. *Med Phys* 2004; 31:2392–2401.

15. Tamponi M, Bona R, Poggiu A, Marini P. A practical tool to evaluate dose distributions using radiochromic film in radiation oncology. *Med Phys* 2015; 31:31–36.

16. del Moral F, Vázquez JA, Ferrero JJ et al. From the limits of the classical model of sensitometric curves to a realistic model based on the percolation theory for GafChromic™ EBT films. *Med Phys* 2009; 36:4015–4026.

17. Lewis DF, Micke A, Yu X. US Patent EP 2825903 A4: Efficient method for radiochromic film dosimetry, Google Patents, 2016.

18. Van Battum LJ, Hoffmans D, Piersma H, Heukelom S. Accurate dosimetry with GafChromic™ EBT film of a 6 MV photon beam in water: What level is achievable? *Med Phys* 2008; 35:704–716.

19. Chan MF, Lewis D, Yu X. Is it possible to publish a calibration function for radiochromic film? *Int J Med Phys, Clin Eng Radiat Oncol* 2014; 3:25–30.

20. Lewis D, Micke A, Yu X, Chan MF. An efficient protocol for radiochromic film dosimetry combining calibration and measurement in a single scan. *Med Phys* 2012; 39:6339–6350.

21. Tamponi M, Bona R, Poggiu A, Marini P. A new form of the calibration curve in radiochromic dosimetry. Properties and results. *Med Phys* 2016; 43:4435–4446.

22. Saur S, Frengen J. GafChromic EBT film dosimetry with flatbed CCD scanner: A novel background correction method and full dose uncertainty analysis. *Med Phys* 2008; 35:3094–3101.

23. Dempsey JF, Low DA, Mutic S et al. Validation of a precision radiochromic film dosimetry system for quantitative two-dimensional imaging of acute exposure dose distributions. *Med Phys* 2000; 27:2462–2475.

24. Gluckman GR, and Reinstein LE. Comparison of three high-resolution digitizers for radiochromic film dosimetry. *Med Phys* 2002; 29:1839–1846.

25. Ranade MK, Li JG, Dubose RS et al. A prototype quantitative film scanner for radiochromic film dosimetry. *Med Phys* 2008; 35:473–479.

26. Rosen BS, Soares CG, Hammer CG et al. A prototype, glassless densitometer traceable to primary optical standards for quantitative radiochromic film dosimetry. *Med Phys* 2015; 42:4055–4068.

27. Bartzsch S, Lott J, Welsch K et al. Micrometer-resolved film dosimetry using a microscope in microbeam radiation therapy. *Med Phys* 2015; 42:4069–4079.

28. Kalef-Ezra J, Karava K. Radiochromic film dosimetry: Reflection vs transmission scanning. *Med Phys* 2008; 35:2308–2311.

29. Lynch BD, Kozelka J, Ranade MK et al. Important considerations for radiochromic film dosimetry with flatbed CCD scanners and EBT GAFCHROMIC® film. *Med Phys* 2006; 33:4551–4556.

30. Menegotti L, Delana A, Martignano A. Radiochromic film dosimetry with flatbed scanners: A fast and accurate method for dose calibration and uniformity correction with single film exposure. *Med Phys* 2008; 35:3078–3085.

31. Schoenfeld AA, Wieker S, Harder D, Poppe B. The origin of the flatbed scanner artifacts in radiochromic film dosimetry—Key experiments and theoretical descriptions. *Phys Med Biol* 2016; 61:7704.
32. Poppinga D, Schoenfeld AA, Doerner KJ et al. A new correction method serving to eliminate the parabola effect of flatbed scanners used in radiochromic film dosimetry. *Med Phys* 2014; 41:021707.
33. Lewis D, Chan MF. Correcting lateral response artifacts from flatbed scanners for radiochromic film dosimetry. *Med Phys* 2015; 42:416–429.
34. Schoenfeld AA, Poppinga D, Harder D et al. The artefacts of radiochromic film dosimetry with flatbed scanners and their causation by light scattering from radiation-induced polymers. *Phys Med Biol* 2014; 59:3575.
35. Paelinck L, De Neve W, De Wagter C. Precautions and strategies in using a commercial flatbed scanner for radiochromic film dosimetry. *Phys Med Biol* 2006; 52:231.
36. McCaw TJ, Micka JA, DeWerd LA. Characterizing the marker-dye correction for Gafchromic® EBT2 film: A comparison of three analysis methods. *Med Phys* 2011; 38:5771–5777.

APPLICATIONS

Kilovoltage X-ray beam dosimetry and imaging

ROBIN HILL AND JOEL PODER

5.1 OVERVIEW OF KILOVOLTAGE X-RAY BEAMS—PROPERTIES

Kilovoltage X-ray beams have a wide application in medicine for diagnostic purposes through various imaging modalities as well as therapeutic applications for cancer treatments and other clinical conditions [1,2]. In addition, these low-energy X-ray units are heavily used for radiobiological research. Kilovoltage X-ray beams are used in CT scanners, mammography X-ray units, diagnostic X-rays, and on-board imagers attached to radiotherapy linear accelerators. The imaging modalities that utilize kilovoltage X-rays include diagnostic X-ray units, CT scanners, and on-board imagers on linear accelerators. The relative dominance of the photoelectric effect and its dependence of the effective atomic number of the different tissues such as bone and soft tissue within this energy range allow clear differentiation between the different tissues.

For therapy applications such as treating skin cancers, the most common X-ray units operate at energies 50–150 kVp. The main dosimetric property of kilovoltage X-ray beams is that the maximum dose occurs close to the surface of the patient, within the first few millimeters, and the dose drops off rapidly with depth due to attenuation and scattering of the beam. In modern days, primary therapeutic use of kilovoltage X-rays is the treatment of skin cancers [3–5] as well as intraoperative therapy. Radiotherapy is often delivered after surgery in which the tumor has been surgically excised, and the radiation is used to eliminate any cancer cells remaining within the tumor bed. Therapeutic kilovoltage X-rays are also used for the treatment of keloids, AIDS-related and non-AIDS-related Kaposi's sarcoma, rectal cancer treatments, electronic brachytherapy systems, intraoperative radiotherapy, and for some palliative radiotherapy cases [6–15].

Intraoperative radiation therapy (IORT) involves the application of high doses of ionizing radiation to the tumor bed at the time of surgery such as for breast cancer patients [9,16]. There are several IORT units that deliver kilovoltage X-ray beams such as the Intrabeam (Carl Zeiss AG, Germany) with X-ray beams energies of up to 50 kVp.

There have been significant advances in these preclinical radiotherapy research platforms that are used for imaging and treating small animals such as mice [17]. These X-ray units combine very high resolution CT imaging with very small X-ray fields down to just a few millimeters and typically operating around 250 kVp for beam delivery down to field sizes of just a few millimeters [17–23]. These units allow experimentation for preclinical radiotherapy techniques such as hypofractionated radiation treatments and combined use of novel medicines with radiotherapy.

For all of these forms of X-ray units, radiochromic film has developed a niche with wide application in dosimetry imaging, and quality assurance (QA). The current chapter will provide an overview of the applications of radiochromic films for these kilovoltage X-ray beams.

5.2 SUITABILITY OF RADIOCHROMIC FILM FOR LOW-ENERGY X-RAYS

Radiochromic films have a number of intrinsic properties that make them desirable for use with low-energy X-rays. Patients treated with low-energy X-rays in radiation therapy are most often prescribed dose at the surface of the skin. Consequently, the characterization of low-energy X-ray beams focuses on measurements at the surface of phantoms. Moreover, in vivo patient measurements are almost exclusively performed at the surface of the patient contour.

The dose of low-energy X-ray beams falls off faster with depth compared with megavoltage photon beams; therefore, the micrometer thickness of radiochromic films makes them attractive for use in measuring surface dose. Radiochromic films are typically 300-μm thick with an active layer of 30 μm, and this property of the film ensures that volume averaging along the depth dose of the low-energy X-ray beam is minimized.

There are a number publications in the literature that show the use of radiochromic film for the characterization of kilovoltage X-ray beams, including the measurement of backscatter factor (BSF) [24–27]. Several of these publications showed that the radiochromic films had favorable agreement in the measurement of BSF compared with other detectors [24] and Monte Carlo simulations [26,27]. There are also a number of publications showing the use of radiochromic films for the measurement of percentage depth doses and beam profiles in the kilovoltage energy range [28–30]. As for the BSF, these publications also show favorable agreement with other detectors [30] and Monte Carlo simulations [28].

An essential property of a dosimeter for use with low-energy X-rays is minimal energy dependence in the kilovoltage energy range. In the kilovoltage X-ray range, the spectrum of X-ray energies can vary considerably with depth in the phantom due to scattering. When measuring absolute dose, the depth of measurement for the calibration films can have a considerable effect on the calculated absolute dose if the X-ray spectrum changes significantly with depth and if the dosimeter has a large energy dependence in this range. Dosimeter response is sensitive to the materials used in its construction. For low-energy X-ray beams, the photoelectric effect is a dominant interaction process, and the photoelectric cross section has a strong dependence on the atomic number of the material.

The most widely used radiochromic films in the field of radiation therapy are the GAFchromic™ external beam therapy (EBT), EBT2, and EBT3 films (Ashland, NJ). A detailed historical information is provided in Chapter 2. The first generation GAFchromic EBT films were first shown to be approximately energy independent over a wide energy range [31,32]. However, the energy dependence of later batches was found to be extremely poor in the kilovoltage energy range, with a 40% decrease in response for a 20-keV beam relative to megavoltage X-rays [33–35]. This variation in energy dependence between batches arose from the difference in chemical composition of the film [31]. A decrease in chlorine content (and therefore effective atomic number) is observed in recent lots of the EBT film that was consistent with the observed decrease response at orthovoltage energies. However, this version of film is no longer manufactured.

The second generation EBT2 films showed improved energy dependence as compared with the first generation product. This was achieved due to an improved chemical composition, and published energy dependence of the films is in the range of 4%–6% between 60 kVp and 10 MeV photons beams [29,33,36]. Lindsay et al. [37] found that the film response, as measured by optical density on the scanner, was up to 14% for the 105 kVp beam as compared with 200 kVp. In comparison, the response for the 200 kVp beam was within 2% as compared with 6 MV beam.

Similar results were also found by Butson et al. [36] for the EBT2 film that had energy response variations of up to 6.5% for X-ray beams with energies from 50 kVp to 10 MV with all being relative to the 6 MV X-ray beams. The maximum response occurred at 100 kVp and from beam energies above 200 kVp; the response was within 1.5% that was well within the uncertainties of the readings. Arjomandy et al. [29] examined the energy dependence of EBT2 film for kilovoltage X-ray beams at energies of 75, 125, and 250 kVp as well as megavoltage X-ray beams, electrons, and proton beams. They found that the energy dependence was relatively small being within measurement uncertainties of $1\sigma = \pm4.5\%$ and concluded that this version of film was very suitable for clinical dosimetry measurements.

The most recently released GAFchromic EBT3 films have been shown to have even further improved energy dependence, with multiple publications showing that the EBT3 films are almost

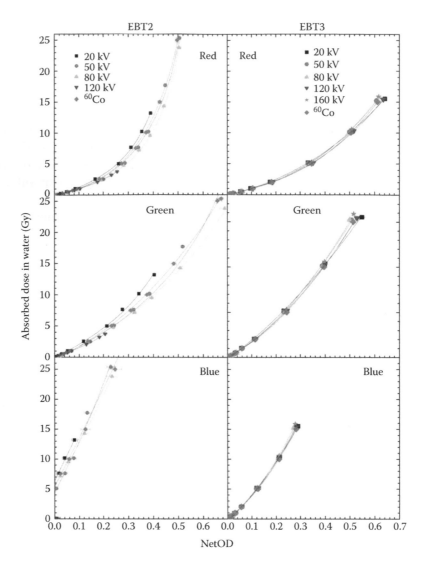

Figure 5.1 The absorbed dose as a function of NetOD using GAFchromic™ EBT3 film for a range of X-ray beam energies ranging from 20 kVp to cobalt 60. (From Guerda, et al., *Bio. Phys. Eng. Exp.*, 2, 045005, 2016.)

completely energy independent down to 20 keV [33,34,38,39]. The study by Massillon-JL et al. [39] evaluated optimal absorbed dose and optical density functions for a range of kilovoltage beam energies and different channels with the results presented in Figure 5.1 showing minimal variation as a function of beam energy. However, an important point from these publications is the recommendation to use the same or a similar energy X-ray beam for calibration of the film as used in the actual measurements in order to minimize any energy response variations and reduce uncertainties. This is particularly recommended for X-ray beams with an energy less than 200 kVp.

5.2.1 ENERGY RESPONSE OF GAFCHROMIC™ FILMS—SUITABILITY FOR kV BEAMS

For the GAFchromic verification-type films such as the XR-QA and similar models, there have been a number of studies performed to check their relative energy response. Butson et al. [40] studied the GAFchromic XR type-R radiochromic film over a range of diagnostic X-ray energies. They showed that the energy response of the GAFchromic XR type-R film was only small over the energy range from 75 to 125 kVp range with a 9% variation. But the film had a larger energy response variation when compared over a larger therapeutic X-ray range with a relative response of more than 10 for 125 kVp X-ray beams as compared with 6-MV beams.

Chiu-Tsao et al. [41] evaluated the energy dependence of the XR-QA film including for low-energy brachytherapy seeds like Palladium-103 with effective energy of 21 keV. In this study, they found that the XR-QA film sensitivity varies with radiation energy by factors of more than 15 depending on which channel of light was used during film scanning and analysis. They concluded that the XR-QA film was suitable for kilovoltage sources with a narrow energy spectrum.

The study by Rampado et al. [42] examined the response of the GAFchromic XR-QA film to a range of X-ray beam energies (28–145 kVp and with various filtrations) in the dose range from 0 to 100 mGy. They found that the response of the film varied according to X-ray beam energy with the maximum response occurring for X-ray beams in the energy range of 80–140 kVp. However, they found that the uncertainty in the dose measurements increased when used for the lower X-ray beam energies.

More recently, Ebert et al. [16] examined the response characteristics of three commercially available radiochromic films when exposed to very low-energy (50 kVp) X-rays from an IORT device. They determined dose–response relationships for the GAFchromic EBT, XR-RV2, and XR-QA films in water and evaluated the effect of changes in X-ray spectrum on response. They found that for all three films, there was significant energy dependence and nonlinear response within the limits of measurement uncertainty.

5.3 RADIOTHERAPY APPLICATIONS

5.3.1 RELATIVE DOSIMETRY

Relative dosimetry data of kilovoltage X-ray beams are measured for the commissioning of the X-ray unit, beam data collection for QA testing, reference dosimetry measurements, and treatment planning calculations [1,43–48]. These data typically include depth doses, profiles, relative output factors (ROFs) both for applicators and lead cutouts, and backscatter factor. Ionization chambers are often considered the gold standard dosimeter for kilovoltage X-ray beam dosimetry in the radiotherapy [49–55]. There are, however, many situations when other dosimeters are required for these measurements. A number of studies have investigated the use of radiochromic films for relative dosimetry of therapeutic kilovoltage X-ray beams with promising results. The high spatial resolution of radiochromic film makes it an ideal dosimeter for measuring lateral profiles of kilovoltage X-ray beams.

5.3.1.1 DEPTH-DOSE AND LATERAL PROFILE MEASUREMENTS

Fletcher and Mills [56] measured depth doses for 50 and 100 kVp X-ray beams using a Wellhoefer small volume ionization chamber, PTW soft X-ray parallel-plate chamber type (N23342), a Wellhoefer photon field diode (PFD) diode as well as GAFchromic EBT, and MD55 films. A Gulmay Medical D3000 DXR unit (Gulmay Limited, Surrey, UK) was used to generate the 50 kV and 100 kV X-rays used in the study for a variety of field sizes. In addition, the BEAMnrc Monte Carlo was used to calculate depth doses in water using an analytical program to generate the primary X-ray beam spectra.

For the film measurements, the film pieces were exposed parallel to the central axis in a solid-water phantom in direct contact with the applicator. The film was held in a central area of the phantom and the whole assembly was held together tightly to minimize the effects of any air gaps as shown in Figure 5.2.

The measurements from all the dosimeters showed agreement with BEAMnrc dose calculations at depths greater than 10-mm, whereas near the surface GAFchromic film and PFD diodes gave the best agreement to Monte Carlo values. The results of the comparison for the 100-kVp X-ray beam are shown in Figure 5.3. The measurements with the two types of GAFchromic film agree well with each other with a maximum difference less than ±4.5%. In comparison, the film depth-dose measurements show an overall agreement with BEAMnrc predicted values, to within ±8% for both energies and at all depths.

Aspradakis et al. [48] used GAFchromic EBT3 film for verification of depth doses of X-ray beams in the energy range 30–200 kVp X-ray unit. They were particularly interested in validating the relative depth doses and within the first 10 mm. For comparison, depth doses were measured with an IC-15 in water and with a PTW Advanced Markus plane-parallel ionization chamber (PTW-Freiburg) and with GAFchromic EBT3 film in a Plastic Water DT® (PWDT) solid phantom (CIRS, Norfolk, VA). Measured depth doses were also compared against values interpolated from the British Journal of Radiology (BJR) Report 25 [57]. Films were calibrated at the surface for low-energy beams (<100 kVp) and at the depth of 2-cm in PWDT for medium-energy beams (>100 kVp). Calibration films were used for each of the X-ray beams used. The depth doses were measured in PWDT with films positioned at different depths vertical to the X-ray beam central axis. After the films were scanned, they used the red channel only for the dose analysis.

For the medium-energy X-ray beam energies greater than 100 kVp [HVL (half-value layer) of 4.52 mm Al], the measurements with the IC-15 in water and with the Advanced Markus and EBT3 film in PWDT agreed within ±2%. The differences from the data published in BJR-25 [57] were

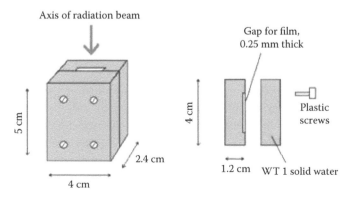

Figure 5.2 The solid-water phantom used to hold the GAFchromic EBT film for the depth-dose measurements. (From Fletcher, C.L. and Mills, J.A., *Phys. Med. Biol.*, 53, N209–N218, 2008.)

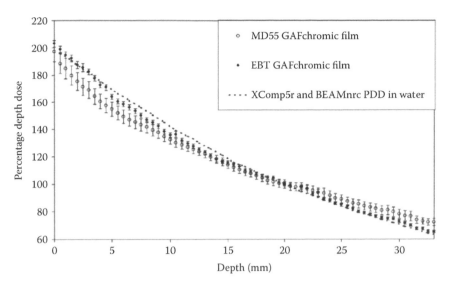

Figure 5.3 A comparison of the percentage depth-dose curves for 100-kV photons measured with EBT film as well as Monte Carlo predictions. (From Fletcher, C.L. and Mills, J.A., *Phys. Med. Biol.*, 53, N209–N218, 2008.)

within ±5% at this energy range. For X-ray beams with energies between 50 and 100 kVp (corresponding to HVLs of 1.2–4.52 mm Al), the differences between doses in the water and PWDT increased with decreasing beam energy but with a maximum with deviations not exceeding 2.5%.

For the 30 and 40 kVp X-ray beams (half value layer [HVL] of 0.45 and 0.84 mm Al, respectively), the differences between depth doses determined between the ionization chamber measurements and the film were greater than 10%. It was noted that for these low X-ray beam energies, the suitability of the detectors in terms of energy dependence and the water equivalence of PWDT was not known and therefore led to a greater uncertainty in the dose measurements.

The work by Sheu et al. [58] summarizes their experience in commissioning two Sensus SRT-100 (Sensus Healthcare, Boca Raton, FL) therapeutic X-ray units that generate X-ray beams with energies of 50–100 kVp. Percentage depth–dose curves were measured with a PTW parallel-plate chamber (N23342) in a solid-water phantom over a range of depths. They used GAFchromic EBT3 film to validate depth doses for several combinations of beam energies and field sizes. They found that depth doses agreed to within 4% over the full range of depths for the 50-kVp beam with a 2.5-cm diameter applicator. They also measured profiles using the GAFchromic EBT3 film and were compared with profiles measured with a parallel-plate ionization chamber. They found that the uniformity of the beam over a circle of 2-cm diameter was within 4% in all four directions. The ionization chamber readings agreed with film measurements to within 1%–3%.

Steenbeke et al. [59] performed an extensive study to see if GAFchromic EBT3 film is suitable as a dosimeter for a Dermopan 2 radiotherapy X-ray unit (Siemens Industry, Berlin, Germany) that generates 50-kVp X-ray beams (HVL of 0.92 mm Al). They used GAFchromic EBT3 films to measure both absolute and relative doses, and these doses were compared with a PTW parallel-plate ionization chamber (type 23342). The relative dosimetry measurements consisted of depth doses, lateral beam profiles, and ROFs. All dose measurements were performed within a polymethyl methacrylate (PMMA) phantom. For the depth-dose measurements, films were placed perpendicular to the beam axis at depths from 0 to 30-mm.

The agreement in the absolute dose measurements at the surface using the ionization chamber and EBT3 film was very good and with differences of less than 1% using both the red and green channels. For the depth-dose measurement, they found an average difference of 1.2% between the ionization chamber and GAFchromic EBT3 film readings with a maximum difference of 3.5%. For output factors, the difference between IC and film doses using the red channel was within 2% for applicator field sizes of 1 and 2 cm. Greater differences were found when using the green channel.

5.3.1.2 RELATIVE OUTPUT FACTORS

The ROF is required to specify the dose output for a particular applicator and/or lead cutout for a patient treatment as compared with the dose output for the reference applicator [2]. In most cases, the ROF is usually specified at the surface of a water phantom and given by the following equation:

$$ROF = \frac{\text{Dose in a particular field size}}{\text{Dose in the reference applicator}}$$

The ROF is typically measured using a suitable ionization chamber such as a parallel-plate chamber [50,52,60]. However, some clinics may perform the ROF measurements at depth and then use depth-dose corrections to determine the dose at the surface using either water or a solid-water equivalent phantom [44,46]. If a solid phantom is used, it should be one that is radiologically water equivalent particularly for lower energy photons such as RMI457 Solid Water (RMI Gammex, USA) [50,61–63].

The GAFchromic EBT3 film has been found to be generally suitable for the measurement of ROFs for X-ray beams in the energy range from 50 to 125 kVp [24]. It is found that the agreement between output factors measured with EBT3 film, a parallel-plate ionization chamber and ratios of published BSFs was generally better than 2%. The larger differences were up to 3.3% that occurred for the smallest field size of 2-cm diameter used in that study.

An alternative method for determining ROFs is by performing GAFchromic film measurements in air and then applying BSFs to determine dose at the water surface [20,44]. The use of published BSF and depth-dose data for the calculation of output factors leads to an increase in dose uncertainties [64,65]. There can be significant differences between measured output factors and factors calculated using the ratio of published BSFs [64,66].

For lead cutouts, it has been shown that there is a dose enhancement due to increased electron scatter from the edge of the lead Lye et al. [67] as shown in Figures 5.4 (a) and (b) for no additional shielding and with 200 μm plastic wrap respectively. This dose enhancement at the edge of the lead cutout can be minimized by wrapping the lead with thin layers of plastic wrap. This means that ROF measured with a lead cutout should use plastic wrap around the lead cutout.

5.3.1.3 BACKSCATTER FACTOR

For radiotherapy purposes, the BSF in water, B_w, is defined as the ratio of the water collision kerma at a point on the beam axis at the surface of a full-scatter water phantom to the water collision kerma at the same point in the primary beam with no phantom present [68,69]. The BSFs for a range of beam qualities are available in various published kilovoltage X-ray beam reference dosimetry protocols [70–72]. It has been shown that the B_w varies as a complex function of the X-ray beam energy, field size, focus to surface distance (FSD), and also with different materials in the phantom [51,73–77].

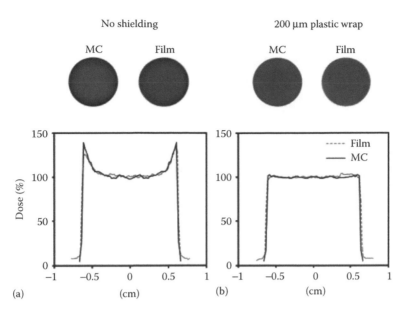

Figure 5.4 Electron contamination from lead cutout calculated using Monte Carlo and measured with GAFchromic EBT2 film. (a) No shielding and (b) 200 μm plastic wrap.

The determination of B_w requires the measurement of dose at the surface of a phantom and in-air at the same point. For low-energy X-ray beams, there may be problems with the accuracy of absorbed dose measurements with ionization chambers or other dosimeters [78,79]. The size and shape of the dosimeter may cause photon fluence perturbations and the energy response of the detector may change if the X-ray spectrum changes between the reference and measuring points [80].

The size and dimensions of the various GAFchromic films make them ideal for the direct measurement with an effective depth corresponding to the International Commission on Radiation Units and Measurements (ICRU) skin depth. GAFchromic EBT and EBT2 film have been used to measure BSF in the energy range from 50 to 280 kVp. In two studies, measured BSFs using EBT film were compared with published BSF and Monte Carlo calculations, giving differences of up to 2.5% [77,81]. A later study by Smith et al. [27] showed a good agreement between BSFs measured using GAFchromic EBT2 film and the published data with a maximum difference of 3%. This level of agreement was similar to the estimated uncertainties in the film measurements.

A summary of BSF measured with GAFchromic EBT and EBT2 films, Monte Carlo calculations, and published data is shown in Table 5.1. Based on these studies, determination of BSF for kilovoltage X-ray beams can be achieved in the clinic by measurements using TLD chips, GAFchromic film, or Monte Carlo calculations.

Several studies have used GAFchromic film to measure changes in the dose due to changes in BSF for various clinical applications. For example, the use of nasal shields, Peter et al. [82] used EBT2 film to measure changes in the dose on the surface of the nose due to having nasal shields inserted during kV treatment. In a similar study, Butson et al. [83] used EBT film to verify doses on the nose in order to verify changes in dose due variations to the geometry and presence of different tissues. There are other areas such as eye shields in which film has been successfully used. Eye shields are used in radiotherapy treatments to reduce dose to the lens which has dose limits [84,85]. The work by Butson et al. [86] used EBT film to determine the dose underneath the eye shields.

Table 5.1 A comparison of BSFs determined by radiochromic film measurements, Monte Carlo calculations, and published values

50 kVp X-ray beam (cm diameter)	EBT film	EBT2 film	BEAMnrc	AAPM TG-61
2	1.091	1.064	1.098	1.096
4	1.132	1.113	1.146	1.141
6	1.188	1.188	1.172	1.169
100 kVp X-ray beam (cm diameter)	**EBT film**	**EBT2 film**	**BEAMnrc**	**AAPM TG-61**
2	1.124	1.098	1.120	1.122
4	1.218	1.197	1.186	1.211
6	1.269	1.234	1.246	1.258
280 kVp X-ray beam (cm diameter)	**EBT film**	**EBT2 film**	**BEAMnrc**	**AAPM TG-61**
2	1.064	1.051	1.059	1.052
4	1.096	1.080	1.095	1.097
6	1.139	1.100	1.134	1.137

Source: Smith, L. et al., *Austral. Phys. Eng. Sci. Med.*, 34, 261–266, 2011; Ma, C.M. and Seuntjens, J.P., *Phys. Med. Biol.*, 44, 131–143, 1999; Kim, J. et al., *Phys. Med. Biol.*, 55, 783–797, 2010.

5.3.1.4 VERY SMALL FIELDS

Several studies have determined the relative dosimetry data for kilovoltage X-rays used for IORT and animal irradiators often with very small field sizes [9,17,20,21,87,88]. The IORT X-ray units typically deliver X-rays with peak potentials of 50 kVp. By comparison, animal irradiators use kilovoltage X-ray tubes generally at higher voltages of around 225 kVp but deliver X-ray beams with small field sizes ranging from 40-mm diameter down to 1-mm diameter.

Depth-dose, lateral profile, and relative output factor data have been determined in some of these studies using GAFchromic film with comparison with both Monte Carlo calculations and various dosimeters. The study by Newton et al. [20] characterized the XRad225Cx (Precision X-Ray Inc) that operates at 225 kVp and has X-ray beams with field sizes down to 1-mm diameter. The agreement in ROFs measured with PRESAGE and EBT2 film was better than 2.8%. They found that there was good consistency in the agreement in depth doses and profiles measured using the ionization chambers, GAFchromic EBT2 film, and the PRESAGE dosimeter for fields down to a 1-mm diameter. Figure 5.5 shows depth-dose and lateral profile data for two fields of 1 and 15 mm diameter.

5.3.1.5 2D DOSIMETRY

The radiation oncologist when choosing particular beam energy for treatment usually considers the beam penetration and the depth of the 95% or 90% relative dose. Further planning information is provided by using 2D isodoses that allow one to check the lateral spread of doses. This has two important purposes—the volume contained by the higher relative dose such as at the 95% or 90% to ensure appropriate dose coverage. In addition, one may be interested in the coverage of the lower dose regions particularly if there is any low dose coverage to the critical structures such as the eye.

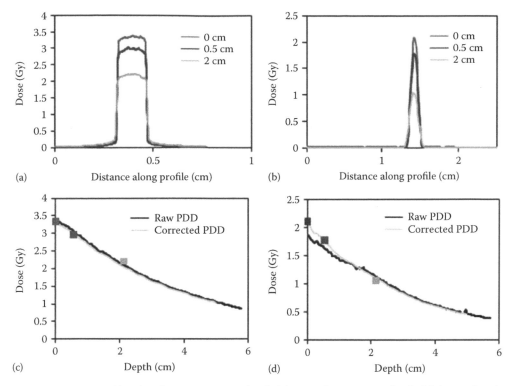

Figure 5.5 Beam profiles for the 15-mm circular field (a) and 1-mm circular field (b) at depths of 0-cm, 0.5-cm and 2-cm in solid water. (c) The PDD curve for the 15-mm circular field and (d) for the 1-mm field.

The measurement of isodose maps is typically performed in a scanning water tank by direct tracking with a small-volume ionization chamber. However, this process can take a long time if there are a large number of X-ray beam energies and treatment applicators. A quicker method involves using the scanning water tank software to calculate isodoses based on a set of depth doses and profile measurements that can be achieved by films. The radiochromic film should be positioned vertically in blocks of suitable solid phantoms. The edge of the film should be flush with the edge of the solid phantoms. The only issue with this sort of measurement is dealing with edge effects of the film [48,56].

5.4 QUALITY ASSURANCE TESTS

5.4.1 LEAKAGE TESTS

At the time of commissioning of a new kilovoltage X-ray unit or a new X-ray tube, it is important to test for any leakage from the tube. This can be readily achieved by wrapping the whole X-ray tube with high sensitivity radiochromic film, for example, GAFchromic RTQA film [48]. The main purpose of this test is to find any hot spots of leakage radiation around the tube and then use a calibrated survey meter to determine actual leakage dose rates.

Radiochromic film may also be used for testing for leakage from the treatment applicators on the kV unit. Aspradakis et al. [48] performed the applicator leakage tests using GAFchromic RTQA film.

The film was wrapped around the applicator walls and results indicated some leakage radiation through the screws as shown on the film placed at the end of the applicator. The absolute doses were then verified by measurements using a Farmer-type ionization chamber.

5.4.2 HALF VALUE LAYER MEASUREMENTS

Knowledge of the beam quality is a precursor to any reference dosimetry (or absolute dose calibration) of kilovoltage X-ray beams. In current dosimetry protocols, the beam quality is specified in terms of the HVL in combination with the peak generating voltage (kVp) [70–72]. The HVL is defined in terms of the thickness of absorber (typically high purity aluminum or copper) that reduces the air-kerma rate by a factor of one half [72]. Many of the parameters used in the reference dosimetry calculations are defined as a function of the HVL. During beam commissioning, it is common to measure both the first and second HVLs for each radiation beam for QA purposes. Radiochromic film can be used to ensure good practice during the HVL measurement. This is done by placing a piece of radiochromic film behind the ionization chamber to ensure that it is positioned at the center of the narrow beam field.

5.4.3 FOCAL SPOT SIZE

The focal spot size and position can have a major impact on the profile shape of X-rays from an X-ray tube [89,90]. This can lead to larger beam penumbra and asymmetry as required for clinical purposes. Measurements for the focal spot can readily be performed using radiochromic film as suggested by Aspradakis et al. [48] using RTQA film. Using film, it was found that the spot size was 7.5×7.5 mm^2 for all X-ray tube potentials on the WOMed therapeutic X-ray unit.

5.4.4 FIELD-SIZE CHECKS

Field-size checks were performed by Aspradakis et al. [48] for all applicators examined from exposures at 150 kV using RTQA film placed on a 5-cm thick plastic block at the end of the applicator. The width of the exposed image was visually compared against the nominal field size as defined by the applicator aperture and found to be in good agreement. Such process with film provides quick and unique opportunity for applicator QA.

5.4.5 IN VIVO DOSIMETRY

Radiochromic film has been shown to be a suitable dosimeter for in vivo dosimetry within kilovoltage X-ray beams [6,59,82,91]. Details can also be found in Chapter 17. This allows verification of doses to the patient and comparison with the prescription dose as specified by the radiation oncologist. This is particularly important in regions where there will be greater dose uncertainties due to tissue inhomogeneities and variations in the FSD of the patient due to contours. For example, in the region around the nose, the presence of cartilage, bone, air gaps, and the geometry of the nose will cause changes in the surface dose.

There are a number of conditions that will allow radiochromic film to be used and provide accurate doses for in vivo dosimetry [1,6,59,66]. These include having a reproducible film scanning processes, using the same X-ray beam energy for the calibration as used for the treatment and not stacking films on top of each other. We have successfully used pieces of film down to film sample sizes of 7×7 mm^2 with uncertainties of about 3%.

Film dosimetry measurements have also been performed to verify changes in surface dose due to changes in backscatter due to the underlying bone or shielding materials. This is particularly dominant for larger field sizes and when the bone is located only a few millimeters below the skin surface. We have found that the film measured the change in surface dose to within 2% as compared with the predicted change in surface dose [51,83,92,93].

5.5 IMAGING SYSTEMS

Imaging is an important component of the planning and treatment for radiotherapy patients. Although CT imaging was traditionally used primarily for diagnostic purposes, it is now a routine process for the treatment planning of all radiotherapy patients [94]. More recently, there have been significant advances in the use of low-energy X-ray imaging prior to and during the treatment process for radiotherapy patients. This has led to greater accuracy for patient setup and the ability to use smaller margins for the planning target volume (PTV) allowing greater doses to the tumor volume and reducing doses to organs at risk. The process of Image-Guided Radiotherapy is now widespread with most linear accelerators and other treatment units having some form of low-energy X-ray imaging [95–97]. This section discusses the use of radiochromic film for QA testing and dosimetry of X-ray imagers that utilize kilovoltage X-ray beams.

5.5.1 X-RAY UNITS

Radiochromic film is being used for QA of X-ray units as well as estimations of skin doses for patients during interventional radiological procedures. Several groups have used radiochromic film for dosimetry studies in interventional radiology procedures including verification of skin doses [98–101] and for the dosimetry of mammography units [102]. However, it was shown that there can be large differences in film doses as compared with those determined by ionization chamber measurements [99]. In one study, the use of radiochromic XR film was investigated for patient skin dose monitoring in a cardiac catheterization laboratory [100]. They found that radiochromic film was a suitable method to identify skin regions at risk from high X-ray doses and could be used for estimation of the localized skin dose to those patients who are susceptible to radiation-induced skin injury.

Gotanda et al. [103,104] evaluated the use of radiochromic XR film for QA of diagnostic X-ray units. This included testing HVL of the X-ray instead of using an ionization chamber. Film measurements were performed with an aluminum step wedge and the HVL determined using from the optical density information. This was performed for X-ray beams ranging in energy from 80, 100, and 120 kVp and found the differences of 10.0%, 7.6%, and 1.2%, respectively, the ionization chambers and GAF-R flim. This suggests that film is potentially suitable for the higher beam energy.

5.5.1.1 CT SCANNERS

A number of studies have investigated the use of radiochromic film for quality assurance testing and dose measurements on CT scanners. Several groups used radiochromic film to determine full width at half maximum of the X-ray beam profiles from a CT scanner. Duan et al. [105] performed a study on X-ray spectrum estimation from a CT scanner using transmission measurements. They verified the width of the X-ray beams in the CT scanner using GAFchromic XR-CT dosimetry film.

Gorny et al. [106] used the GAFchromic HXR film for CT dosimetry due to its higher sensitivity as compared with the GAFchromic XR film. They found that this HXR film could be used for measurement of dose profiles for very low doses below 100 mGy and with a precision of better than 1%. Rampado et al. [107] investigated using GAFchromic XR-QA film for determining the computed tomography dose index (CTDI). They found differences of nearly 50% in the response of the film depending on its orientation relative to the radiation beam, integrating the film. However, when the radiation dose was integrated over 360 degrees corresponding to the full rotation of the CT scanner, the difference in CTDI measured with the film and an ionization chamber was within two standard deviations of uncertainty. As such, it was concluded that this film could be used to estimate values of CTDI for CT scanners.

5.5.2 ON-BOARD IMAGER UNITS

On-board imager (OBI) units are low-energy X-ray units mounted onto linear accelerators that use kilovoltage energies, which are typically in the range from 40 to 125 kVp for accurate patient positioning [95,108,109]. These OBI units operate in two different modes as follows: 2D imaging with planar X-rays and 3D imaging via cone beam CT (CBCT) modes [108]. A number of studies have used a range of dosimeters for determining dosimetry of these OBI units including radiochromic film for the dosimetry of these X-ray units [107,109–118].

Hu and McLean [112] compared three different dosimetry systems for CBCT dosimetry as follows: pencil ionization chambers, a Farmer-type thimble ionization chamber, and GAFchromic XR QA2 film in various phantoms. They found good agreement to within 3% for the three dosimeters for the scanning protocols studies. However, they noted that the measured CBCT doses were up to 40% higher compared with the manufacturer doses. As such, this showed the suitability and importance of independent dose measurements that can be performed in the local radiotherapy department using radiochromic film. The Figure 5.6 shows the beam profiles using the high quality head scanning protocol.

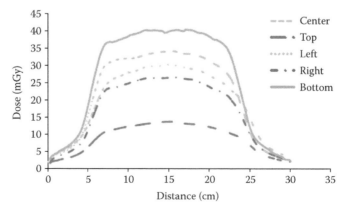

Figure 5.6 Profiles measured with the GAFchromic XR QA2 films using high-quality head scan on a Varian linear accelerator. (From Hu, N. and McLean, D., *Austral. Phys. Eng. Sci. Med.*, 37, 779–789, 2014.)

5.6 SUMMARY

Radiochromic film has found wide application in the dosimetry and QA of therapeutic kilovoltage X-ray units for treating cancer patients as well as in imaging systems that use low-energy X-rays. The near tissue equivalence, high spatial resolution, and relative ease of use of radiochromic film has shown it to be very useful in radiotherapy and imaging applications. It is anticipated that the use of radiochromic film will increase for these reasons and find further applications.

REFERENCES

1. Hill R, Healy B, Holloway L et al. Advances in kilovoltage X-ray beam dosimetry. *Phys Med Biol* 2014;59:R183.
2. Mayles P. Kilovoltage X-rays. In: Mayles P, Nahum AE, Rosenwald J (Eds.). *Handbook of Radiotherapy Physics*. Boca Raton, FL: CRC Press; 2007. pp. 439–449.
3. Amdur RJ, Kalbaugh KJ, Ewald LM et al. Radiation therapy for skin cancer near the eye: Kilovoltage X-rays versus electrons. *Int J Radiat Oncol Biol Phys* 1992;23:769–779.
4. Locke J, Karimpour S, Young G et al. Radiotherapy for epithelial skin cancer. *Int J Radiat Oncol Biol Phys* 2001;51:748–755.
5. Poen JC. *Clinical Applications of Orthovoltage Radiotherapy: Tumours of the Skin, Endorectal Therapy and Intraoperative Radiation Therapy*. Madison, WI: Medical Physics Publishing; 1999.
6. Avanzo M, Rink A, Dassie A et al. In vivo dosimetry with radiochromic films in low-voltage intraoperative radiotherapy of the breast. *Med Phys* 2012;39:2359–2368.
7. Clausen S, Schneider F, Jahnke L et al. A Monte Carlo based source model for dose calculation of endovaginal TARGIT brachytherapy with INTRABEAM and a cylindrical applicator. *Z Med Phys* 2012;22:197–204.
8. D'Alimonte L, Sinclair E, Seed S. Orthovoltage energies for palliative care in the 21st century: Is there a need? *Radiography* 2011;17:84–87.
9. Eaton DJ. Quality assurance and independent dosimetry for an intraoperative X-ray device. *Med Phys* 2012;39:6908–6920.
10. Eaton DJ, Barber E, Ferguson L et al. Radiotherapy treatment of keloid scars with a kilovoltage X-ray parallel pair. *Radiother Oncol* 2012;102:421–423.
11. Eaton DJ, Best B, Brew-Graves C et al. In vivo dosimetry for single-fraction targeted intraoperative radiotherapy (TARGIT) for breast cancer. *Int J Radiat Oncol Biol Phys* 2012;82:e819–e824.
12. Gérard JP, Myint AS, Croce O et al. Renaissance of contact X-ray therapy for treating rectal cancer. *Exp Rev Med Dev* 2011;8:483–492.
13. Rong Y, Welsh JS. Surface applicator calibration and commissioning of an electronic brachytherapy system for nonmelanoma skin cancer treatment. *Med Phys* 2010;37:5509–5517.
14. Schneider F, Fuchs H, Lorenz F et al. A novel device for intravaginal electronic brachytherapy. *Int J Radiat Oncol Biol Phys* 2009;74:1298–1305.
15. Fletcher C, Mills J, Baugh G et al. Comparison of 50 kV facilities for contact radiotherapy. *Clin Oncol* 2007;19:655–660.
16. Ebert MA, Asad AH, Siddiqui SA. Suitability of radiochromic films for dosimetry of low energy X-rays. *J Appl Clin Med Phys* 2009;10:232–240.
17. Verhaegen F, van Hoof S, Granton PV et al. A review of treatment planning for precision image-guided photon beam pre-clinical animal radiation studies. *Z Med Phys* 2014;24:323–334.
18. Bazalova M, Zhou H, Keall PJ et al. Kilovoltage beam Monte Carlo dose calculations in submillimeter voxels for small animal radiotherapy. *Med Phys* 2009;36:4991–4999.
19. Chow JCL. Depth dose dependence of the mouse bone using kilovoltage photon beams: A Monte Carlo study for small-animal irradiation. *Rad Phys Chem* 2010;79:567–574.

20. Newton J, Oldham M, Thomas A et al. Commissioning a small-field biological irradiator using point, 2D, and 3D dosimetry techniques. *Med Phys* 2011;38:6754–6762.
21. Verhaegen F, Granton P, Tryggestad E. Small animal radiotherapy research platforms. *Phys Med Biol* 2011;56:R55–R83.
22. Aldelaijan S, Nobah A, Alsbeih G et al. Dosimetry of biological irradiations using radiochromic films. *Phys Med Biol* 2013;58:3177.
23. Wong J, Armour E, Kazanzides P et al. High-resolution, small animal radiation research platform with X-ray tomographic guidance capabilities. *Int J Rad Oncol Biol Phys* 2008;71:1591–1599.
24. Gill S, Hill R. A study on the use of Gafchromic™ EBT3 film for output factor measurements in kilovoltage X-ray beams. *Australas as Phys Eng Sci Med* 2013;36:465–471.
25. Butson MJ, Cheung T, Yu PKN. Radiochromic film for verification of superficial X-ray back-scatter factors. *Aust ralas Phys Eng Sci Med* 2007;30:269–273.
26. Kim J, Hill R, Mackonis EC et al. An investigation of backscatter factors for kilovoltage X-rays: A comparison between Monte Carlo simulations and Gafchromic EBT film measurements. *Phys Med Biol* 2010;55:783.
27. Smith L, Hill R, Nakano M et al. The measurement of backscatter factors of kilovoltage X-ray beams using Gafchromic™ EBT2 film. *Australas as Phys Eng Sci Med* 2011;34:261–266.
28. Claire LF, John AM. An assessment of GafChromic film for measuring 50 kV and 100 kV percentage depth dose curves. *Phys Med Biol* 2008;53:N209.
29. Arjomandy B, Tailor R, Anand A et al. Energy dependence and dose response of Gafchromic EBT2 film over a wide range of photon, electron, and proton beam energies. *Med Phys* 2010;37:1942–1947.
30. Arjomandy B, Tailor R, Zhao L et al. EBT2 film as a depth-dose measurement tool for radiotherapy beams over a wide range of energies and modalities. *Med Phys* 2012;39:912–921.
31. Rink A, Vitkin IA, Jaffray DA. Energy dependence (75 kVp to 18 MV) of radiochromic films assessed using a real-time optical dosimeter. *Med Phys* 2007;34:458–463.
32. Chiu-Tsao S-T, Ho Y, Shankar R et al. Energy dependence of response of new high sensitivity radiochromic films for megavoltage and kilovoltage radiation energies. *Med Phys* 2005;32:3350–3354.
33. Brown TAD, Hogstrom KR, Alvarez D et al. Dose-response curve of EBT, EBT2, and EBT3 radiochromic films to synchrotron-produced monochromatic X-ray beams. *Med Phys* 2012;39:7412–7417.
34. Bekerat H, Devic S, DeBlois F et al. Improving the energy response of external beam therapy (EBT) GafChromicTM dosimetry films at low energies (≤100 keV). *Med Phys* 2014;41:022101.
35. Richter C, Pawelke J, Karsch L et al. Energy dependence of EBT-1 radiochromic film response for photon (10 kVp–15 MVp) and electron beams (6–18 MeV) readout by a flatbed scanner. *Med Phys* 2009;36:5506–5514.
36. Butson MJ, Yu PKN, Cheung T et al. Energy response of the new EBT2 radiochromic film to X-ray radiation. *Radiat Meas* 2010;45:836–839.
37. Lindsay P, Rink A, Ruschin M et al. Investigation of energy dependence of EBT and EBT-2 Gafchromic film. *Med Phys* 2010;37:571–576.
38. Sorriaux J, Kacperek A, Rossomme S et al. Evaluation of Gafchromic® EBT3 films characteristics in therapy photon, electron and proton beams. *Phys Med* 2013;29:599–606.
39. Massillon-JL G, Iván DM-M, Porfirio D-A. Optimum absorbed dose versus energy response of Gafchromic EBT2 and EBT3 films exposed to 20–160 kV X-rays and 60 Co gamma. *Bio Phys Eng Exp* 2016;2:045005.
40. Butson MJ, Cheung T, Yu PKN. Measurement of energy dependence for XRCT radiochromic film. *Med Phys* 2006;33:2923–2925.

41. Chiu-Tsao ST, Ho Y, Shankar R et al. Energy dependence of response of new high sensitivity radiochromic films for megavoltage and kilovoltage radiation energies. *Med Phys* 2005;32:3350–3354.

42. Rampado O, Garelli E, Deagostini S et al. Dose and energy dependence of response of Gafchromic® XR-QA film for kilovoltage X-ray beams. *Phys Med Biol* 2006;51:2871–2881.

43. Butson MJ, Mathur J, Metcalfe PE. Dose characteristics of a new 300 kVp orthovoltage machine. *Aust ralas Phys Eng Sci Med* 1995;18:133–138.

44. Evans PA, Moloney AJ, Mountford PJ. Performance assessment of the Gulmay D3300 kilovoltage X-ray therapy unit. *Br J Radiol* 2001;74:537–547.

45. Hill R, Mo Z, Haque M et al. An evaluation of ionization chambers for the relative dosimetry of kilovoltage X-ray beams. *Med Phys* 2009;36:3971–3981.

46. Jurado D, Eudaldo T, Carrasco P et al. Pantak therapax SXT 150: Performance assessment and dose determination using IAEA TRS-398 protocol. *Br J Radiol* 2005;78:721–732.

47. Podgorsak EB. *Radiation Physics for Medical Physicists*. Berlin, Germany: Springer-Verlag; 2006.

48. Aspradakis MM, Zucchetti P. Acceptance, commissioning and clinical use of the WOmed T-200 kilovoltage X-ray therapy unit. *Br J Radiol* 2015;88:20150001.

49. Das IJ, Akber SF. Ion recombination and polarity effect of ionization chambers in kilovoltage X-ray exposure measurements. *Med Phys* 1998;25:1751–1757.

50. Healy BJ, Gibbs A, Murry RL et al. Output factor measurements for a kilovoltage X-ray therapy unit. *Aust ralas Phys Eng Sci Med* 2005;28:115–121.

51. Healy BJ, Sylvander S, Nitschke KN. Dose reduction from loss of backscatter in superficial X-ray radiation therapy with the Pantak SXT 150 unit. *Aust ralas Phys Eng Sci Med* 2008;31:49–55.

52. Hill R, Holloway L, Baldock C. A dosimetric evaluation of water equivalent phantoms for kilovoltage X-ray beams. *Phys Med Biol* 2005;50:N331–N344.

53. Li XA, Ma CM, Salhani D. Measurement of percentage depth dose and lateral beam profile for kilovoltage X-ray therapy beams. *Phys Med Biol* 1997;42:2561–2568.

54. Li XA, Salhani D, Ma CM. Characteristics of orthovoltage X-ray therapy beams at extended SSD for applicators with end plates. *Phys Med Biol* 1997;42:357–370.

55. Klevenhagen SC, D'Souza D, Bonnefoux I. Complications in low energy X-ray dosimetry caused by electron contamination. *Phys Med Biol* 1991;36:1111–1116.

56. Fletcher CL, Mills JA. An assessment of GafChromic film for measuring 50 kV and 100 kV percentage depth dose curves. *Phys Med Biol* 2008;53:N209–N218.

57. British Journal of Radiology Report 25. Central axis depth dose data for use in radiotherapy. London, UK: British Institute of Radiology; 1996.

58. Sheu R-D, Powers A, Lo Y-C. Commissioning a 50–100 kV X-ray unit for skin cancer treatment. *J Appl Clin Med Phys* 2015;16:161–174.

59. Steenbeke F, Gevaert T, Tournel K et al. Quality assurance of a 50-kV radiotherapy unit using EBT3 GafChromic film a feasibility study. *Tech Cancer Res Treat* 2016;15:163–170.

60. Williams JR, Thwaites DI. *Radiotherapy Physics in Practice*. Oxford, UK: Oxford University Press; 2000.

61. Hill R, Holloway L, Baldock C. A dosimetric evaluation of water equivalent phantoms for kilovoltage X-ray beams. *Phys Med Biol* 2005;50:N331.

62. Meigooni AS, Li Z, Mishra V et al. A comparative study of dosimetric properties of plastic water and solid water in brachytherapy applications. *Med Phys* 1994;21:1983–1987.

63. Reniers B, Verhaegen F, Vynckier S. The radial dose function of low-energy brachytherapy seeds in different solid phantoms: Comparison between calculations with the EGSnrc and MCNP4C Monte Carlo codes and measurements. *Phys Med Biol* 2004;49:1569–1582.

64. Chica U, Anguiano M, Lallena AM. Study of the formalism used to determine the absorbed dose for low-energy X-ray beams. *Phys Med Biol* 2008;53:6963–6977.

65. Munck AF, Rosenschold P, Nilsson P, Knoos T. Kilovoltage X-ray dosimetry—an experimental comparison between different dosimetry protocols. *Phys Med Biol* 2008;53:4431–4442.

66. Gill S, Hill R. A study on the use of Gafchromic™ EBT3 film for output factor measurements in kilovoltage X-ray beams. *Aust ralas Phys Eng Sci Med* 2013;36:1–7.

67. Lye JE, Butler DJ, Webb DV. Enhanced epidermal dose caused by localized electron contamination from lead cutouts used in kilovoltage radiotherapy. *Med Phys* 2010;37:3935–3939.

68. Grosswendt B. Backscatter factors for X-rays generated at voltages between 10 and 100 kV. *Phys Med Biol* 1984;29:579–591.

69. Ma CM, Seuntjens JP. Mass-energy absorption coefficient and backscatter factor ratios for kilovoltage X-ray beams. *Phys Med Biol* 1999;44:131–143.

70. Andreo P, Burns DT, Hohlfield K et al. Absorbed dose determination in external beam radiotherapy, an international code of practice for dosimetry based on standards of absorbed dose to water. Technical Report Series No. 398. Vienna, Austria: International Atomic Energy Agency; 2000.

71. Klevenhagen SC, Aukett RJ, Harrison RM et al. The IPEMB code of practice for the determination of absorbed dose for X-rays below 300 kV generating potential (0.035 mm Al-4 mm Cu HVL; 10-300 kV generating potential). *Phys Med Biol* 1996;41:2605–2625.

72. Ma CM, Coffey CW, DeWerd LA et al. AAPM protocol for 40-300 kV X-ray beam dosimetry in radiotherapy and radiobiology. *Med Phys* 2001;28:868–893.

73. Grosswendt B. Dependence of the photon backscatter factor for water on source-to-phantom distance and irradiation field size. *Phys Med Biol* 1990;35:1233–1245.

74. Grosswendt B. Dependence of the photon backscatter factor for water on irradiation field size and source-to-phantom distances between 1.5 and 10 cm. *Phys Med Biol* 1993;38:305–310.

75. Johns HE, Cunningham JR. *The Physics of Radiology*. Springfield, IL: Charles C. Thomas; 1983.

76. Klevenhagen SC. The build-up of backscatter in the energy range 1 mm Al to 8 mm Al HVT (radiotherapy beams). *Phys Med Biol* 1982;27:1035–1043.

77. Kim J, Hill R, Claridge ME et al. An investigation of backscatter factors for kilovoltage X-rays: A comparison between Monte Carlo simulations and Gafchromic EBT film measurements. *Phys Med Biol* 2010;55:783–797.

78. Klevenhagen SC. Experimentally determined backscatter factors for X-rays generated at voltages between 16 and 140 kV. *Phys Med Biol* 1989;34:1871–1882.

79. Eaton DJ, Doolan PJ. Review of backscatter measurement in kilovoltage radiotherapy using novel detectors and reduction from lack of underlying scattering material. *J Appl Clin Med Phys* 2013;14:5–17.

80. Patrocinio HJ, Bissonnette JP, Bussière MR et al. Limiting values of backscatter factors for low-energy X-ray beams. *Phys Med Biol* 1996;41:239.

81. Butson MJ, Cheung T, Yu PKN. Radiochromic film for verification of superficial X-ray backscatter factors. *Aust Phys Eng Sci Med* 2007;30:269–273.

82. Peter K, Butson MJ. Measurement of effects of nasal and facial shields on delivered radiation dose for superficial X-ray treatments. *Phys Med Biol* 2013;58:N95.

83. Butson MJ, Cheung T, Yu PKN. Measurement of dose reductions for superficial X-rays backscattered from bone interfaces. *Phys Med Biol* 2008;53:N329–N336.

84. Baker CR, Luhana F, Thomas SJ. Absorbed dose behind eye shields during kilovoltage photon radiotherapy. *Br J Radiol* 2002;75:685–688.

85. Wang D, Sobolewski M, Hill R. The dosimetry of eye shields for kilovoltage X-ray beams. *Australas Phys Eng Sci Med* 2012;35:491–495.

86. Butson MJ, Cheung T, Yu PK et al. Measurement of radiotherapy superficial X-ray dose under eye shields with radiochromic film. *Phys Med* 2008;24:29–33.

87. Eaton DJ, Duck S. Dosimetry measurements with an intra-operative X-ray device. *Phys Med Biol* 2010;55:N359–N369.

88. Pidikiti R, Stojadinovic S, Speiser M et al. Dosimetric characterization of an image-guided stereotactic small animal irradiator. *Phys Med Biol* 2011;56:2585–2599.

89. Baldwin Z, Fitchew R. The influence of focal spot size, shape, emission profile and position on field coverage in a Gulmay D3300 kilovoltage X-ray therapy unit. *Australas Phys Eng Sci Med* 2014;37:515–523.

90. Oliveira A, Fartaria M, Cardoso J et al. The determination of the focal spot size of an X-ray tube from the radiation beam profile. *Radiat Meas.*, 2015;82:138–145.

91. Currie M, Bailey M, Butson M. Verification of nose irradiation using orthovoltage X-ray beams. *Aust ralas Phys Eng Sci Med* 2007;30:105–110.

92. Hill R, Healy B, Holloway L et al. An investigation of dose changes for therapeutic kilovoltage X-ray beams with underlying lead shielding. *Med Phys* 2007;34:3045–3053.

93. Chow JCL, Grigorov GN. Effect of the bone heterogeneity on the dose prescription in orthovoltage radiotherapy: A Monte Carlo study. *Rep Pract Oncol Radiother* 2012;17:38–43.

94. Van DJ (Ed.). *The Modern Technology of Radiation Oncology*. Madison, WI: Medical Physics Publishing; 1999.

95. Ding GX, Munro P. Radiation exposure to patients from image guidance procedures and techniques to reduce the imaging dose. *Radiother Oncol* 2013;108:91–98.

96. Kim J, Wen N, Jin J-Y et al. Clinical commissioning and use of the Novalis Tx linear accelerator for SRS and SBRT. *J Appl Clin Med Phys* 2012;13:124–151.

97. Solberg TD, Medin PM, Ramirez E et al. Commissioning and initial stereotactic ablative radiotherapy experience with Vero. *J Appl Clin Med Phys* 2014;15:205–225.

98. Chu RY, Thomas G, Maqbool F. Skin entrance radiation dose in an interventional radiology procedure. *Health Phys* 2006;91:41–46.

99. Delle CS, Carosi A, Bufacchi A et al. Use of GAFCHROMIC XR type R films for skin-dose measurements in interventional radiology: Validation of a dosimetric procedure on a sample of patients undergone interventional cardiology. *Phys Med* 2006;22:105–110.

100. Giles ER, Murphy PH. Measuring skin dose with radiochromic dosimetry film in the cardiac catheterization laboratory. *Health Phys* 2002;82:875–880.

101. Herron B, Strain J, Fagan T et al. X-ray dose from pediatric cardiac catheterization: A comparison of materials and methods for measurement or calculation. *Pediatr cardiol* 2010;31:1157–1161.

102. Soliman K, Bakkari M. Examination of the relevance of using radiochromic films in measuring entrance skin dose distribution in conventional digital mammography. *Rad Prot Dosim* 2015:ncv126.

103. Gotanda T, Katsuda T, Gotanda R et al. Half-value layer measurement: Simple process method using radiochromic film. *Aust ralas Phys Eng Sci Med* 2009;32:150–158.

104. Gotanda T, Katsuda T, Gotanda R et al. Evaluation of effective energy for QA and QC: Measurement of half-value layer using radiochromic film density. *Aust ralas Phys Eng Sci Med* 2009;32:26–29.

105. Duan X, Wang J, Yu L et al. CT scanner X-ray spectrum estimation from transmission measurements. *Med Phys* 2011;38:993–997.

106. Gorny KR, Leitzen SL, Bruesewitz MR et al. The calibration of experimental self-developing Gafchromic® HXR film for the measurement of radiation dose in computed tomography. *Med Phys* 2005;32:1010–1016.

107. Rampado O, Garelli E, Ropolo R. Computed tomography dose measurements with radiochromic films and a flatbed scanner. *Med Phys* 2010;37:189–196.

108. Song WY, Kamath S, Ozawa S et al. A dose comparison study between XVI® and OBI® CBCT systems. *Med Phys* 2008;35:480–486.

109. Ding GX, Coffey CW. Radiation dose from kilovoltage cone beam computed tomography in an image-guided radiotherapy procedure. *Int J Rad Oncol Biol Phys* 2009;73:610–617.

110. Giaddui T, Cui Y, Galvin J et al. Characteristics of Gafchromic XRQA2 films for kV image dose measurement. *Med Phys* 2012;39:842–850.

111. Giaddui T, Cui Y, Galvin J et al. Comparative dose evaluations between XVI and OBI cone beam CT systems using Gafchromic XRQA2 film and nanoDot optical stimulated luminescence dosimeters. *Med Phys* 2013;40:062102.
112. Hu N, McLean D. Measurement of radiotherapy CBCT dose in a phantom using different methods. *Australas Phys Eng Sci Med* 2014;37:779–789.
113. Jaffray D, Siewerdsen J. Cone-beam computed tomography with a flat-panel imager: Initial performance characterization. *Med Phys* 2000;27:1311–1323.
114. Jaffray DA, Siewerdsen JH, Wong JW et al. Flat-panel cone-beam computed tomography for image-guided radiation therapy. *Int J Rad Oncol Biol Phys* 2002;53:1337–1349.
115. Létourneau D, Wong JW, Oldham M et al. Cone-beam-CT guided radiation therapy: Technical implementation. *Radiother Oncol* 2005;75:279–286.
116. Nobah A, Aldelaijan S, Devic S et al. Radiochromic film based dosimetry of image-guidance procedures on different radiotherapy modalities. *J Appl Clin Med Phys* 2014;15:229–239.
117. Scandurra D, Lawford CE. A dosimetry technique for measuring kilovoltage cone-beam CT dose on a linear accelerator using radiotherapy equipment. *J Appl Clin Med Phys* 2014;15:4658.
118. Smith L, Haque M, Morales J et al. Radiation dose measurements of an on-board imager X-ray unit using optically-stimulated luminescence dosimeters. *Australas as Phys Eng Sci Med* 2015;38:665–669.

Application of radiochromic film for dosimetric and quality assurance of the brachytherapy sources

ALI S. MEIGOONI AND SHARIFEH A. DINI

6.1 INTRODUCTION OF BRACHYTHERAPY

6.1.1 DEFINITION OF BRACHYTHERAPY

The term brachytherapy refers to short-distance therapy as compared with teletherapy which means long-distance therapy. In this treatment strategy, the source of radiation is placed either directly into the tumor or adjacent to that to deliver the prescribed radiation. This treatment modality was introduced after December 26, 1898 with discovery of ^{226}Ra by Marie and Pierre Curie [1]. Traditionally, the brachytherapy sources were made of different sealed radioactive materials, such as ^{226}Ra, ^{137}Cs, ^{192}Ir, ^{60}Co, and ^{125}I [2,3]. However, recently, different electronic brachytherapy sources have been introduced using miniature X-ray tubes [4]. These sources provide similar dosimetric characteristics as the radioactive sources. First use of radiation for the treatment of prostate cancer was reported by Pasteau and Degrais in 1914 [5]. Selection of the source type with different type of radiation emission (gamma ray, X-ray, or beta ray) and different energies of the radiation created the variations in the field of brachytherapy treatment [6]. Aside from the differences in attenuation of the radiation emitted by these sources and their interactions with different tissues, the concept of inverse-square law is the main foundation of the dose fall-off in the brachytherapy treatments.

6.1.2 DIFFERENT CLASSIFICATIONS OF BRACHYTHERAPY SOURCES

The optimum method of the brachytherapy treatment technique and radiation source type may vary from one treatment site to another. Therefore, brachytherapy treatments are classified in different categories based on (i) the energy of the radiation source, (ii) dose rate, and (iii) type of the radiation, as shown in the following sections.

6.1.2.1 CLASSIFICATION IN TERMS OF ENERGY OF THE SOURCE
6.1.2.1.1 Low energy

Sources emitting photons less than 50 keV are considered as low-energy photon emitters. With these photons, the majority of the interaction of photons with matter would be through photoelectric interactions [7]. Therefore, these photons are locally absorbed, and they are good for interstitial brachytherapy treatments for tumors such as prostate. Moreover, these sources are used for treatment of ocular melanoma [8]. Sources listed in this category are ^{125}I, ^{103}Pd, and ^{131}Cs [9–11].

6.1.2.1.2 High energy

Sources emitting photons greater than 50 keV are considered high-energy emitters [7]. With these photons, the main method of the interaction would be Compton scattering. Therefore, these photons may travel to some distances before they are completely absorbed. These sources are good for temporary intracavitary and interstitial treatments. Sources in this category are ^{137}Cs, ^{60}Co, and [192]Ir [7].

6.1.2.2 CLASSIFICATION IN TERM OF DOSE RATE OF THE SOURCE

6.1.2.2.1 Low dose rate

Brachytherapy treatments performed with a dose rate of less than 100 cGy/h are known as low dose rate (LDR) implants [12,13]. Traditional intracavitary implants of the gynecological patients with ^{137}Cs and [226]Ra sources and prostate seed implants with ^{125}I, [103]Pd, and [131]Cs are among this group [11,13].

6.1.2.2.2 High dose rate

Brachytherapy treatments performed with a dose rate of greater than 200 cGy/min (1200 cGy/h) is known as high dose rate (HDR) implants [14]. Intracavitary of the gynecological patients and interstitial implants of the prostate and breast implants using high dose ^{192}Ir source is among this group. Recently, HDR ^{60}Co has been introduced as an alternative to the ^{192}Ir HDR source [15,16].

6.1.2.3 CLASSIFICATION IN TERMS OF TYPE OF THE SOURCE

6.1.2.3.1 Photon emitter

Brachytherapy sources emitting gamma ray or X-ray are used for treatment of tumors with using interstitial or intracavitary treatments [9–11]. [226]Ra, ^{137}Cs, ^{60}Co, ^{125}I, [103]Pd, and [131]Cs are among the sources of this type [9–11].

6.1.2.3.2 Beta emitter

Brachytherapy sources emitting β-ray are used for treatment of superficial tumors, such as cardiovascular restenosis, or pterygium disease [17,18]. Sources such as ^{90}Sr and ^{32}P are among these sources [19].

6.2 DOSIMETRIC CHARACTERISTICS

6.2.1 TG-43

6.2.1.1 FORMALISM

In 1995, Task Group 43 (TG-43) of the American Association of Physicists in Medicine (AAPM) introduced a worldwide recommendation for dosimetry of brachytherapy sources [9]. The updated version of TG-43 formalism (known as TG-43U1) was published in 2004 [10] to eliminate shortcomings in the original formalism, to clarify some of the definitions. For example, the active lengths of the sources with different geometry were clearly defined. This and some subsequent reports introduced the dosimetry of the sources that were not available during the preparation of the original report [10,11].

According to the recommendations of TG-43 protocol [9,10], the absorbed dose rate distribution around a sealed brachytherapy source, for line source approximation, can be determined using the following formalism:

$$\dot{D}(r,\theta) = \Lambda\, S_K\, \frac{G_L(r,\theta)}{G_L(r_0,\theta_0)}\, g_L(r)\, F(r,\theta) \tag{6.1}$$

where:

Λ is the dose-rate constant

$G_L(r, \theta)$ is the geometry function

$g_L(r)$ is the radial dose function

$F(r, \theta)$ is the 2D anisotropy function

$(r_0 = 1$ cm, $\theta_0 = \pi/2)$ is the reference point.

The above-mentioned quantities are defined and discussed in detail in TG-43reports [9,10]. The subscript L has been added in TG-43U1[10] to denote the line source approximation used for the geometry function. S_K is the air kerma strength of the brachytherapy source. The geometry function, $G_L(r, \theta)$, takes into account the effect of the activity distribution within the source and the distance between the source and point of interest. The geometry function is defined in the AAPM TG-43 [9,10] as

$$G_L(r,\theta) = \begin{cases} \dfrac{\beta}{Lr\sin\theta} & \text{if } \theta \neq 0° \\[2mm] (r^2 - L^2/4)^{-1} & \text{if } \theta = 0° \end{cases}$$

where β is the angle in radians, subtended by the point of interest, $P(r, \theta)$, to the tips of the active length of a hypothetical line.

6.2.1.2 DOSE-RATE CONSTANT

In the TG-43 report [9,10], the quantity of dose-rate constant is analogues to the output of the linear accelerator, and it is defined as a ratio of the dose rate at a reference point ($r_0 = 1$ cm, $\theta_0 = 90°$) to the air kerma strength of the source (S_K) as

$$\Lambda = \frac{\dot{D}\left(1\,\text{cm}, \pi/2\right)}{S_k} \tag{6.2}$$

6.2.1.3 RADIAL DOSE FUNCTION

The radial dose function, $g_L(r)$, describes the attenuation in tissue of the photons emitted from the brachytherapy source. This quantity is analogous to the tissue-maximum ratio of the external beam therapy. The radial dose function is defined as

$$g_L = (r) = \left[\frac{\dot{D}\left(r,\pi/2\right)G_L\left(r_0,\pi/2\right)}{\dot{D}\left(r_0,\pi/2\right)G_L\left(r,\pi/2\right)}\right] \tag{6.3}$$

where $\dot{D}\,(r, \pi/2)$ and $\dot{D}\,(r_0, \pi/2)$ are the dose rates measured at distances of r and r_0, respectively, along the transverse axis of the source.

6.2.1.4 2D ANISOTROPY FUNCTION

2D anisotropy function, $F(r, \theta)$, presenting the angular nonuniformity of dose distributions relative to the longitudinal axis of the source. This quantity has been introduced because of the fact that the brachytherapy sources are normally cylindrical shape with different thicknesses of the attenuator at different orientations, presence of a cylindrical X-ray marker within the source that is made of a high-z metal, and finally, distribution of the activity within the source geometry. The 2D anisotropy function is defined as

$$F(r,\theta) = \left[\frac{\dot{D}(r,\theta) G_L\left(r,\frac{\pi}{2}\right)}{\dot{D}\left(r,\frac{\pi}{2}\right) G_L(r,\theta)} \right] \tag{6.4}$$

6.2.2 TG-60

6.2.2.1 CHARACTERISTICS OF INTRAVASCULAR BRACHYTHERAPY SOURCES

Using gamma- and beta-emitter intravascular brachytherapy sources, radiation doses in the range of 15–30 Gy can be delivered for reduction of the restenosis in coronary arteries [20]. This technique creates minimal normal tissue toxicity because of the high localization of dose delivery to the immediate vicinity of radioactive brachytherapy sources. It has been reported that with this technique, the restenosis rate may drop from roughly 40% to well below 10%, if radiation is delivered to the obstruction site during or after angioplasty.

In traditional brachytherapy, dosimetry at distances of the order of a few millimeters from radioactive sources is poorly known. However, in intravascular brachytherapy, the entire lesion may be in the order of 1–3 mm in thickness. In late 1995, AAPM formed TG-60 [20] was developed to investigate and report the current state of intravascular brachytherapy physics. One of the responsibilities of this task group was recommending a dose-prescription site (or region) for selected intravascular brachytherapy procedures (e.g., 1-mm depth from the arterial lumen surface or 2 mm from the center of the lumen, etc.). This task group adapted a modified TG-43 formalism for the dosimetric characteristics of the intravascular brachytherapy sources. The main modification that was introduced by this task group on the TG-43 recommendations was on the reference point located at a distance of 2 mm instead of 1 cm from the source center, along the transverse axis of the source. This modification was also implemented on determination of dose-rate constant, radial dose function, geometry function, and 2D anisotropy function. Therefore, care must be taken in application of the appropriate dosimetric information for a given source for a brachytherapy treatment. For example, there are two different sets of dosimetric parameters for a given model of ^{192}Ir source: one set based on the TG-43 recommendation for general brachytherapy and another set based on the TG-60 recommendation for the intravascular brachytherapy.

6.3 DETECTORS FOR BRACHYTHERAPY DOSIMETRY

Radiation dosimetry in brachytherapy is challenging due to small distances and low energy. Dosimetry can be performed either by experimental or theoretical techniques. Experimental methods are performed using several different detectors such as thermoluminescent dosimeter (TLD), diode, ion-chambers, and films. However, due to spatial resolution, energy, dose, dose rate, angle, and water equivalency, most detectors fail in one or other parameters. The theoretical calculations are accomplished using different Monte Carlo (MC) simulation codes such as monte-carlo n-particle (MCNP) and electron gamma shower (EGS). A brief review of these dosimetry methods and devices is provided in the following.

6.3.1 THERMOLUMINESCENT DOSIMETER

The term luminescence is referred to the emission of light from some kind of solid material, such as phosphors [21–23]. This emission is the release of energy stored within the solid material, through some type of prior excitation of the solid electronic system (i.e., by visible, infrared, or ultraviolate light and ionization radiation). The wavelength of the emitted light is a characteristic of the luminescent material. The ability to store the radiation energy is important in luminescence dosimetry and is generally associated with the presence of activator (i.e., impurity atoms and structural defects). The wide variety of TLD materials and their different physical forms (Figure 6.1) allow the determination of different radiation qualities at dose levels from μGy to kGy [26]. Some of the major advantages of TLD dosimeters over other detectors are their small physical size and that no cables or auxiliary equipment is required during the dose measurement. This makes them well suited for a wide range of applications in medicine. However, although some of the larger institutions with extensive experiences in TLD dosimetry commonly achieve quite good results, TLD appears very difficult to work with and have reproducible results [24,25]. Energy dependence of the TLD response strongly depends on its chemical composition [24]. For example, LiF or CaF2 TLDs have very different energy-dependent characteristics (Figure 6.2). Finally, it is mostly recommended to use the values from several TLD measurements for determination of dose at each point to achieve a more accurate dosimetry. Most of the published data are based on LiF doped with magnesium and titanium (LiF:Mg, Ti). Other materials such as $CaSO_4$:Dy are also widely used. More recent TLD materials such as LiF:Mg, Cu, and P show greater potential for radiation dosimetry.

6.3.2 DIODES

Silicon diode dosimeters consist of p–n junction between n-type and p-type materials [27]. The n-type material is doped with impurities of a free of pentavalent element (e.g., phosphorous) called a donor. The p-type silicon is doped with impurities of a trivalent element (e.g., boron) called

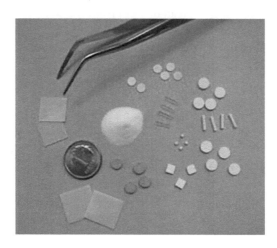

Figure 6.1 Different forms of TLD detectors. (From http://www.ifj.edu.pl/ccb/en/badania/projekt4.php.)

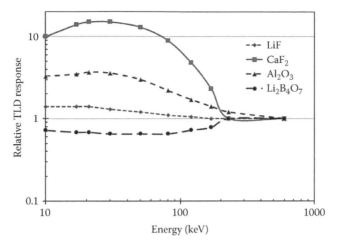

Figure 6.2 Energy dependence of TLD response.

acceptor. Each donor can contribute 32 free electrons to the silicon, and each acceptor can accept an electron. The diodes are produced by taking n-type or p-type silicon and counterdoping the surface to produce the opposite type material [27,28]. Both n- and p-type diodes are commercially available. An n-type diode is formed by doping acceptor impurities into n-type silicon. A p-type diode is formed by doping donor impurities into p-type silicon. In either case, a spatially doped matter creates a region where p- and n-type silicones are in direct contact.

Diodes are used in the short-circuit mode, as this mode exhibits a linear relationship between the measured charge and the absorbed dose. Unlike the ion-chambers, diodes are usually operated without an external bias voltage to reduce leakage current. Diodes are more sensitive and smaller in size compared with typical ionization chambers. However, they are not used as absolute dosimeters as they cannot be calibrated by the standard laboratory. Diode responses are generally energy dependent, particularly at low-energy radiation level [10,23]. The commercially available diodes are in different geometrical shapes and sizes (Figure 6.3) [28]. Generally, cylindrical diodes are used for brachytherapy dosimetry. The sensitivity of diodes depends on their radiation history, so the calibration has to be repeated periodically [29]. Diodes responses are commonly temperature dependent, dose-rate dependent, and angular (directional) dependent [29,30].

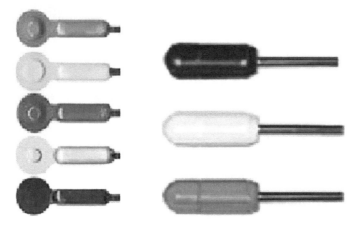

Figure 6.3 Different models of the diode detectors. (From http://www.rpdinc.com/diodes-155.)

6.3.3 ION CHAMBERS

Ionization chambers are made of a cavity filed with regular air (i.e., it is open to air in the room) or pressurized gas (i.e., argon) to create ion-pairs due to radiation interactions. These ions are then collected by an electrode using electric voltage (known as bias voltage). Ionization chambers have various shapes and sizes for different applications (Figure 6.4) [31–33].

The wall and the collecting electrode are separated with a high-quality insulator to reduce the leakage current when a polarizing voltage is applied to the chamber. These detectors are the most consistent measuring devices and present the least energy dependence and dose-rate dependence for measurements of radiation. Ion-chambers are normally calibrated by one of the national calibration laboratories (i.e., National Institute of Standard and Technology (NIST) or Accredited Dosimetry Calibration Laboratory (ADCL)); thus, they are used as absolute dosimeters. However, as these detectors are commonly made of air with a low density (about 1/1000 of water), it produces lower signals per unit dose of radiation relative to diode and TLD detectors. Ionization chambers are generally used for radiation dosimetry in external beam therapy, because of the high dose rate (HDR) radiation to create high signal-to-noise ratio (i.e., high statistical accuracy). Therefore, in brachytherapy with relatively low dose rate (LDR), particularly at larger distances from the source center, the ionization chambers do not play much role.

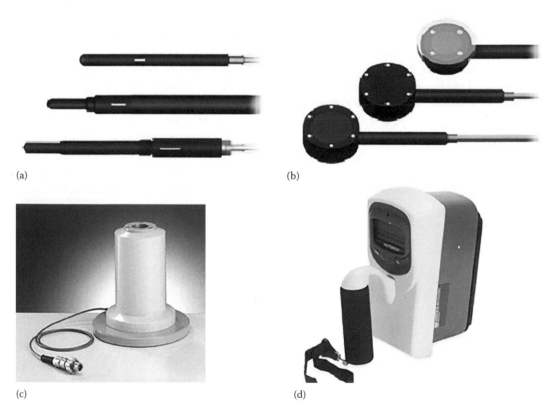

(a)

(b)

(c)

(d)

Figure 6.4 (a) Cylindrical ionization chambers (From http://www.standardimaging.com/exradin/), (b) parallel plate ionization chambers (From http://www.standardimaging.com/exradin/), (c) well type ionization chamber (From http://www.ptw.de/well-type_chamber.html), and (d) ion-chamber survey meter (From http://en-us.fluke.com/products/hvac-iaq-tools/fluke-481-desi.html).

6.3.4 MONTE CARLO SIMULATION

Although the experimental dosimetry of the brachytherapy sources may be considered as the gold standard and they are used for validation of the theoretical calculations, they may be very complicated and time consuming [10,34,35]. Particularly, accurate experimental setup in a high-dose gradient area around the brachytherapy sources is very crucial for these procedures [23]. An alternative to the experimental technique is the application of the modern MC codes. These codes enable us to create a very difficult source and detector geometry and create 2D and 3D dose distributions from any source geometry. MC-based dosimetry has been validated by different investigators as a valuable tool in brachytherapy source dosimetry. There are several different commercially available MC codes such as PTRAN, ETRAN, ITS, MCNP, PENELOPE, GEANT4, and EGS4. Description of all these codes and their application for various brachytherapy sources is available in the literature. Aside from the source and phantom geometry, in these codes, the applications of the accurate and appropriate cross-section libraries are very important in derivation of the source dosimetric characteristics. Significantly, multiple investigators were able to demonstrate that they were able to create a very detailed, accurate, and reproducible data by using a MC code that has been validated using experimental data [36]. Moreover, these codes provided a means to determine the radiation distribution at some points that experimental dosimetry was either very difficult or impossible, such as the area near the surface of the source, or dose distribution around any heterogeneity located at the vicinity of the source.

6.4 RADIOCHROMIC FILM'S PROPERTIES FOR BRACHYTHERAPY DOSIMETRY

Within past several years, different models of GAFchromic™/Radiochromic films (RCFs) have been introduced by International Specialty Products (Wayne, NJ). These film models have been examined by different investigators for dosimetry of various types of radiations, particularly for brachytherapy sources [37–39]. Applications of early models of this film type in radiation dosimetry have been discussed in detail in AAPM Task Group 55 (TG-55) [40]. Several investigators had shown that the characteristic of this film type, such as its high spatial resolution, tissue-equivalent base, and no requirement for processing after irradiation, are ideal for their applications as discussed in Chapter 3 and various references [37–41]. This detector type reduces some of the difficulties of the experimental setups in high-dose gradient area, particularly for dosimetry at short distances relative to the sources. However, the low sensitivities of these film types have created limitations for their applications in low-dose radiation fields such as dosimetry around the LDR brachytherapy sources. For example, dose rate at 1-cm distance from an ^{125}I source is approximately 1 cGy/h per U (unit of air kerma strength) at 1-cm distance. Therefore, for a clinically used source with a maximum activity of 2U, it will take about 250 h (about 10 days) to create 500 cGy, which is the threshold dose for some of the RCF films. With this example, neglecting the tissue attenuation, it would take 6250 h (260 days) exposure to perform a measurement at 5 cm from the source. Lowering the threshold dose to 100 cGy, in the newer film model, the measured time for these two locations will reduce to 2 days and 52 days, respectively.

Energy dependence of the RCF response is another subject that has been evaluated by many investigators [24,38–44]. Dose response of the newer models of RCFs has been shown to have minimal energy dependence in the megavoltage energy range. However, there are data to indicate variation with energy for kilovoltage X-rays. Butson et al. [45,46] had shown a variation in sensitivity of external beam therapy (EBT2) films of up to 7% for energies 50–250 kVp. More recently, Dini et al. [43] have evaluated the XR-type R and T films and compared the results with MD-55 films. These results show larger sensitivity (i.e., OD/Gy) for the new films than the MD-55-2 films. However, they have observed a much larger energy dependence of the XR-type R and T than MD-55-2 film. Moreover, the color of XR type R film turns from amber to dark greenish-black, whereas XR type T films turn from orange to brownish-black, depending on the level of exposure. In 2009, Richter et al. [47] published their results regarding the investigation of the energy dependence of the EBT-1 film for photon beams ranging from 10 kVp to15 MVp and electron beams ranging from 6 to 18 MeV. They found strong energy dependence for EBT-1 RCFs. In a detailed investigation, Sutherland and Rogers [48] evaluated the energy dependence of the EBT and EBT2 films using MC calculation technique. They have demonstrated that the absorbed-dose energy dependence $f(Q)$ of the EBT film to be approximately constant ($\pm0.6\%$) for monoenergetic photon beams ranging from 100 keV to 18 MeV but varied for low-energy photons ranging from 3 to 100 keV. The low-energy variation of $f(Q)$ was attributed to ratio of the mass-energy absorption coefficient of water to the EBT active material below 10 keV and due to the increase in photon interactions in the surface layer between 10 and 100 keV.

Another limitation of the application of the RCF in brachytherapy dosimetry was its nonuniformity of the film response [49]. Meigooni et al. [38] had shown the impact of the film nonuniformity on the radial dose function of HDR ^{192}Ir dosimetry by comparison of the film dosimetry with TLD data. Zhu et al. [41] and Chiu-Tsao et al. [50] had also independently examined the nonuniformity of the RCF response for different models of the films.

6.5 ROLE OF RADIOCHROMIC FILM ON SOURCE CHARACTERIZATION

6.5.1 LOW DOSE RATE; LOW-ENERGY PHOTON EMITTER SOURCES

6.5.1.1 DOSIMETRY OF ^{125}I, [103]Pd, AND [131]Cs BRACHYTHERAPY SOURCES

Chiu-Tsao et al. [50] have examined the application of the MD-55 RCF for dosimetric evaluation of [125]I. They have measured the optical densities of the irradiated films using a Macbeth spot densitometer (Model TG502) with 1-mm aperture, an He–Ne laser scanning microdensitometer (Pharmacia LKB, Model 2222-010, with a red light filter 632.8 nm and 100 µm apertures). Additional detail on scanning system can be found in Chapter 4. Morrison et al. [51] have reported their assessment of the calibration technique of the EBT3 RCF for low-energy brachytherapy sources. They have designed a custom phantom to hold a single ^{125}I seed. Film pieces were scanned with an Epson 10000XL flatbed scanner, and the resulting 48-bit RGB TIFF images were analyzed using both FilmQA Pro software and MATLAB. They have concluded that the energy dependence between 6 MV and ^{125}I photons is significant such that film calibrations should be done with an appropriately low-energy source when performing low-energy brachytherapy dose measurements. Their results indicate that earlier doses of 1 Gy, absolute dose measurements can be made with an accuracy of 1.6% for 6 MV beams and 5.7% for ^{125}I seed exposures if using the ^{125}I source for calibration, or 2.3% if using the 75 kVp photon beam for calibration. Aldelaijan et al. [52] evaluated the use of RCF for the dosimetry of ^{192}Ir source.

6.5.2 LOW DOSE RATE; HIGH-ENERGY PHOTON-EMITTER SOURCES

6.5.2.1 DOSIMETRY OF ^{192}Ir, ^{137}Cs, AND ^{60}Co BRACHYTHERAPY SOURCES

As noted in the text, due to the low sensitivity of the RCF responses, there are limited investigations in their applications for LDR brachytherapy. In 2006, Le et al. [53] evaluated the possibility of 2D dosimetry of LDR brachytherapy sources using RCF. They have measured the precision and accuracy of the RCF dosimetry of the ^{137}Cs sources using the MD-55-2 films. They have read the irradiated films with an He–Ne laser scanner. The measured data were compared with the MC simulated data. They reported an agreement of within 4% between measurement and MC simulated data in the dose range of 3–60 Gy.

6.5.3 HIGH DOSE RATE; PHOTON-EMITTER SOURCES

6.5.3.1 DOSIMETRY OF ^{192}Ir AND ^{60}Co HDR BRACHYTHERAPY SOURCES

In 1997, Meigooni et al. [38] utilized Model MD-55-2 RCF for dosimetric characterization of the ^{192}Ir HDR remote after loading system. A comparison between their measured radial dose function of ^{192}Ir using RCF and LiF TLD values shows poor agreements between the two data sets due to the non-uniformity of the film response. However, Uniyal et al. [54] investigated the accuracy and suitability of using GAFchromic EBT2 film for a dosimetry method in the transverse plane of HDR ^{192}Ir brachytherapy source. They measured the radial dose function and dose-rate constant using GAFchromic EBT2 film and TLD. They had shown that the dose-rate constant and radial dose function measured by GAFchromic EBT2 film are in agreement with their TLD data within 3.9% and 2.8%, respectively. Moreover, they had shown the agreement of the data with the MC-simulated values by Williamson et al. [55] (Figure 6.5). Therefore, they demonstrated the suitability of using GAFchromic EBT2 film dosimetry in characterization of dose distribution in the transverse plane of HDR ^{192}Ir source.

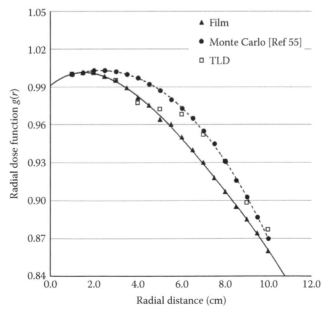

Figure 6.5 Comparison of the measured radial-dose function of ^{192}Ir with TLD, GAFchromic film, and Monte Carlo simulated data. (Reproduced from Uniyal, S.C. et al., *Phys. Med.*, 28, 2, 129–133, 2012.)

Schumer et al. [56] reported their evaluation of brachytherapy dosimetry with RCF for high dose rate ^{192}Ir source from a nucletron brachytherapy planning system. The optical density of the exposed films was determined with a modified Scanditronix film scanner, and the film was calibrated with ^{192}Ir using manually calculated exposure times. They had demonstrated that the dose distribution calculated by the nucletron brachytherapy planning system (v. 13.3), at a distance of 1.0 cm, is validated to within ±4% of the measured dose distribution. The advantages and limitations of RCF as a dosimetry tool are also addressed in their work.

Ghorbani et al. [57] examined the application of EBT film for dosimetric evaluation of HDR GZP6 ^{60}Co brachytherapy sources. They have utilized a Microtek color ScanMaker (1000XL) for their project and compared their measured data with MC-simulated values. Their results indicated that RCF measurements are in good agreement with the MC calculations (4%).

6.5.4 BETA-EMITTER SOURCES

6.5.4.1 DOSIMETRY OF ^{90}Sr/^{90}Y AND ^{106}Ru BRACHYTHERAPY SOURCES

Duggan et al. [58] measured dose distribution around HDR beta emitter source for the intravascular brachytherapy. They concluded that the maximum and minimum of the dose rates along the axis, at a radial distance of 2 mm from the axis, over the centered 24.5 mm, are within ±10% of the average. Moreover, Kirisitsa et al. [59] obtained the reference isodose lines by MC calculations and also GAFchromic film dosimetry for three endovascular brachytherapy devices currently in clinical use (^{192}Ir seed ribbon, ^{32}P wire source, and ^{90}Sr seed train). They found that the agreement between the measured and calculated is better for beta emitter sources than the photon emitters. Song et al. [60] also presented the results of their investigations for application of MD55-2 film, for dosimetric evaluation of the Cordis Checkmate™ ^{192}Ir sources used in the intravascular brachytherapy system.

6.6 RADIOCHROMIC FILM IN SPECIAL APPLICATIONS

6.6.1 EYE PLAQUE

Acar et al. [61] have published their experimental data for measuring the absolute dose distributions in eye phantom for collaborative ocular melanoma study (COMS) eye plaques with ^{125}I seeds (model I25.S16) using radiochromic EBT film dosimetry (Figure 6.6). They had to correct their measured data for the scattered radiation from the gold-plaque, which was determined using a MC-simulation technique. Their measurements along the plaque's central axis were performed using the EBT film for (a) uniformly loaded plaques (14–20 mm in diameter) and (b) a 20-mm plaque with single seed. In addition, they have measured the dose values in the off-axis direction at depths of 5 and 12 mm for all four plaque sizes. The EBT film calibrations were performed at ^{125}I photon energy. MC calculations were performed using MCNP5 code for a single seed at the center of a 20-mm plaque in homogeneous water and polystyrene medium. The EBT film-measured absolute dose rate values (film) were compared with those calculated using plaque simulator (PS) with homogeneous assumption (PS Homo) and heterogeneity correction (PS Hetero). Their results indicated the central axis depth–dose rate values for a single seed in 20-mm plaque measured using EBT film and calculated with MCNP5 code were compared, and agreement within 9% was found. They concluded that the calculated doses for uniformly loaded plaques using PS with heterogeneity correction option enabled were corroborated by the EBT film-measurement data. Radiochromic EBT

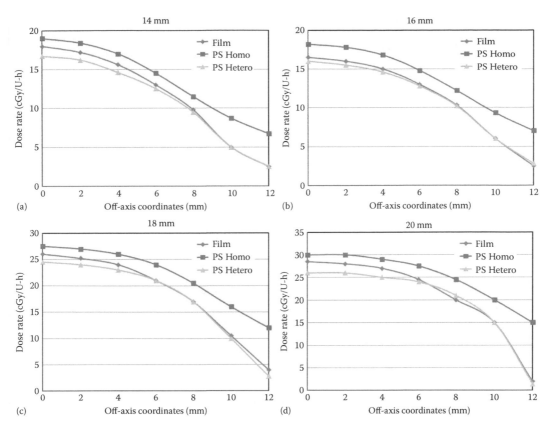

Figure 6.6 Comparison of the off-axis dose rates at a depth of 5 mm that was measured for different ^{125}I eye-plaques (a: 14 mm, b: 16 mm, c: 18 mm, and d: 20 mm plaques) using EBT film and calculated with plaque simulator program (v5.3.9) with homogeneous water (PS Homo) and with heterogeneity correction (PS Hetero). (Reproduced from Acar, H. et al., *Med. Phys.*, 40, 1, 011708-1-13, 2013.)

film dosimetry is feasible in measuring absolute dose distributions in eye phantom for COMS eye plaques loaded with single or multiple ^{125}I seeds. AAPM task group 129 reports present a detailed discussions regarding eye-plaque dosimetry, including the application of the RCF for dosimetry of various source types as well as the plaque assembly [8].

Heon et al. [62] have examined the dosimetry of ^{32}P ophthalmic applicator as an alternative to ^{90}Sr/^{90}Y irradiation devices. They have used RCF to measure the reference dose, axial depth dose distributions, and trans-axial dose profiles at various depths in water. They recommended measuring dose profiles as well as the reference dose rate for every new applicator.

6.6.2 SURFACE IMPLANT

The nonmalignant disease known as pterygium is generally developed on the conjunctiva, or clear membrane that covers the eye. Pterygium is a red growth of excessive blood vessels on white part of the eye that can also grow onto the cornea. This disease is treated with surgery or radiation using a β-emitter contact applicator such as ^{90}Sr [63,64]. These applicators are generally calibrated by the ADCL using the RCF to measure the dose rate as well as the uniformity of the activity distributions.

6.6.3 OTHER APPLICATIONS

RCFs are often used for quality assurance (QA) purposes in various brachytherapy procedures. For example, Elekta has developed a précised design device that uses a precisely marked RCF strip to verify the dwell position of the HDR ^{192}Ir sources. These QA are performed during the daily QA or QA during the source exchange and Annual QA.

6.7 SUMMARY

6.7.1 ADVANTAGES OF RCF

Higher special resolution, lower energy dependence of the RCF relative to other radiation detectors such as TLD and diodes are among the advantages of these detectors for brachytherapy sources. These films enable us to measure the radiation distribution at the vicinity of the surface of the brachytherapy sources that is not possible with other detectors. In additions, no requirement of the processing after irradiation provides a great advance of these films. Although the investigators had shown that the film response is not linear relative to the absorbed dose, nonlinearity can easily be corrected by calibration of samples of films prior to the dosimetric procedure.

6.7.2 DISADVANTAGES OF RCF

Low film response with the absorbed dose, particularly for earlier models of the films, was one of the disadvantages of this film for their application in brachytherapy dosimetry. Presently, the major application of the RCF is concentrated in HDR brachytherapy dosimetry. Moreover, nonuniformity of the film response created another obstacle for expansion of the application of this film for radiation dosimetry.

REFERENCES

1. Curie E. *Madame Curie*. Garden City, NJ: Doubleday, Doran and Company; 1937.
2. Pierquin B, Wilson F, Chassagne D. *Modem Brachytherapy*. New York: Masson Publishing; 1987.
3. Pierquin B, Marinello G. *A Practical Manual of Brachytherapy*. Madison, WI: Medical Physics Publishing; 1997.
4. Liu D, Poon E, Bazalova M et al. Spectroscopic characterization of a novel electronic brachytherapy system. *Phys Med Biol* 2007;53:61–75.
5. Pasteau O, Degrais P. The radium treatment of cancer of the prostate. *Arch Roentgen Ray* 1914;18:396–410.
6. Awan SB, Hussain M, Dini SA, Meigooni AS. Historical review of interstitial prostate brachytherapy. *Iran J Radiat Res* 2008;5:153–168.
7. Perez-Calatayud J, Das RK, DeWerd LA et al. Dose calculation for photon-emitting brachytherapy sources with average energy higher than 50 keV: Report of the AAPM and ESTRO radiotherapy. *Med Phys* 2012;39:2904–2929.
8. Chiu-Tsao ST, Astrahan MA, Finger PT et al. Dosimetry of ^{125}I and [103]Pd COMS eye plaques for intraocular tumors: Report of task group 129 by the AAPM and ABS. *Med Phys* 2012;39:6161.
9. Nath R, Anderson LL, Luxton G et al. Dosimetry of interstitial brachytherapy sources report of AAPM radiation therapy committee task group 43. *Med Phys* 1995;22:209–234.

10. Rivard MJ, Coursey BM, DeWerd LA et al. Update of AAPM task group no. 43 report: A revised AAPM protocol for brachytherapy dose calculations. *Med Phys* 2004;31:633–674.

11. Rivard MJ, Butler WM, DeWerd LA et al. Supplement to the 2004 update of the AAPM task group no. 43 report. *Med Phys* 2007;34:2187–2205.

12. Amoush A, Wilkinson A. Low dose rate brachytherapy for prostate cancer at the Cleveland Clinic: A technical review. *Appl Radiat Oncol* 2014;10–13. Available from: http://cdn.agilitycms.com/applied-radiation-oncology/ARO_02-14_Wilkinson.pdf.

13. Patel FD, Sharma SC, Negi PS et al. Low dose rate vs. high dose rate brachytherapy in the treatment of carcinoma of the uterine cervix: A clinical trial. *Int J Radiat Oncol Biol Phys* 1994;28:335–341.

14. Kubo DH, Glasgow GP, Pethel TD et al. High dose-rate brachytherapy treatment delivery: Report of the AAPM radiation therapy committee task group no. 59. *Med Phys* 1998;25:375–403.

15. Ghorbani M, Bahreyni TMT, Mowlavi AA et al. A Application of a color scanner for ^{60}Co high dose rate brachytherapy dosimetry with EBT radiochromic film. *Radiol Oncol* 2012;46:363–369.

16. Bahreyni TMT, Ghorbani M, Mowlavi AA et al. Air kerma strength characterization of a GZP6 Cobalt-60 brachytherapy source. *Rep Pract Oncol Radiother* 2010;15:190–194.

17. Choi CH, Han HS, Son KJ et al. Dosimetry of a new P-32 ophthalmic applicator. *Med Phys* 2011;38:6143–6451.

18. Bloch P, Bonan R, Wallner P, Lobdell J. Dosimetry for an Sr90/Y90 source train used for intra-vascular radiation of a hemodialysis graft. *Cardio Radiat Med* 2003;4:90–94.

19. Supe SJ, Mallikarjuna RS, Sawant SG. Dosimetry of spherical 90Sr/90Y9 β-ray eye applicators. *Am J Roentgenol Radium Ther Nucl Med* 1975;123:36–41.

20. Nath R, Amols H, Coffey C et al. Intravascular brachytherapy physics: Report of the AAPM radiation therapy committee task group no. 60. *Med Phys* 1999;26:119–152.

21. Nieto JA, Thermoluminescence dosimetry (TLD) and its application in medical physics, *AIP Conf Proc* 2004;724:20.

22. DeWerd LA, Liang Q, Reed JL, Culberson WS. The use of TLDs for brachytherapy dosimetry. *Radiat Measure* 2014;71:276–281.

23. Kirov AS, Williamson JF, Meigooni AS, Zhu Y. TLD, diode and Monte Carlo dosimetry of an ^{192}Ir source for high dose-rate brachytherapy. *Phys Med Biol* 1995;40:2015–2036.

24. Muench PT, Meigooni AS, Nath R, McLaughlin WL. Photon energy dependence of the sensitivity of radiochromic film compared with silver halide and LIF TLDs used for brachytherapy dosimetry. *Med Phys* 1991;18:769–775.

25. Meigooni AS, Meli JA, Nath R. Influence of the variation of energy spectra with depth in the dosimetry of ^{192}Ir using LiF TLD. *Phys Med Biol* 1988;33:1159–1170.

26. The Bronowice Cycltron Center. Available from: http://www.ifj.edu.pl/ccb/en/badania/projekt4.php.

27. Piermattei A, Azario L, Monaco G et al. p-type silicon detector for brachytherapy dosimetry. *Med Phys* 1995;22:835–839.

28. Radiation Products and Design, Inc. Available from: http://www.rpdinc.com/diodes-155.

29. Saini AS, Zhu TC. Energy dependence of commercially available diode detectors for in-vivo dosimetry. *Med Phys* 2007;34:1704–1711.

30. Kertzscher G, Rosenfeld A, Beddar S et al. In vivo dosimetry: Trends and prospects for brachytherapy. *Br J Radiol* 2014;87: (accessed on February 6, 2014).

31. Standard imaging. Available from: http://www.standardimaging.com/exradin/.

32. PTW. Available from: http://www.ptw.de/well-type_chamber.html.

33. Fluke. Available from: http://en-us.fluke.com/products/hvac-iaq-tools/fluke-481-desi.html.

34. Bohm TD, DeLuca Jr. PM, DeWerd LA. Brachytherapy dosimetry of ^{125}I and ^{103}Pd sources using an updated cross section library for the MCNP MC transport code. *Med Phys* 2003;30:701–710.

35. Taylor REP, Rogers DWO. EGSnrc MC calculated dosimetry parameters for ^{192}Ir and ^{169}Yb brachytherapy sources. *Med Phys* 2008;35:4933–4944.
36. Chandola RM, Tiwari S, Kowar MK, Choudhary V. Review article: Monte Carlo and experimental dosimetric study of the mHDR-v2 brachytherapy source. *J Cancer Res Ther* 2010;6:421–426.
37. Saylor MC, Tamargo TT, McLaughlin WL et al. A thin film recording medium for use in food irradiation. *Radiat Phys Chem* 1988;31:529–536.
38. Meigooni AS, Kleiman MT, Johnson JL et al. Dosimetric characteristics of a new high-intensity ^{192}Ir source for remote afterloading. *Med Phys* 1997;24:2008–2013.
39. Meigooni AS, Sanders MI, Ibbott GS, Szeglin SR. Dosimetric characteristics of an improved radiochromic film. *Med Phys* 1996;23:1883–1888.
40. Niroomand-Rad A, Blackwell CR, Coursey BM et al. Radiochromic film dosimetry. Recommendations of AAPM radiation therapy committee task group 55. *Med Phys* 1998;25:2093–2115.
41. Zhu Y, Kirov AS, Mishra V et al. Quantitative evaluation of radiochromic film response for two-dimensional dosimetry. *Med Phys* 1997;24:223–231.
42. Brown TAD, Hogstrom KR, Alvarez D et al. Dose-response curve of EBT, EBT2, and EBT3 radiochromic films to synchrotron-produced monochromatic X-ray beams. *Med Phys* 2012;39:7412–7417.
43. Dini SA, Koona RA, Ashburn JR, Meigooni AS. Dosimetric evaluation of GAFCHROMIC® XR type T and XR type R films. *J Appl Clin Med Phys* 2005;6:114–134.
44. Sutherland JGH, Rogers DWO. Monte Carlo calculated absorbed-dose energy dependence of EBT and EBT2 film. *Med Phys* 2010;37:1110–1116.
45. Butson MJ, Cheung T, Yu PKN. Weak energy dependence of EBT gafchromic film dose response in the 50 kVp–10 MVp X-ray range. *Appl Radiat Isot* 2006;64:60–62.
46. Butson MJ, Yu PKN, Cheung T, Alnawaf H. Energy response of the new EBT2 radiochromic film to X-ray radiation. *Radiat Meas* 2010;45:836–839.
47. Richter C, Pawelke J, Karsch L, Woithe J. Energy dependence of EBT-1 radiochromic film response for photon (10 kVp–15 MVp) and electron beams (6–18 MeV) readout by a flatbed scanner. *Med Phys* 2009;36:5506–5514.
48. Sutherland JGH, Rogers DWO. Monte Carlo calculated absorbed-dose energy dependence of EBT and EBT2 film. *Med Phys* 2010;37:1110–1116.
49. Niroomand-Rad A, Chiu-Tsao S, Soars C et al. Comparison of uniformity of dose response of double-layer radiochromic film (MD-55-2) measured at 5 institution. *Phys Med* 2005;21:15–21.
50. Chiu-Tsao ST, de la Zerda A, Lin J, Ho KJ. High sensitivity GafChromic film dosimetry for ^{125}I seed. *Med Phys* 1994;21:651–657.
51. Morrison H, Menon G, Sloboda RS. Radiochromic film calibration for low-energy seed brachytherapy dose measurement. *Med Phys* 2014;41:072101-1-11.
52. Aldelaijan S, Mohammed H, Tomi N et al. Radiochromic film dosimetry of HDR ^{192}Ir source radiation fields. *Med Phys* 2011;38:6074–6083.
53. Le Y, Ali I, Dempsey JF, Williamson JF. Prospects for quantitative two-dimensional radiochromic film dosimetry for low dose rate brachytherapy sources. *Med Phys* 2006;33:4622–4634.
54. Uniyal SC, Sharma SD, Naithani UC. A dosimetry method in the transverse plane of HDR Ir-192 brachytherapy source using gafchromic EBT2 film. *Phys Med* 2012;28:129–133.
55. Williamson JF, Thomadsen BR, Nath R (Eds.). *Brachytherapy Physics*. Madison, WI: Medical Physics Publishing; 1995.
56. Schumer W, Fernando W, Carolan M et al. Verification of brachytherapy dosimetry with radiochromic film. *Med Dosimet* 1999;24:197–203.
57. Ghorbani M, Bahreyni TMT, Mowlavi AA et al. A Application of a color scanner for ^{60}Co high dose rate brachytherapy dosimetry with EBT radiochromic film. *Radiol Oncol* 2012;46:363–369.

58. Duggan DM, Coffey CW, Lobdell JL, Schell MC. Radiochromic film dosimetry of a high dose rate beta source for intravascular brachytherapy. *Med Phys* 1999;26:2461–2464.

59. Kirisitsa C, Georga D, Wexbergb P et al. Determination and application of the reference isodose length (RIL) for commercial endovascular brachytherapy devices. *Radiother Oncol* 2002;64:309–315.

60. Song H, Roa DE, Yue N et al. Application of Gafchromic® film in the dosimetry of an intravascular brachytherapy source. *Med Phys* 2006;33:2519–2524.

61. Acar H, Chiu-Tsao ST, Özbay İ et al. Evaluation of material heterogeneity dosimetric effects using radiochromic film for COMS eye plaques loaded with ^{125}I seeds (model I25.S16). *Med Phys* 2013;40:011708-1-13.

62. Heon CC, Soo HH, Son KJ et al. Dosimetry of a new P-32 ophthalmic applicator. *Med Phys* 2011;38:6143–6151.

63. Smitt MC, Donaldson SS. Radiation therapy for benign disease of the orbit. *Sem Radiat Oncol* 1999;9:179–189.

64. Alaniz-Camino F. The use of postoperative beta radiation in the treatment of pterygia. *Ophthal Surg* 1982;13:1022–1025.

Megavoltage beam commissioning

IORI SUMIDA, DAVID BARBEE, AND INDRA J. DAS

7.1 INTRODUCTION

The field of radiation therapy has revolutionized with the introduction of linear accelerator-based megavoltage beams and characterized by their increased skin sparing and deeper penetration depth relative to kilovoltage and isotope-generated (teletherapy) beams. These beam characteristics led to improved outcome in radiation treatment of most diseases by improved sparing of skin and other organs at risk while allowing for more uniform dose distributions. Innovation and advancement of linear accelerator technology rapidly progressed with the development of multileaf collimators (MLCs) for designing dynamic beam's eye view blocking needed for three-dimensional conformal radiation therapy (3DCRT), intensity-modulated radiotherapy (IMRT), and volumetric-modulated arc therapy (VMAT). The evolution of radiation treatment from 2D to 3DCRT and beyond is only possible with advances in radiation technology with highly complex motion and dose delivery providing hypofractionation with very high dose rate (2400 cGy/min). This chapter details the use of radiochromic film (RCF) in the routine quality assurance of medical accelerators, collection of beam data for commissioning, and treatment-planning system (TPS) quality assurance and validation.

However, before these sophisticated radiation emitting machines and techniques can be used for patient's treatment, they must go through acceptance testing and proper commissioning and validation by qualified personnel [1] before being used clinically. The commissioning of accelerators is a time-consuming process requiring significant data acquisition, analysis, and processing, as discussed by Das et al. in TG-106 [2]. The majority of beam-data collection is performed using a single detector in a scanning water phantom, wherein the beam is static and the chamber response changes as the detector position changes within the water phantom. However, in the case of dynamically changing beam systems (e.g., MLC movement in IMRT/VMAT or jaw movement in virtual/soft wedges), multiple measurements are required to fully characterize beam properties in space. Measurement of such fields using a single detector would require repeated beam deliveries for each new measurement position, which significantly increases acquisition time and is an inefficient measurement technique. This has led to the development of 1D and 2D measurement systems that allow for rapid collection of dynamically changing beam data in a single beam delivery. Such systems specific to commissioning beam data include 1D and 2D diode and chamber arrays (SNC Waterproof Profiler, PTW LA48) as well as radiographic and RCF. Ion chamber and diode array systems are primarily used for beam data commissioning allowing for rapid collection of data; however, the upfront cost of such systems is prohibitively expensive considering their limited clinical utilization. Moreover, spatial resolution is typically on the order of millimeters and requires software and hardware interfaces, and array calibration for each beam energy to normalize detector response.

In contrast, film techniques are cheaper, can be utilized for a wide range of clinical applications in addition to commissioning, and offer submillimeter resolution. RCF, specifically, requires no response calibration, is near tissue equivalent, can be cut to specific application, and requires no processor or development. The primary drawbacks to film techniques involve the need for digitization (film scanning) for advanced quantitation and analysis and calibration curves to relate dose to film response. Radiographic film and RCF provide unique opportunity to collect 2D data in a single exposure. With disappearance of radiographic films and emergence of RCF having many advantages (self-developing, no dark room, energy dependency, high dose range, etc.), as discussed in Chapter 3, it provides opportunity for beam data collection and have beam explored by many investigators [3–5].

7.1.1 HISTORICAL VIEW OF 2D DOSIMETER

Point dosimetry allows us to verify the absolute dose, whereas 2D dosimetry of dose distributions requires higher dimensional measurements. Currently, there are several options for 2D dosimeters such as radiographic film, RCF, diode arrays, and ionization chamber arrays. Historically, the majority of beam data collection techniques were done using radiographic film in the megavoltage energy range. Film is convenient to use, providing permanent records of integrated dose distributions in a single exposure from which a set of isodose curves can be obtained in the film plane [6]. However, radiographic film has several disadvantages, such as an energy-dependent response due to oversensitivity at low photon energies, a complex development process, and requires a dark room and processor [7,8]. Innovations in RCF have removed many of these drawbacks; the complex development process has been eliminated, and a dark room is not necessary. With the appropriate film-scanning technique [9] for the conversion process from the film density to dose, it is possible to perform the absolute dosimetry as well as the evaluation of dose distributions. Two-dimensional diode and ionization chamber arrays are also available to use for the same purpose. The repeatability of dose measurements is superior in these detectors compared with those in film detectors [10,11]; however, the diode has an energy, dose rate, and temperature dependence [12–14], and the ionization chamber has a volume averaging effect with larger cavity [15–17]. Some of these problems are not encountered when using RCF; thus, some prefer to use RCF with comfort.

7.1.2 RADIOCHROMIC PROPERTIES

Film dosimetry in radiotherapy plays an important role for qualitative geometrical verification and quantitative dose measurements. As a quality control (QC) of radiotherapy machines, coincidence of light and radiation fields, MLC leaf positions, dose profile at depth in phantom, and the coincidence of rotational axes of gantry, collimator, and couch with respect to the isocenter, so-called *star-shot* irradiation, are verified by the use of film measurement. Historically, radiographic film has been used for the above QC items; however, RCF has recently begun replacing radiographic film for these tasks. Development of RCF and its characteristics can be found in Chapters 2 and 3 of the current book.

7.1.3 COMPLEX GEOMETRY WITH HETEROGENEITY

Patients' bodies contain many structures with different densities and composition, such as soft tissue, lung, cartilage, bone and air cavity, and occasionally implanted materials and alloys including titanium, steel, gold, and dental amalgam. Dose measurements incorporating such heterogeneity changes are challenging due to changes in electronic equilibrium and low energy scatter through various heterogeneity interfaces. Dose can be measured after the beam passed through the inhomogeneity medium for both ionization chamber and film; however, in case of measuring at the interface between homogeneous and inhomogeneous mediums, film measurement is better suited to know the dose changes extending over the interface in phantom. RCF film measurement is especially advantages in cases in which the dose measurement is conducted to know the scatter radiation from the surface of high-density materials such as metals or alloys. In these situations, the film should be adhered to the metal surface as closely as possible to measure dose. Backscatter dose from metallic interfaces can be challenging due to short distances as described by Das and Khan [18]. RCF film has been successfully used for complex geometry as shown by Shimamoto et al. [19]. Similarly, RCF could be used in many complex situation including lungs.

7.2 RCF AND QUALITY ASSURANCE IN MEGAVOLTAGE

7.2.1 COMPONENT OF QUALITY ASSURANCE TG-142

Quality-assurance programs are essential to the safety and accuracy of radiotherapy delivery. Task Group 142 [20] provides recommendations on machine QA parameters, some of which can be performed using RCF. The QA items such as the coincidence of light field and radiation field, star shots for rotation axes, MLC tests for leaf characteristics, and soft wedge could be easily performed using RCF.

7.2.2 COINCIDENCE OF LIGHT FIELD AND RADIATION FIELD

Ensuring that the light field is commensurate with the radiation field is especially important for treatment cases in which the light field is used to clinically define or match the radiation field to be delivered. Such cases include electron–photon field matching, light-field entry points relative to tattoos or patient anatomy, and clinical electron setups. Errors in light field radiation field alignment may result from positioning errors in mirror or bulb positioning or beam steering.

Field size is defined as the size of collimator opening at source–axis-distance (SAD) 100 cm. Figure 7.1 shows light and radiation fields from 5×5 to 20×20 cm^2 film on a single film as well as an independent jaw test. Independent jaw tests consist of bringing each jaw to a half beam block and irradiating a single RCF. This test is useful in determining symmetry and gap width of half beam blocks as well as diagnosing jaw skewness.

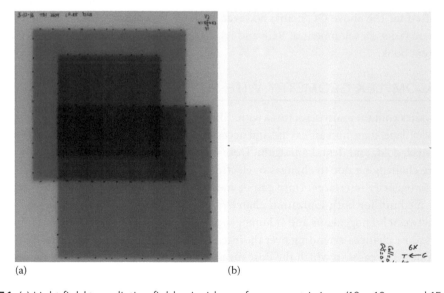

(a)   (b)

Figure 7.1 (a) Light field to radiation field coincidence for symmetric jaws (10 × 10 cm and 15 × 15 cm) and asymmetric jaws (X1 = 5, X2 = 10, Y1 = 15, and Y2 = 0 cm) delivered on a single 8 × 10″ RCF. (b) Independent jaw comprises four quadrant irradiation, in which each jaw is brought to half beam block and irradiated. Note that there is no overdose at junctions.

7.2.3 STAR SHOTS

Star-shot analysis determines the coincidence of radiation isocenter and mechanical isocenter with respect to the rotation axes of gantry, collimator, and couch. Treatment room laser should be matched to the mechanical isocenter before performing star-shot irradiation. As an example, during gantry star-shot irradiation, the RCF is placed on the treatment couch and parallel to the beam axis in transverse (lateral–vertical) plane and is sandwiched between solid water phantoms. Then, slit beams made by the X-jaw at several gantry angles (e.g., 0°, 45°, 90°, and 135°) irradiate the RCF to create the star-shot pattern. As with the procedure of the coincidence of light field and radiation field, the vertical and lateral marks with respect to the treatment room laser are marked on the RCF by a marker pen. The radiation isocenter would be a center of gravity calculated from the slit fields. The coincidence of radiation isocenter to mechanical isocenter could be calculated by the displacement between the center of gravity from slit beams and the cross point of the four marks as shown in Figure 7.2.

7.2.4 MLC TESTS

MLC leaf evaluation can be performed using RCF. Properties such as leaf-position accuracy and leaf transmission can be performed for mechanical and dosimetric checks, respectively. These evaluations are necessary before clinical implementation of MLC programs and routinely as QC checks for accurate beam modeling and to ensure accurate dose calculation and delivery. According to the TG-142 report, the procedures of MLC tests are distinguished in non-IMRT and IMRT. In this section, the procedures for non-IMRT are presented; IMRT and VMAT are discussed in Chapter 8 of this book. For the verification of field shape using MLC blocking, the planned field should be compared with the radiation field. For example, the RCF is placed on the treatment couch and perpendicular to the beam axis in coronal (lateral-long) plane and is sandwiched between solid water phantoms. By irradiation for getting enough film density, the appropriate monitor units (MUs) would be necessary. To evaluate the difference between the planned and irradiated dose distribution, distance-to-agreement evaluation is useful for quantitative check.

For MLC leaf-transmission measurement, films must be irradiated for both MLC banks such that the RCF is sufficiently exposed, followed by a third film with no MLC blocking to serve as a normalization scan to remove beam nonuniformity. The irradiated field will contain some nonuniformity due to the absence or presence of the flattening filter as well as the amount of buildup used. Sufficient and equivalent buildup is required for the three measurements to ensure that electron contamination does not provide additional exposure to the open field, causing reduced transmission measurements after open-field normalization. In addition, all films should be marked by laser or crosshair for purposes of spatial registration to normalize the MLC-blocked films with the open-field film. The ratio of the closed-field film to the open-field film represents the transmission and should be measured for interleaf and intraleaf transmission. MLC transmission films will typically require a factor of 5–10 greater number of MUs than the open-field MUs for sufficient exposure for comparison, due to the leaves' expected 1%–2% transmissions. This factor should be included during the open-field normalization. In addition, transmission may be obtained from MLC leaves from alternating banks to investigate the interleaf leakage between alternating leaves. When an ionization chamber is used to measure the MLC leaf transmission, the size of the detector is relatively large to the leaf width. Therefore, average leaf transmission including interleaf and intraleaf would be measured. As the spatial resolution for film dosimetry

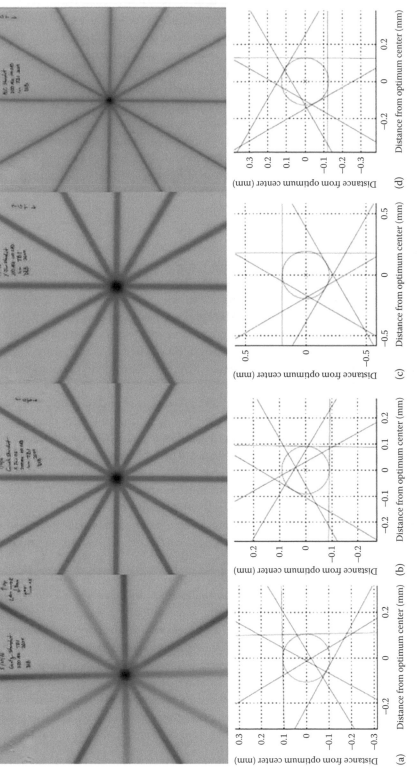

Figure 7.2 Star-shot patterns delivered on RCF for (a) gantry, (b) couch, (c) collimator, and (d) MLC radiation isocenters (top row). The corresponding evaluation of each digitized RCF star shot using Dose Lab Pro (Mobius Medical) wherein the diameter of the largest circle contained within the largest triangle formed from intersecting spokes measures the radiation isocenter (bottom row).

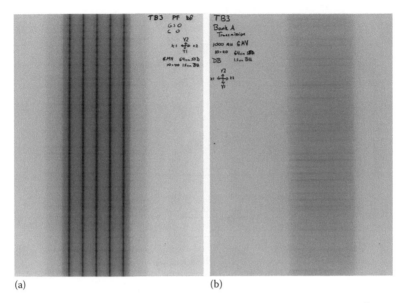

(a)   (b)

Figure 7.3 (a) A picket fence pattern delivered on RCF, testing MLC positional accuracy. (b) MLC transmission measured using RCF. The transmission film has two registration points that allow for registration to the open-field film for normalization. Both films were delivered on a Varian millennium 120 MLC using 6 MV with 1.5-cm build-up at source–film distance of 64 cm to include all 60 leaf pairs.

is high, it is possible to distinguish these leaf transmissions that are necessary for MLC modeling in some TPSs (Figure 7.3).

Implementation of VMAT programs also requires measurement of variable MLC speed, dose rate, and gantry speed that can be performed using RCF and is reviewed in Chapter 8.

7.2.5 SOFT WEDGE (DYNAMIC AND VIRTUAL)

Soft wedges are created using modulation of jaws in various ways based on the concept developed by Petti et al. [21]. This technology is called commercially as enhanced dynamic wedge and virtual wedge in Varian and Siemens accelerators, respectively. Movement of the collimating jaw during irradiation results in the beam modulation. To verify the slope of dose distribution, two-dimensional detectors such as film, diode array, and electronic portal imaging device are necessary. RCF is more useful to perform modulated 2D dose distribution for soft wedge fields, especially when the dose distribution in the transverse plane is investigated, which will be discussed later in Section 7.3.1.4. RCF also provides additional advantages of measuring dose in parallel and perpendicular directions relative to the beam axis and can be performed at any depth in water, solid water, or heterogenous phantoms.

7.3 RCF APPLICATIONS IN 3DCRT COMMISSIONING

This section provides the application of RCF for commissioning of 3D conformal radiotherapy systems. Before continuing, it is important to note that this section deals solely with large field commissioning. Please see the following chapters for detailed discussions of RCF in other sub-components of commissioning: IMRT/VMAT (Chapter 8) small field dosimetry (Chapter 13), and buildup (Chapter 16). The application of RCF consists of the following two primary phases: data collection for beam modeling and validation of beam models. In both situations, RCF is irradiated in a specific beam geometry, digitized, converted to dose, and then either input into the TPS for

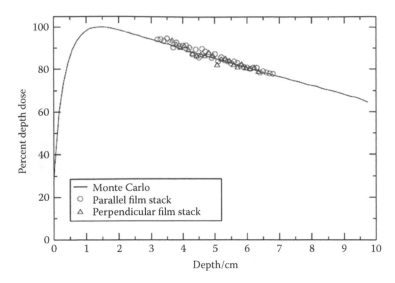

Figure 7.4 Stacked phantom approach for PDD. Data for parallel and perpendicular stacking are shown along with Monte Carlo generated 6-MV beam used for dose evaluation and the TPS commissioning. (Adapted from McCaw, T.J. et al., *Med. Phys.*, 41, 052104, 2014.)

purposes of beam modeling or compared with the expected doses from the TPS. In most cases, the same irradiated film can be used for both purposes.

The beam parameters acquired during commissioning provide the foundation for treatment planning. These can be regularly checked using RCF. The most common parameters are percent depth dose (PDD) and beam profiles. These two 2D parameters could be easily acquired using RCF. McCaw et al. [22] have provided a stacked phantom approach for the depth–dose measurement as shown in Figure 7.4. The stacking phantom provides minimum air column that is known for dosimetric error. Similarly, beam profiles at different depths can be acquired, which has been shown to be satisfactory with 1.5% [22]. Instead of such phantom, film can be exposed in a water phantom with the help of plastic holders or book self-clamps to keep film in a single plane. A simple device for film exposure is shown by Arjomandy et al. [23]. The advantage of water is that it is homogeneous and has no air/medium perturbation. Several investigators have provided measurements in water [24–27]. In case film is placed in water, it should be kept only during exposures with minimum duration as water could seep through the film layers causing peripheral damage [27]. The films should be dried out uniformly and individually after taken out of water. Various technical information about irradiation in water are provided by Van Battum et al. [24] and claimed that accuracy ±1.5% is easily achievable.

RCF provides dosimetrically accurate measurement in homogeneous media; however, in 3DCRT, nearly all clinical planning situations in the human body involve inhomogeneities. The calculation algorithms should provide a high degree of accuracy in complex geometries and should be verified by measurements, wherein RCF is a very useful tool. During commissioning, it is possible to simplify the irradiation method used clinically and to investigate the accuracy of the individual dose calculation to estimate the uncertainty of the irradiation accuracy of the dose. In most cases, some simplification can be noted: the use of a rectangular irradiation fields, the use of a wedges, the tilt of the surface of the body with respect to the beam, and the presence of an inhomogeneous region (lung, air, and bone) in the irradiation field. The calculated dose profile depends on the method of dose calculation, for example, the Clarkson method, the convolution/superposition, collapsed-cone method, or the Monte Carlo method. The accuracy of some of the algorithms differs significantly as noted by Akino et al. [28] can be easily performed in phantom with RCF. Furthermore, clinical irradiation comprises a combination of many physical parameters. Therefore, it is important to investigate each of these parameters in which RCF plays an important role due to its characteristics as described earlier in Chapters 2 and 3.

7.3.1 MEASURED DOSE DISTRIBUTION COMPARED WITH THE CALCULATED DOSE PLANE

The equipment required for dose investigations include an ionization chamber for the absolute dose measurement and films, a scanner, and dose profile analysis software for the relative dose-profile measurement. First, a treatment plan for commissioning on a TPS and outputting a point dose and dose profile is proposed. Water-equivalent phantoms, into which the ionization chamber and films can be inserted and virtual water-equivalent phantoms, created within the TPS, are used as the irradiation targets. An example of such practice was clearly introduced in ESTRO Booklet No. 7 [29], for commissioning regarding the dose investigation for a TPS. Detailed information can be viewed from this booklet about the process and need for verification in PDD and profiles that care essentials for the beam data.

7.3.1.1 OPEN SQUARE AND RECTANGULAR FIELDS

This basic verification consists of delivering square and rectangular fields ranging from 5×5 to 40×40 cm^2 to a measurement device. It is possible to check the output factor by changing the irradiation field size and to check the extent of the collimator exchange effect by changing the sizes in the X and Y-directions that is also energy dependent [30,31]. Many investigators have studied the use of RCF for beam data, such as PDDs, and dose profiles, using RCF and compared with various detectors including ion chambers, diode, TLD, and Monte Carlo calculations. As high-resolution data from film can be acquired, many types of analysis can be made such as shown in Figure 7.5. It shows the distributions of (a) dose difference and (b) distance to agreement between TPS and RCF for a simple 10×10 cm^2 square field at 100-cm SAD and 10-cm depth. Dose profiles for comparison in horizontal and vertical directions are shown in subfigures of (c) and (d), respectively. Both profiles are normalized at the beam center. It was found that the measured dose was a little higher than the planned dose outside the field in subfigures of (a) and (c). It should be noted that such systematic discrepancy might be derived from the TPS modeling for MLC transmission factor and jaw transmission factor. In contrast, the dose profiles outside the field are well matched between TPS and RCF in vertical direction (d). Although the dose profiles in TPS and RCF are well matched and visually confirmed in (c) and (d), distance-to-agreement approach is suitable to evaluate quantitatively shown in (b). As the spatial resolution for RCF is high (e.g., 0.3 mm/pixel in 75 dpi), it is a suitable way for the evaluation of dose distribution especially at high dose gradient region.

Various investigators have proven that RCF can be used in beam modeling in Monte Carlo or commissioning, such as penumbra region matching in both electron and photon-intensity profiles [32,33]. The accuracy of RCF in photon beam measurement can be relatively good up to 2.8% [34] and even better as shown by McCaw et al. [22] up to 1.5%. The accuracy is dependent on the exposure conditions, care of the film, and reading processes that have been described in Chapter 2 through 4. An analysis of RCF data in water was presented with three sets of film and compared with an ion chamber. Figure 7.6 shows data from Van Battum [24] indicating accuracy within ±1.8% across the entire profile.

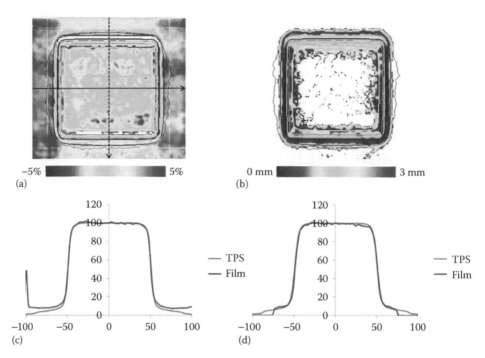

Figure 7.5 Analysis of beam data using radiochromic film in a solid water phantom exposed in a reference condition. (a) Dose difference between TPS and RCF, (b) distance to agreement (DTA) between TPS and RCF, (c) beam profile comparison in MLC direction, and (d) comparison of TPS and film with primary jaws indicating a good agreement.

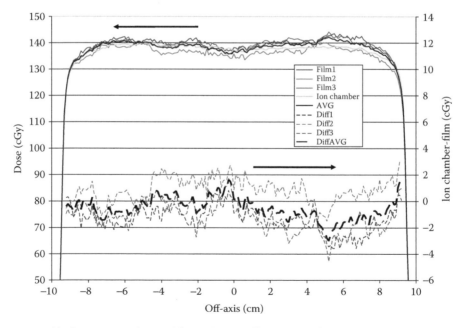

Figure 7.6 Profile from 6 MV derived from three different RCF films in water. Solid line indicates average values for profile and difference from the ion chamber. (Adapted from Van Battum, L.J. et al., *Med. Phys.*, 35, 704–716, 2008.)

7.3.1.2 PERCENT DEPTH DOSE

In general, PDD curves are measured using an ionization chamber in a water tank. Alternatively, this can be acquired using RCF either in a phantom or water tank. As mentioned by McCaw et al. [22], when using a solid water phantom, dose perturbation due to air should be taken into account. Figure 7.7 depicts PDD measurement using RCF sandwiched between water-equivalent slabs and exposed parallel to the beam axis for a 4-MV beam. Comparison between the RCF and ion chamber measurements demonstrates that the two methods are commensurate to within 2% for these conditions. The PDD measured by the RCF is useful for quick periodic QA without a 3D water tank, thus saving significant amount of time in a busy department.

7.3.1.3 OFF-AXIS FIELD

Half beam either by block or by independent jaws is required in many clinical cases such as breast, head and neck, and spine irradiation. Dosimetry of such fields can easily be verified using RCF in a phantom. In addition, off-axis data can be acquired using film. Many clinical situations like medulloblastoma may require matching fields in 3D. In these situations, fields can be visually verified with RCF, whereas doses can be quantitatively compared with TPS provided doses. Similar approaches in photon and proton beams have been demonstrated with radiographic film [35,36] that can be achieved with RCF in a complex geometry.

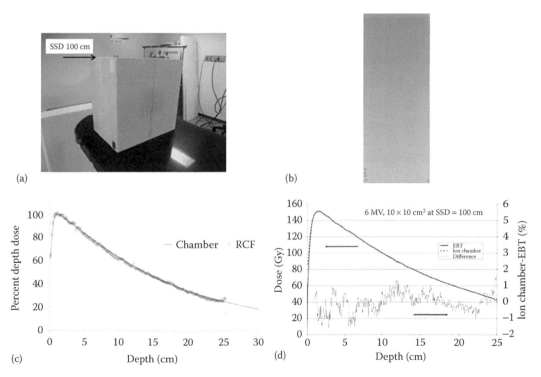

Figure 7.7 PDD measurements with RCF: (a) setup arrangement, (b) 2D data set for PDD on film, and (c) PDD comparison between chamber (solid line) and data point from RCF. Similar and more precise data taken in water are shown in (d). Such quick data acquisition with film may be very useful for periodic QA or after machine repair. (Adapted from Van Battum, L.J. et al., *Med. Phys.*, 35, 704–716, 2008.)

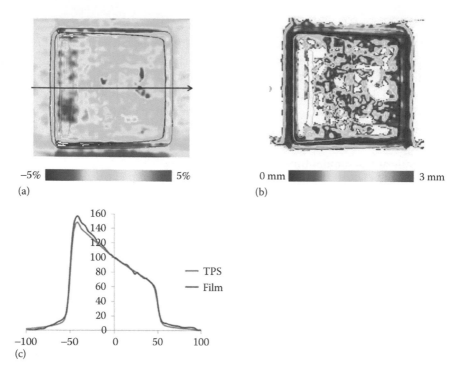

Figure 7.8 Comparison of TPS and RCF data acquired in a phantom for a 45° wedge angle at 10-cm depth. (a) Dose difference between TPS and RCF, (b) DTA between two data sets, and (c) actual dose profiles from TPS and film at the black arrow in (a). The discrepancies at heel could be because of poor modeling of TPS.

7.3.1.4 WEDGE FIELD

Commissioning of soft wedges requires 2D detector that is not available in many institutions. This can be easily verified using film in a phantom or water. Figure 7.8 shows the distributions of (a) dose difference and (b) distance to agreement between TPS and RCF. Wedge field irradiation was performed for a 10×10-cm^2 field at 100-cm SAD and at 10-cm depth. Dose profiles along the dose gradient for comparison in horizontal direction are shown in subfigure (c). Both profiles are normalized at the beam center. It is found that the measured dose at heel side was a little higher than the planned dose in subfigures of (a) and (c). It should be noted that such systematic discrepancy might be derived from the TPS modeling for wedged shape. Distance-to-agreement distribution shows the near agreement between TPS and RCF as shown in (b).

7.3.1.5 INHOMOGENEITY

In addition to homogenous system, 3DCRT requires validation of inhomogeneous media that can be tested in an appropriate phantom. For the inhomogeneous section, a low-density region simulating lung and a high-density region for simulating bone can be used. The accuracy of the dose calculation for the absorbed dose of a beam passing through an inhomogeneous section is evaluated by comparing the calculation algorithms available in the TPS. The dose inside the inhomogeneous part and the transmitted dose vary significantly depending on the irradiation field size, beam energy, and the dose-calculation algorithm used [37]. Point-dose validations using ion chamber can be very tedious and in some instances not feasible due to placement (drilling) within the phantom. RCF provides a great opportunity to measure dose in a plane with inhomogeneity.

To verify the dose calculation corrections in heterogeneous medium, the low- and high-density materials (lung and bone) are prepared before the TPS commissioning. Inhomogeneity corrections require the conversion of CT numbers to electron density. If a correction model is unable to deal with inhomogeneity corrections, one should delineate manually the specific region and assign the appropriate CT number or electron density to calculate the dose accurately. For head-and-neck regions, metal artifact from the dental crown would be present and invalidate CT numbers in soft tissue. In addition to the dose verification at the beam central axis below the inhomogeneity medium, one should recognize the dose calculation accuracy for whether the TPS is able to calculate the effect of electric disequilibrium at the field edges and interfaces. As film is a 2D detector, it is suitable to use in these regions.

7.3.1.6 OBLIQUE INCIDENCE

Megavoltage 3DCRT uses oblique incidences that perturb the buildup dose due to electron transport moving toward surface as described by Jackson [38]. Details of the obliquity factors have been also provided by Gerbi et al. [39] in a homogeneous medium. However, in a clinical setting with a complexity of patients, obliquity is frequently problematic. Such difficulty could be resolved with RCF measuring dose distribution accurately with a selection of proper phantom or in a water phantom. The dose-calculation accuracy of the TPS could also be evaluated with oblique beam incidence very similar to the study conducted by Akino et al. [40].

7.3.1.7 ABSORPTION AND SCATTERING FROM THE TREATMENT COUCH

For commissioning of 3DCRT, in a case with multiple beams irradiated from different angles, the beams could pass through the treatment couch, and the subsequent transmitted beams would have a decreased dose as described by Olch et al. [41]. Furthermore, if the irradiation target is close to the couch side of the irradiated body, it is possible that the dose could be increased by the scattered radiation from the treatment couch and that the dose could then be higher than the calculated dose. Therefore, considering such uncertainties, it is possible to reduce the effect of scattered photons from the treatment couch by increasing the distance between the couch and irradiated body with absorbing and ignorable materials, such as Styrofoam. However, such techniques in a clinical setting are not always feasible. If dose verification is needed at the interface, modeling the treatment couch shape with an accurate relative electron density (physical density) and making a dose calculation for the integrated model comprising the body contour and the modeled treatment couch could be used for calculation. Verification of the combined couch and body model or estimated dose at interface can be measured using RCF.

7.4 QUALITY ASSURANCE OF A TREATMENT-PLANNING SYSTEM

Before a TPS can be used clinically, the accuracy of the dose calculation needs to be verified with all its parameters, such as PDD, off-center ratio, slope of beam penumbra, and transmission factor of jaw and MLC. These must be modeled correctly to accurately match calculated data to acquired measurements. Beam data are commissioned using water phantom as recommended by TG-106 [2]; however, quick verification can be achieved for quality-assurance purpose by two-dimensional detectors similar to film. RCFs have become popular as they are self-developing, tissue equivalent, and can be placed in phantom. Many of the characteristics have been described in Chapter 3 and need not to be elaborated here. However, its utility in megavoltage beam dosimetry is discussed in the current chapter.

7.4.1 TPS COMMISSIONING

There are many guidelines for TPS commissioning. In modern times, some of the older references are obsolete. Central axis data in most cases are in agreement including RCF; however agreement could be a problem due to steep dose gradient in the beam penumbra regions. The deviation between the calculated and measured doses is not expressed as the relative error; rather the distance expresses the difference in the dose distribution. Homogeneous phantoms correspond to water phantoms and water-equivalent phantoms. For inhomogeneous phantoms, the acceptable dose criteria are shown for the transmission of a beam through a substance, the density of which is different from that of water, with lung-density equivalent or bone-density equivalent. Commissioning of the dose-calculation accuracy of the TPS has been provided by Venselaar et al. [42]. Table 7.1 can be used as a standard for the comparison using ion chamber or film. Based on data provided and from various references, it is clear that RCF will provide data within acceptable limit with precautions as recommended in the literature.

7.4.2 COMPLEX PLANS

Details of complex techniques such as IMRT, VMAT, stereotactic radiosurgery (SRS), stereotactic body radiation therapy (SBRT) can be reviewed in their respective chapters in this book. The commissioning of TPS dose calculation is conducted using a single beam in anterior–posterior direction. This is a convenient way to assess the accuracy of dose calculation to determine which parameters have the greatest dosimetric impact or deviation from TPS calculation. Complex plans consisting of multiple beams (e.g., four fields box irradiation) are used to evaluate the overall dose calculation accuracy along with combinations of wedged and MLC-shaped beams. RCF placed in the appropriate phantom can be used to validate these complex plans. Furthermore, orientation of the RCF within the phantom can often be arranged to investigate dose gradients in multiple

Table 7.1 Acceptable criteria for external beam-dose calculations

Situation	Absolute dose at normalization point (%)	Beam central axis (%)	Inner beam (%)	Beam penumbra (mm)	Outer beam (%)	Buildup region (%)
Homogeneous phantoms						
Square fields	0.5	1	1.5	2	2	20
Rectangular fields	0.5	1.5	2	2	2	20
Asymmetric fields	1	2	3	2	3	20
Blocked fields	1	2	3	2	5	50
MLC-shaped fields	1	2	3	3	5	20
Wedged fields	2	2	5	3	5	50
External surface variations	0.5	1	3	2	5	20
SSD variations	1	1	1.5	2	2	40
Inhomogeneous phantoms[a]						
Slab inhomogeneities	3	3	5	5	5	–
3D inhomogeneities	5	5	7	7	7	–

[a] Excluding regions of electronic disequilibrium.

directions and in planning target regions, organs at risk, and heterogeneity or buildup regions. For example, in cases of beam attenuation due to beams passing through the treatment couch, film could be inserted in a transverse plane to investigate scatter from the couch and additional dose deposition at the couch-phantom interface. As for another complex geometry to verify the dose calculation accuracy, the complex surfaces (e.g., sharp point, square corner, convexities, etc.) in patient shape such as head-and-neck region can be investigated using RCF.

7.4.3 DOSE CALCULATION

Whenever the accuracy of a TPS dose-calculation algorithm is evaluated, care should be taken in regards to the resolution of the dose-calculation grid, which is known to affect calculation accuracy [40,43]. The finer grid resolution used, the slower the dose calculation, but the more accurate the calculation. In particular, the region of steep dose gradient is subject to the grid resolution in terms of the visualization of the isodose line. For dose-volume evaluation, small volume organs, such as lens, optic nerves, and optic chiasms, are also subject to the grid resolution as the number of voxels for these organs is relatively smaller than those of other organs Srivastava et al. [43]. If RCF data for dose calculation do not agree with TPS-dose calculation for high gradient or small-volume regions, the grid resolution should be investigated and the grid size possibly reduced.

7.5 SUMMARY

The 3DCRT is still a major treatment modality in which commissioning, continuous verification, and QA data are critical for patient treatment. RCF provides 2D data set that can be acquired in any simple and complex geometry. Even though data can be acquired with other devices such as ion chamber or other reliable detector, RCF provides unique opportunity to a clinical physicist for independent verification QA. It can provide accurate relative dose distribution in buildup region and PDD as well as TMR, and profiles. In addition, use of RCF for QA and commissioning and quick 2D verification of treatment algorithm is extremely valuable. The RCF film can be used in solid phantom as well as in water, thus reducing some of the disadvantages of RCF as discussed in this book.

REFERENCES

1. Das IJ, Anderson A. Equipment and software: Commissioning from the quality and safety perspective. In: Dicker AP, Williams TR, Ford E (Eds.). *Quality and Safety in Radiation Oncology: Implementing Tools and Best Practices for Patients, Providers, and Payers.* New York: Demos Medical; 2016. pp. 73–84.
2. Das IJ, Cheng CW, Watts RJ et al. Accelerator beam data commissioning equipment and procedures: Report of the TG-106 of the therapy physics committee of the AAPM. *Med Phys* 2008;35:4186–4215.
3. Dempsey JF, Low DA, Mutic S et al. Validation of a precision radiochromic film dosimetry system for quantitative two-dimensional imaging of acute exposure dose distributions. *Med Phys* 2000;27:2462–2475.
4. McLaughlin WL, Yun-Dong C, Soares CG et al. Sensitometry of the response of a new radiochromic film dosimeter to gamma radiation and electron beams. *Nucl Instrum Meth Phys Res* 1991;A302:165–176.
5. McLaughlin WL, Soares CG, Sayeg JA et al. The use of a radiochromic detector for the determination of stereotactic radiosurgery dose characteristics. *Med Phys* 1994;21:379–388.

6. Williamson JF, Khan FM, Sharma SC. Film dosimetry of megavoltage photon beams: A practical method of isodensity-to-isodose curve conversion. *Med Phys* 1981;8:94–98.

7. Pai S, Das IJ, Dempsey JF et al. TG-69: Radiographic film for megavoltage beam dosimetry. *Med Phys* 2007;34:2228–2258.

8. Das IJ. Radiographic film. In: Rogers DWO, Cyglar JE (Eds.). *Clinical Dosimetry Measurements in Radiotherapy.* Madison, WI: Medical Physics Publishing; 2009. pp. 865–890.

9. Niroomand-Rad A, Blackwell CR, Coursey BM et al. Radiochromic film dosimetry: Recommendations of AAPM radiation therapy committee task group 55. American Association of Physicists in Medicine. *Med Phys* 1998;25:2093–2115.

10. McKenzie EM, Balter PA, Stingo FC et al. Toward optimizing patient-specific IMRT QA techniques in the accurate detection of dosimetrically acceptable and unacceptable patient plans. *Med Phys* 2014;41:121702.

11. McKenzie EM, Balter PA, Stingo FC et al. Reproducibility in patient-specific IMRT QA. *J Appl Clin Med Phys* 2014;15:241–251.

12. Saini AS, Zhu TC. Dose rate and SDD dependence of commercially available diode detectors. *Med Phys* 2004;31:914–924.

13. Saini AS, Zhu TC. Temperature dependence of commercially available diode detectors. *Med Phys* 2002;29:622–630.

14. Létourneau D, Gulam M, Yan D et al. Evaluation of a 2D diode array for IMRT quality assurance. *Radiother Oncol* 2004;70:199–206.

15. Low DA, Parikh P, Dempsey JF et al. Ionization chamber volume averaging effects in dynamic intensity modulated radiation therapy beams. *Med Phys* 2003;30:1706–1711.

16. Fenwick JD, Kumar S, Scott AJ, Nahum AE. Using cavity theory to describe the dependence on detector density of dosimeter response in non-equilibrium small fields. *Phys Med Biol* 2013;58:2901–2923.

17. Bouchard H, Kamio Y, Palmans H et al. Detector dose response in megavoltage small photon beams. II. Pencil beam perturbation effects. *Med Phys* 2015;42:6048–6061.

18. Das IJ, Khan FM. Backscatter dose perturbation at high atomic number interfaces in megavoltage photon beams. *Med Phys* 1989;16:367–375.

19. Shimamoto H, Sumida I, Kakimoto N et al. Evaluation of the scatter doses in the direction of the buccal mucosa from dental metals. *J Appl Clin Med Phys* 2015;16:233–234.

20. Klein EE, Hanley J, Bayouth J et al. Task group 142 report: Quality assurance of medical accelerators. *Med Phys* 2009;36:4197–4212.

21. Petti PL, Siddon RL. Effective wedge angles with universal wedge. *Phys Med Biol* 1985;30:985–991.

22. McCaw TJ, Micka JA, DeWerd LA. Development and characterization of a three-dimensional radiochromic film stack dosimeter for megavoltage photon beam dosimetry. *Med Phys* 2014;41:052104.

23. Arjomandy B, Tailor R, Zhao L, Devic S. EBT2 film as a depth-dose measurement tool for radiotherapy beams over a wide range of energies and modalities. *Med Phys* 2012;39:912–921.

24. Van Battum LJ, Hoffmans D, Piersma H, Heukelom S. Accurate dosimetry with GafChromic EBT film of a 6 MV photon beam in water: What level is achievable? *Med Phys* 2008;35:704–716.

25. Aldelaijan S, Devic S, Mohammed H et al. Evaluation of EBT-2 model GAFCHROMIC film performance in water. *Med Phys* 2010;37:3687–3693.

26. Arjomandy B, Tailor R, Zhao L, Devic S. EBT2 film as a depth-dose measurement tool for radiotherapy beams over a wide range of energies and modalities. *Med Phys* 2012;39:912–921.

27. Butson MJ, Cheung T, Yu PK. Radiochromic film dosimetry in water phantoms. *Phys Med Biol* 2001;46:N27–N31.

28. Akino Y, Das IJ, Cardenes HR, Desrosiers CM. Correlation between target volume and electron transport effects affecting heterogeneity corrections in stereotactic body radiotherapy for lung cancer. *J Radiat Res* 2014;55:754–760.

29. Mijnheer B, Olszewska A, Fiorino C et al. ESTRO Booklet #7: Quality assurance of treatment planning systems practical examples for non-IMRT photon beam. Brussels, Belgium: ESTRO; 2004.

30. Meeks SL. Clinical implications of collimator exchange effect, relative collimator and phantom scatter. *Med Dosim* 1996;21:27–30.

31. Sternick ES. Treatment aim and degree of accuracy required. In: Wright AE, Boyer AL (Eds.). *Advances in Radiation Therapy Treatment Planning*. New York: AAPM; 1983. pp. 11–15.

32. Almberg SS, Frengen J, Kylling A, Lindmo T. Monte Carlo linear accelerator simulation of megavoltage photon beams: Independent determination of initial beam parameters. *Med Phys* 2012;39:40–47.

33. Chan MF, Chiu-Tsao S, Li J et al. Confirmation of skin doses resulting from bolus effect of intervening alpha cradle and carbon fiber couch in radiotherapy. *Technol Cancer Res Treat* 2012;11:571–581.

34. Aland T, Kairn T, Kenny J. Evaluation of a Gafchromic EBT2 film dosimetry system for radiotherapy quality assurance. *Australas Phys Eng Sci Med* 2011;34:251–260.

35. Cheng CW, Das IJ, Chen DJ. Dosimetry in the moving gap region in craniospinal irradiation. *Br J Radiol* 1994;67:1017–1022.

36. Cheng CW, Das IJ, Srivastava SP et al. Dosimetric comparison between proton and photon beams in the moving gap region in cranio-spinal irradiation (CSI). *Acta Oncol* 2013;52:553–560.

37. Papanikolaou N, Battista JJ, Boyer AL et al. Tissue inhomogeneity correction for megavoltage photon beams: Report of the task group no. 65 of the radiation therapy committee of the American Association of Physicist in Medicine, AAPM Report No 85. Madison, WI: AAPM; 2004.

38. Jackson W. Surface effects of high-energy X-rays at oblique incidence. *Br J Radiol* 1971;44:109–115.

39. Gerbi BJ, Meigooni AS, Khan FM. Dose buildup for obliquely incident photon beams. *Med Phys* 1987;14:393–399.

40. Akino Y, Das IJ, Bartlett GK et al. Evaluation of superficial dosimetry between treatment planning system and measurement for several breast cancer treatment techniques. *Med Phys* 2013;40:011714.

41. Olch AJ, Gerig L, Li H et al. Dosimetric effects caused by couch tops and immobilization devices: Report of AAPM Task Group 176. *Med Phys* 2014;41:061501.

42. Venselaar J, Welleweerd H, Mijnheer B. Tolerances for the accuracy of photon beam dose calculations of treatment planning systems. *Radiother Oncol* 2001;60:191–201.

43. Srivastava SP, Cheng CW, Das IJ. The dosimetric and radiobiological impact of calculation grid size on head and neck IMRT. *Pract Radiat Oncol* 2016;7:209–217.

Intensity-modulated radiotherapy and volumetric-modulated arc therapy

TANYA KAIRN

8.1 INTRODUCTION

Modulated radiotherapy treatments such as intensity-modulated radiotherapy (IMRT) and volumetric-modulated arc therapy (VMAT, sometimes called IMAT) involve the use of moving multileaf collimator (MLC) leaves to deliver beams with a spatially varying intensity. These treatments are planned using inverse planning, where a sophisticated computerized fluence optimization routine within the radiotherapy treatment-planning system is used to produce beams capable of delivering complex dose distributions within the patient (Figure 8.1).

Modulated techniques are especially useful for treating convoluted or concave treatment targets [1], or targets that are close to or abutting one or more organs-at-risk (OAR) (such as the volume in Figure 8.1). For example, for treating the prostate, in which the target abuts the radiosensitive rectum and is surrounded by other OAR, IMRT and VMAT treatments have been shown to be capable of delivering similar doses to the prostate with substantially reduced OAR doses compared with conventional (static, 3D conformal) radiotherapy [2]. This improved OAR sparing can lead to reduced gastro-intestinal toxicities when conventional prostate prescriptions are used [3] or can enable the use of escalated prescription doses [4]. Similarly, modulated radiotherapy treatments have been shown to produce significantly lower rates of severe side effects (grade $\geq$ 3 toxicities) from head and neck treatments (including esophagus [4], oropharynx, paranasal sinus, and nasopharynx [5]) while achieving similar control and survival rates to conventional radiotherapy treatments [4–6]. Modulated radiotherapy techniques are being used with an increasing frequency to improve dose conformity and potentially reduce radiation side effects for a growing range of anatomical sites [7–10].

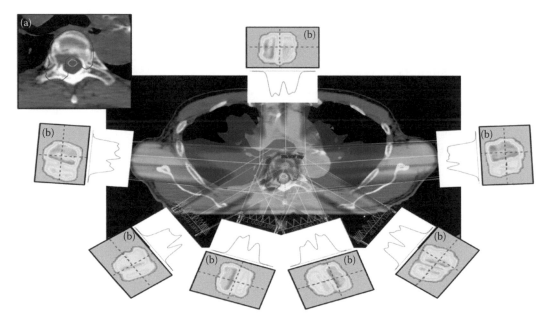

Figure 8.1 Example of a dose distribution produced by an IMRT treatment plan: Modulated beams combine to produce a concave high-dose region, covering the target (vertebra) while sparing a critical organ-at-risk (spinal cord). Main image: transverse slice through CT of patient anatomy, showing dose distribution as heat map (red = higher dose, blue = lower dose). Inset (a) Contours indicating the treatment target (red) and the spinal cord (blue). Insets (b) Two-dimensional dose distributions for each of the beams contributing to the treatment, with profiles showing the relative intensity across the center of each beam.

The planning and delivery of modulated treatments involve complex processes, with numerous opportunities for inaccuracies to be introduced and compounded. However, the manifold clinical benefits of treating with modulated radiotherapy justify the additional time and effort required to establish and verify that the treatments are delivered accurately.

To ensure accurate modulated radiotherapy delivery, thorough commissioning and ongoing testing of modulated treatment-planning and delivery systems are widely recommended [11–14]. In particular, it is important to verify the accuracy of the beam data used by the treatment-planning system (including MLC transmission and interleaf leakage); establish the accuracy of the treatment-planning system's dose calculations for modulated treatments and identify any necessary limitations that must be applied during planning (including modulation limits, field size limits, and MLC speed limits); and test the accuracy with which treatments can be delivered—both generally (in terms of factors including MLC motion and positioning accuracy) and specifically (including pretreatment quality assurance of patient treatments). Further testing may also be required whenever an existing modulated radiotherapy-treatment program needs to be broadened to include additional anatomical sites, whenever particularly challenging one-off clinical cases are encountered or whenever ongoing review or auditing of the modulated treatment process is undertaken.

The current chapter describes how radiochromic film dosimetry techniques can be used to achieve all of these important clinical goals and thereby contribute to the ongoing quality improvement of a modulated radiotherapy program.

8.2 QUALITY ASSURANCE FOR LINACS DELIVERING MODULATED TREATMENTS

8.2.1 MULTILEAF COLLIMATOR QUALITY ASSURANCE

The extra tests and tighter tolerances that are needed for quality assurance of Linacs that deliver modulated radiotherapy treatments have been developed and investigated extensively, beginning when practical systems for delivering modulated radiotherapy treatments were first developed in the early 1990s [15–17].

A useful summary of recommended tests and tolerances for Linacs delivering modulated radiotherapy treatments has been provided by Task Group 142 and subsequently discussed by Task Group 100 of the American Association of Physicists in Medicine (AAPM), as part of their respective reports on routine Linac-quality assurance [11] and radiotherapy quality management [18]. These reports, known colloquially as TG-142 and TG-100, specifically recommend detailed testing and ongoing quality assurance of the Linac's MLC system, especially if it is used to deliver modulated radiotherapy, because the performance and characteristics of the MLC system can have substantial effects on the dose delivered during modulated treatments [19–22].

8.2.2 LEAF-SIDE AND LEAF-GAP EFFECTS

Film measurements have been used to demonstrate that when there are small changes in patient position over the course of treatment (intra- and interfraction motion), the effective MLC transmission dose is substantially affected by the amount of radiation passing between adjacent MLC leaves (interleaf leakage), as well as by the radiation transmitted through the MLC leaf body [20].

Although MLC transmission and leakage affect the doses delivered during all the treatments that use MLCs, the related property, known as the *tongue-and-groove effect* [23–25], affects modulated treatments in particular. The tongue-and-groove effect has been shown to produce small regions of underdosage, due to dynamic overlapping of the steps (tongues or grooves) in the sides of MLC leaves, which are designed to minimize interleaf leakage when adjacent leaves are aligned.

The TG-142 report notes that it is important to verify, "leaf body, side and end characteristics do not change over time, the most vulnerable being the leaf side rigidity due to leaf inter-digitation, as it may affect interleaf leakage [11]." It is therefore important to measure interleaf leakage and the tongue-and-groove effect during the process of commissioning a Linac for modulated radiotherapy treatment delivery and to verify the constancy of these properties at least annually thereafter [11,13,26,27].

As radiochromic film provides high-resolution measurements of relatively large, two-dimensional dose planes, radiochromic film is very well suited to use in evaluating MLC leaf-side and leaf-gap effects (Figure 8.2), which may be underestimated by scanning ionization chambers (due to volume averaging), overestimated by scanning silicon diodes or amorphous silicon electronic portal imaging devices (EPIDs) (due to detector overresponse), or barely detected at all by array-based dosimeters (due to inadequate resolution). Quantitative measurements of these properties require the use of carefully calibrated radiochromic dosimetry film. For example, the leakage measurement requires a calibration extending down to as low a dose as can be reliably measured. However, routine verification may be performed qualitatively (via visual examination or conversion to relative net optical density), by irradiating a less costly type of radiochromic film, designed for use in qualitative examination of radiotherapy beams (such as GAFchromic™ RT-QA2 film).

8.2.3 LEAF-END EFFECTS

For MLC leaves with rounded ends (as are used in many Linacs), transmission of radiation and not light through the narrowest part of each leaf means that the size of the light field used in patient setup and optical field checking does not exactly match the size of the radiation field used in the treatment [28–30]. This effect is important for modulated treatments, in which projections of MLC leaf ends may be overlying any part of the treated volume, rather than simply surrounding it, as each beam is delivered.

The ability to directly inscribe radiochromic film with markings (including beam landmarks such as light field edges and cross-hair positions, as well as field orientation and other beam

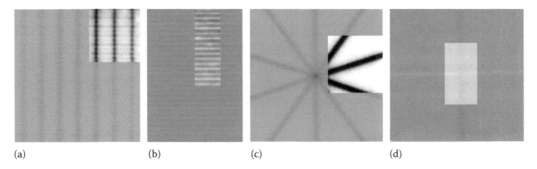

(a)   (b)   (c)   (d)

Figure 8.2 Sample radiographic film results from MLC tests: (a) abutting-leaf transmission test, also showing interleaf leakage, (b) tongue-and-groove test (lighter stripes show tongue and groove overlap regions), (c) MLC star shot, and (d) quadrant fields (the lighter, horizontal stripe shows tongue and groove overlap region, the darker vertical stripe shows leaf-end transmission region, and the presence and orthogonality of both regions shows MLC leaf banks are correctly aligned). Yellow-background images show the visual appearance of RT-QA2 film after irradiation, grayscale insets show regions from contrast-boosted red-channel images, used to examine fine details.

delivery records) makes radiochromic film especially useful for evaluating the coincidence of the MLC-defined radiation field with the MLC-defined light field and to verify the accuracy of MLC positioning, for simple and complex static fields. For modulated fields, film may be marked and irradiated using a series of apertures across the MLC travel range.

Similarly, the dose transmitted through abutting (closed) leaf ends may be measured either by using a static abutting-leaf field (Figure 8.2) or by dynamically using a moving slit field [26]. Boyer et al. have suggested, "Any deviation of the net optical density along the match line of about 20% above or below the average net optical density indicates a positioning problem" [26].

These measurements should be completed and the results should be analyzed quantitatively, when the Linac is commissioned for modulated radiotherapy delivery. In particular, it is important to establish that light-radiation field offset and abutting leaf leakage are constant across the MLC travel range and do not vary with gantry angle (due to gravitational effects on leaf banks) [27,31].

8.2.4 LEAF BANK ALIGNMENT

The measurements required to confirm the alignment of the MLC leaf banks with each other and with the collimator axis or rotation can also be performed effectively using radiochromic film. Both the MLC spoke-shot test [11,27,31] and the *quadrant field* MLC leaf bank alignment test [32] may be completed efficiently by delivering all fields required by each test to one piece of film (Figure 8.2). Analysis can then be performed via full calibration and conversion to dose, at commissioning, and by visual examination or conversion of relative net optical density, during routine Linac quality-assurance testing.

8.2.5 LEAF POSITIONING ACCURACY

The accuracy of MLC leaf positioning during modulated beam delivery can be quantified using a *picket-fence* test [11,12,33] in which a sheet of film is irradiated using a dynamic MLC field in which all leaves in both banks travel simultaneously, from one side of the field to the other, with several predetermined stops at regular distances during beam delivery. All pairs of leaves are separated by the same, small, constant distance, so the resulting film shows dark lines (where the MLC apertures stopped) against a lighter background (similar to the abutting leaf transmission test shown in Figure 8.2(a). This test was designed specifically to enable MLC positioning errors to be observed quickly, via visual examination, without the need for film to be digitized or further analyzed [33]. This test can enable very small (submillimeter) offsets in MLC position to be easily observed and quickly corrected (usually via reinitialization but potentially via MLC leaf motor replacement). The picket-fence test is, by definition, performed using a two-dimensional dosimeter, with radiochromic film usually being the preferred choice due to its high resolution and ability to provide geometric as well as dosimetric information.

The consistency of MLC performance can also be evaluated by regularly completing film measurements of a small sample of modulated treatment beams, comparing the results against the planned dose distributions and investigating variations from baseline results. Sample beams should include beams that are typical of those used locally and may therefore include beams taken from treatments of various anatomical sites (if beam geometry or complexity varies substantially between anatomical sites [34]). At least one beam that unequivocally passes local quality-assurance tests and at least one beam that produces a more borderline (or failing) result should be included in the sample. Regular retesting of these beams provides a means to monitor treatment delivery accuracy, observe systematic trends or gradual changes in MLC performance over time, and quantify the effects of MLC positioning variability on the accuracy of locally typical modulated treatments. Repetition of these tests and evaluation of the constancy of the results also provides a means to maintain staff familiarity with, and confidence in, the film dosimetry process.

8.3 COMMISSIONING OF MODULATED RADIOTHERAPY TREATMENT-PLANNING, DELIVERY, AND DOSIMETRY SYSTEMS

8.3.1 COMMISSIONING

Before a treatment-planning or delivery system can be used for modulated radiotherapy, the system must be specifically *commissioned* for that purpose. Commissioning involves a detailed evaluation of the suitability of the system for its intended use as well as the optimization of user-controllable parameters and the provision of local guidelines and limitations on using the system. Radiochromic film has an important role in all of these aspects of the modulated radiotherapy system commissioning process due to its submillimeter precision, which is only limited by the resolution of scanner, as described in Chapter 4.

It should be noted that commissioning of any radiotherapy program is a substantially larger task than simply preparing and verifying the treatment-planning system, measuring dosages and evaluating dosimeters, and establishing whether and which treatments or treatment modalities will be released for clinical use. Commissioning a modulated radiotherapy program involves much more than dosimetry. However, this section does not discuss issues relating to radiation shielding, patient immobilization, target contouring, margins, target localization, image guidance, motion management, staff availability and workloads, or any of the wide range of other important areas in which film dosimetry does not make a substantial contribution. Information on the broader issues surrounding radiotherapy commissioning should be sought elsewhere [35–37].

8.3.2 MEASUREMENT AND OPTIMIZATION OF TREATMENT-PLANNING SYSTEM PARAMETERS

In addition to the extensive set of profiles and factors that must be measured and checked in order for a treatment-planning system to be commissioned for use in conventional (static) radiotherapy treatment planning [36,37], several additional parameters must be investigated before the system can also be used for planning modulated treatments. Depending on the specific treatment-planning system used, these parameters may include the maximum allowed gantry speed or gantry speed variability (maximum acceleration or deceleration per control point), maximum number of monitor units per beam/arc/degree, MLC transmission and leakage, maximum MLC leaf speed, minimum MLC leaf opening, MLC tip radius, MLC tongue and groove width, or MLC penumbra width.

Many of these parameters are related to properties of the MLC and are used by the treatment-planning system to predict the effects of treating the target through moving MLC leaves, rather than using a static MLC aperture in which the target receives no transmission dose. Most of these parameters can be directly measured, although some are supplied by the software vendor based on measurements supplied by the user, and some should be revised or optimized to improve the deliverability of planned MLC motions and maximize the accuracy of modulated radiotherapy-treatment dose calculations.

For example, the *dosimetric leaf gap* is an optimizable parameter in the Varian Eclipse treatment-planning system. This parameter accounts for the effects of rounded MLC leaf ends, as the planning system otherwise treats the MLC as a solid, flat-edged block, with a variable aperture. An initial estimate of the required dosimetric leaf gap can be obtained by using an ionization chamber in water to measure point doses in sliding-slit MLC fields of various widths, and then extrapolating these results (minus the MLC transmission component) to identify the width of the slit field that would result in a zero dose at the chamber [38,39]. This estimate should then be validated or iteratively optimized via thorough testing of treatment-plan dose calculation accuracy using a range of simple geometric test fields and realistic clinical treatment fields, with the delivered doses measured using a high-resolution medium, such as radiochromic film. Careful film measurements may indicate that different values of the dosimetric leaf gap are optimal for use in planning modulated treatments of different anatomical sites [40] or that the conventional, ion chamber-based dosimetric leaf-gap measurement is unsuitable for calculating doses from less-conventional beams (Figure 8.3).

Many of the treatment-planning parameters listed at the start of this section are ideally measured using film. In addition, similar to the dosimetric leaf gap, many of the parameters are best optimized by using film to evaluate their effects on the dosimetric accuracy of simple test plans as well as clinical treatment plans.

Test volumes and constraints that have been specifically designed for evaluating the performance of treatment-planning systems during commissioning for IMRT are provided with the report of the AAPM's Task Group 119 [41]. (For dosimetric testing of the resulting plans, the TG-119 report specifically recommends the use of film.)

8.3.3 VERIFICATION OF DOSE CALCULATION AND TREATMENT DELIVERY ACCURACY

Before a radiotherapy treatment-planning system can be released for clinical use in a modulated radiotherapy program, the accuracy of the delivery of planned MLC motions and the accuracy of treatment dose calculations must be verified in a range of clinically expected and unexpected

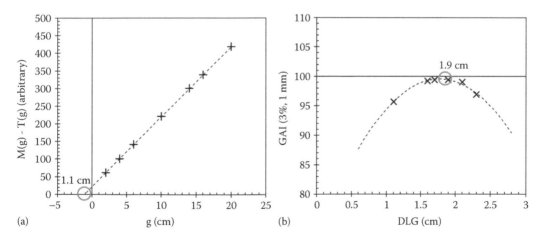

Figure 8.3 Dosimetric leaf gap (DLG) values (circled) obtained via: (a) standard measurement (where g is the sliding gap width, and M and T are the measurement and transmission values for each gap width), and (b) optimization using a film measurement of a hypofractionated spine VMAT treatment plan (where GAI is the gamma agreement index, percentage of points passing a gamma evaluation using 3% dose-difference and 1 mm distance-to-agreement criteria), for the same beam (6 MV, flattening filter free) from the same Linac (a Varian Truebeam). (Courtesy of Bess Sutherland and Nigel D. Middlebrook.)

scenarios. This verification is usually achieved through the use of end-to-end testing. End-to-end testing involves the dose measurements delivered to a phantom using a radiotherapy treatment specifically planned using purpose-scanned 3D-computed tomography (CT) images of that phantom [42–45], using a process similar to the following:

1. Select a phantom for use in the measurement. The phantom may be homogeneous or heterogeneous, simple (rectilinear, cylindrical, spherical) or humanoid (2.5-dimensional [46] or anthropomorphic), commercially produced or constructed in-house, and should be chosen with reference to the purpose of the specific end-to-end test being completed.

2. Set up the phantom for CT simulation, in the treatment position, with the dosimeter in situ (scrap or previously irradiated film may be used for this purpose). Radio-opaque markers should be used to record the phantom orientation and the positions of lasers.

3. Import the phantom CT into the treatment-planning system. Verify the accuracy of the geometry and density of the phantom as represented by the treatment-planning system (TPS) (compared with physical geometry and vendor-specified density) and discontinue the test if the phantom is not represented accurately.

4. For end-to-end testing using an ionization chamber (which is designed to provide a measurement of dose to water by accumulating charge produced by ionization events in a small volume of air), the density (or HU values) of the active volume should be overridden with the density (or HU) of water. This step is not necessary for end-to-end testing using film, where any differences between the film density and the phantom density should be included in the planned dose calculation.

5. Plan the test treatment using the same process, beam arrangement, and inverse-planning objectives that would be used if a patient was treated using the proposed modulated radiotherapy technique. (When commissioning a modulated radiotherapy treatment program, it is advisable to plan a range of treatments, with different targets and OAR geometries, different isocenter positions, and different beam arrangements, to provide an indication of the performance of the treatment-planning system across a range of clinically expected and unexpected scenarios.)

6. Record the planned (calculated) dose to the dosimeter. For an ionization chamber, this would be the point dose at the collecting electrode. For a film measurement, the dose plane at the location of the film should be exported for comparison with the measurement.

7. Set up the phantom for treatment. For simple, coplanar treatments, verification imaging may not be required, and a rigid and well-labeled phantom may be set up with reference to room lasers and field light alone. For more complex treatments, or for tests in which the image-guidance system is under evaluation, verification imaging should be used to correct the positioning of the phantom. Although this X-ray imaging will not affect the measurements provided by dosimeters (including ionization chambers), the additional radiation dose should be taken into account (by either subtracting an estimated imaging dose or adding an additional uncertainty to the measurement) when analyzing film dosimetry results.

8. Deliver the planned treatment to the phantom. If several treatment plans are being evaluated, then the film should be changed between each measurement. If changing the film involves disassembling the phantom, then the phantom position must be reset (including verification imaging, if used) before each measurement. Calibration films should also be irradiated to known doses at this time. Calibration doses must extend above the maximum dose expected in this measurement.

9. Evaluate the degree of agreement between the planned and the measured treatment dose to dosimeter, with reference to published tolerances [42,47] and measurement uncertainties. For film dosimetry, this should involve using an established development time, scanning method, calibration relationship, and analysis technique.
10. Record and report on the results. Provide a local recommendation on whether the proposed modulated radiotherapy method can be used to treat the proposed anatomical site. It is important at this stage to provide the department with an explicit statement of what types of treatment have been tested and approved and what types of treatment will require additional testing before approval. (For example, it may be necessary to recommend minimum and maximum field sizes for accurate IMRT treatment dose calculation, or a limit on allowed collimator angles to minimize the effects of interleaf-leakage on the delivery of VMAT treatments, or it may be necessary to recommend that IMRT/VMAT techniques can be released for use in treating some anatomical sites but not others.)

End-to-end testing is widely recommended as part of periodic treatment-planning system commissioning and quality assurance and should also be completed whenever updates, upgrades, or other changes are made to the treatment-planning system. To ensure accurate planning and delivery of modulated radiotherapy treatments, end-to-end testing should be performed during and after the treatment-planning system parameter verification and optimization process described in Section 8.3.2.

In addition, comprehensive dosimetric end-to-end testing is advisable whenever a new treatment modality is introduced or whenever the use of an existing treatment modality is broadened to cover additional anatomical sites. These additional commissioning measurements are especially important for modulated treatments, as modulated beam geometry and complexity can vary substantially between anatomical sites [34].

Figure 8.4 exemplifies the use of radiochromic film in the commissioning of a radiotherapy treatment-planning system for the delivery of complex, modulated radiotherapy treatments to a specific anatomical site. This example summarizes the general steps involved when film measurements are used for optimizing treatment-planning system configuration data or identifying reliable treatment-planning methods. In this case and in numerous similar instances described elsewhere [48–51], radiochromic film was selected for use due to its ability to provide accurate, high-resolution, two-dimensional dose images of complex radiotherapy treatments, without noticeable angular dependence or energy dependence (in the megavoltage range) [52,53].

As suggested by Figure 8.4, important clinical considerations may affect the specific method by which the end-to-end testing is undertaken. If necessary, aspects of the end-to-end testing should be repeated over several iterations to improve the agreement between the dose calculation provided by the treatment-planning system and the dose measurement provided by the film, as the treatment planning process is refined or the beam-configuration data are optimized. Although initial investigations may involve the use of homogeneous phantoms, it is important to verify the accuracy of the dose calculation using a phantom containing density heterogeneities that are appropriate for the anatomical site under investigation. When comparing the dose planes from the treatment-planning system and from the corresponding film measurements, two-dimensional comparison algorithms (such as Low et al. gamma evaluation [54]) may provide indications of overall agreement as well as useful information about the locations of areas of disagreement, but comparisons between dose profiles and even dose points at specific locations should also be used to investigate regions of specific clinical interest.

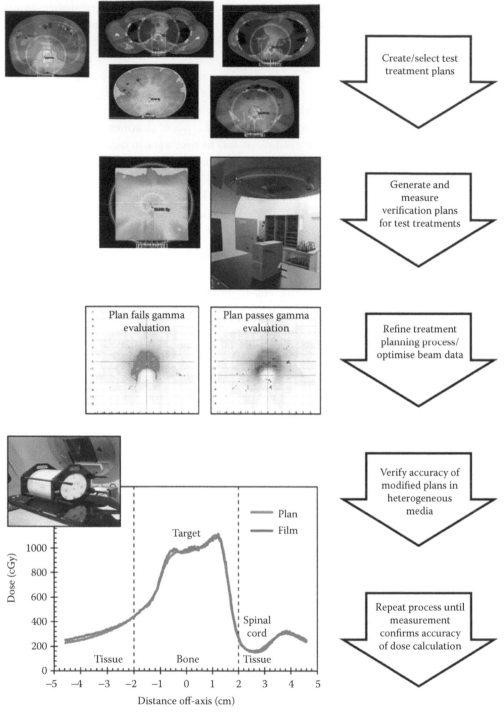

Figure 8.4 General steps in the use of radiochromic film for modulated radiotherapy commissioning. This example shows results from a process of commissioning a VMAT system for treating for vertebral metastases using a hypofractionated protocol. (The treatments shown in this figure were planned by Daniel Papworth. The recalculations of the treatment plans and film measurements in the homogeneous phantom were completed by Bess Sutherland.)

Further precautions and considerations that should be taken into account to produce radiochromic film measurements with sufficient accuracy for use in the commissioning of a modulated radiotherapy program include the following:

- Careful selection and commissioning of the specific film and scanner to be used in the measurements (see Chapter 4)
- Minimizing exposure of the film to optical and ultraviolet light [55]
- Labeling the film with reference marks, including indications of orientation, before cutting [56]
- Careful cutting of the film, to avoid or minimize delamination [57]
- Minimizing or accounting for the effects of film nonuniformity [55,56,58,59]
- Minimizing or accounting for the effects of scanner output variation [55,56,58], scanner nonuniformity [58,59] and maintaining consistent optical scatter conditions within the scanner [55]
- Using a large enough number of warmup scans or a small enough number of films scans to minimize scanner warmup variations [55]
- Using a film or a film scanning setup that is designed to avoid artifacts [57,58]
- Selecting a scanning resolution that is an appropriate compromise between spatial resolution and scanning noise
- Consistently scanning all films with the same side facing the scanning bed and with the same orientation on the scanning bed [56,60]
- Selecting a suitable calibration relationship and verifying the accuracy of the calibration by converting the pixel values, optical densities, or net optical densities from the calibration films into dose, using the calibration relationship, and verifying that these *measured* doses match the known doses delivered during calibration
- Using a range of calibration doses that extends well above and below the range of doses expected in the measurement films
- Investigating the accuracy with which the measurement phantom can be set up for irradiation and evaluating the sensitivity of the resulting measurement to likely setup errors
- Making sure that air gaps around measurement films are avoided (by cutting film to an appropriate size or by filling gaps with spare or previously irradiated film) unless those gaps are also present in the treatment plan [61]
- Subtracting or otherwise accounting for imaging dose, when evaluating agreement between planned and measured film doses
- Considering differences in film response to megavoltage and kilovoltage photons, when accounting for imaging dose or when evaluating out-of-field dose measurements [62]
- Using a local secondary-standard ionization chamber, or a field chamber that has been cross-checked with the local secondary-standard chamber, to acquire a measurement in a high-dose plateau region of the modulated treatment and to verify the accuracy of the film result as an *absolute* measurement in Gy [41,63].

Systematic, film-based, end-to-end testing of a range of clinical plans has the potential to provide confirmation that the treatment-planning system and the treatment-delivery hardware are suited to their intended use, in a modulated radiotherapy program. These tests also have the potential to allow any likely errors or uncertainties to be identified early so that recommendations or limitations on treatment planning techniques, geometries, suitable anatomical sites, and uncertainties may be provided, before the program is adopted.

8.3.4 COMMISSIONING OF ALTERNATIVE DOSIMETERS FOR ONGOING QUALITY ASSURANCE

Despite the advantages of using radiochromic film for routine-modulated radiotherapy quality assurance (addressed in Section 8.4), institutional needs, planning and treatment schedules, or workflows or management decisions may mandate that dosimeters other than film be used for routine patient-specific pretreatment quality assurance. In such cases, film remains a valuable tool for establishing the accuracy, sensitivity, and limitations of other, lower resolution dosimeters.

Radiochromic film measurements may be used to provide the baseline data required to evaluate the angular dependence of other dosimeters and may provide a high-resolution reference for evaluating the suitability of low-resolution arrays or point dosimeters for use in the quality assurance of specific treatment types [64–66].

Radiochromic film measurements may also be used to refine the quality assurance process. Film measurements, which can easily be obtained in multiple planes or at multiple orientations, may be used to identify whether and which broad types of modulated treatments can be effectively verified using two-dimensional or three-dimensional array-based dosimeters. Film measurements, which provide a detailed indication of the variability of the dose delivered by each plan, may be used to inform the selection of comparison criteria (including gamma evaluation criteria) and action levels, so that lower-resolution dosimeters can be used to more-reliably identify the plans that should pass or fail quality-assurance testing.

8.4 PATIENT-SPECIFIC QUALITY ASSURANCE FOR MODULATED RADIOTHERAPY TREATMENTS

8.4.1 PATIENT-SPECIFIC QUALITY ASSURANCE

Patient-specific quality assurance, or the dosimetric verification of the deliverability of individual patient treatment plans, has been recommended for treatments involving modulated radiotherapy [1,21]. The patient-specific quality assurance process involves similar steps to an end-to-end test (see Section 8.3.3) and can be summarized as follows (Figure 8.5):

1. Select a phantom for use in the measurement. Generally, the same phantom will be used for all measurements of the same type of treatment, within a particular patient-specific quality assurance program, and this phantom will have already been selected and imported into the treatment-planning system when the program was commissioned.

2. Copy the patient's treatment plan onto the pre-prepared CT image of the quality assurance phantom and recalculate the dose using the same dose calculation algorithm and resolution that was used to calculate the dose in the patient's clinical-treatment plan. For modulated radiotherapy treatments, it is very important that the treatment plan not be reoptimized at this step, and that all numbers of monitor units and MLC positions used in the clinical treatment plan be preserved in the phantom plan. If the treatment-planning system does not allow this step to be completed easily, without reoptimization or recalculation of MLC positions, then that planning system should not be used for planning modulated radiotherapy treatments.

3. Record the planned dose to the dosimeter. For an ionization chamber, this would be the point dose at the collecting electrode. For a film measurement, the dose plane at the location of the film should be exported for comparison with the measurement.

4. Set up the phantom for treatment, as described at step 5 in Section 8.3.3.
5. Deliver the planned treatment to the phantom and irradiate the calibration films, as described at step 6 in Section 8.3.3. Calibration films should also be irradiated to known doses at this time. Calibration doses must extend above the maximum dose expected in this measurement.
6. Evaluate the degree of agreement between the planned and the measured treatment dose to dosimeter, as described at step 7 in Section 8.3.3.
7. Record and report on the results. This should include an indication of whether the treatment plan has *passed* or *failed* its quality assurance test. A passing result indicates that the planned dose has been delivered to the phantom with sufficient accuracy to justify the assumption that the dose to the patient has been calculated correctly by the treatment-planning system. A failing result indicates that the planned dose has not been delivered accurately to the phantom, either due to an issue with treatment delivery (such as inaccurate MLC positioning) or due to an inaccuracy in the treatment-planning system's calculation of the dose (which may arise due to unpredicted MLC leakage or an excessive number of small beam apertures in the treatment plan [34,67,68]).

For modulated radiotherapy treatments, patient-specific quality assurance measurements are advisable due to the complexity of the treatment-delivery process (with collimators moving in complex patterns, sometimes in combination with gantry rotation and/or dose rate variation) as well as the complexity of the treatment plans themselves (with spatially varying fluences and doses across each beam, requiring accurate calculations of dose from small beam segments and in

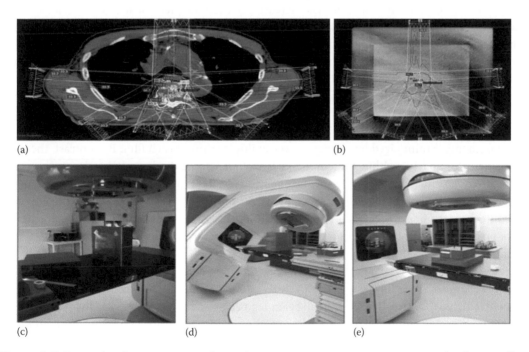

(a) (b) (c) (d) (e)

Figure 8.5 Example of a patient-specific quality assurance measurement using radiochromic film, showing duplication of the patient treatment plan: (a) onto a preexisting CT scan of the quality assurance phantom (b), setup (c), and irradiation (d) of the measurement film (in the transverse plane, in this case), and immediate irradiation (e) of calibration films under local reference conditions. (The treatment shown in (a) and (b) was planned by Daniel Papworth.)

high-dose-gradient regions). It is important to verify that each Linac can deliver each calculated patient's treatment dose as planned.

Point-dose measurements are inadequate for this purpose, as the accuracy of dose calculated at one point in a spatially varying dose distribution cannot be assumed to be indicative of the accuracy of dose calculated at another point. In addition, ionization chambers (or other finite *point* dosimeters) may produce inaccurate point-dose measurements in a modulated radiation field due to volume averaging effects; care must be taken to place the chamber (or dosimeter) in a dose plateau, at a location that does not necessarily represent the region of greatest clinical interest.

Consequently, two-dimensional measurements of modulated beams are widely recommended, with or without an additional point-dose measurement [21,41,52,63], with numerous authors reporting the suitability of radiochromic film for this purpose [52,55,63,69]. For example, an early examination of the usefulness of GAFchromic EBT2 film for modulated radiotherapy quality assurance [52] showed that film measurements were able to provide "sufficient information to verify that (IMRT, VMAT, and TomoTherapy®) treatments could be delivered with an acceptable level of accuracy, while also providing additional information on low-level dose variations (including MLC leakage and TomoTherapy beam threading) that were not predicted by the treatment planning systems." Figure 8.4 in Section 8.3.3 provides an indication of the detailed information that can be obtained by completing radiochromic film-based quality assurance measurements of modulated radiotherapy beams.

8.4.2 SUITABILITY OF RADIOCHROMIC FILM

Since the introduction of reliable and ambient-light insensitive forms of radiochromic film in the early 2000s [70], radiochromic film has increasingly been adopted for use in modulated radiotherapy quality assurance [70,71]. Radiochromic film has several key features that make it highly advantageous for use as a modulated radiotherapy patient-specific quality assurance tool.

The spatial resolution of radiochromic film allows modulated dose distributions to be measured in substantially more detail than any commercial array-based dosimeters. The resolution of film-dose measurements is limited by scanning resolution (rather than the internal polymer structure of the film) as shown in Chapter 4. Common scanning resolutions range from 72 DPI (0.35 mm) to 300 DPI (0.08 mm) [72], in which the noise contribution to the higher-resolution measurements is minimized by averaging over multiple scans of each film. By contrast, the minimum detector spacing within contemporary two-dimensional diode and chamber arrays range from 2.5 to 7.2 mm [65,73–75], and three-dimensional systems are able to measure doses at points separated by 5 mm or more [64,76]. Amorphous-silicon EPIDs, which are a standard component of contemporary linear accelerators, may be calibrated to measure dose [77–79], and generally have pixel spacings from 0.4 to 0.8 mm, with only the latest models providing a resolution similar to film (0.34 mm pixels [80]).

Radiochromic film is not affected by variations in temperature or pressure (unlike chambers and possibly chamber arrays [81]), is approximately tissue equivalent, and therefore does not overrespond when irradiated using small fields or small beam segments (unlike diodes [82–84] and possibly diode arrays [85]) and do not overrespond to low-energy photons and thereby producing questionable results in low-dose, high-scatter, out-of-field regions (unlike EPIDs [86,87]). In addition, radiochromic film does not produce anomalous results when irradiated at high instantaneous dose rates, unlike electronic systems, which can become saturated by excessive electron fluences [88,89].

Radiochromic film is also lightweight and flexible, can be cut to size, can easily be marked for reproducible alignment and registration with planned-dose planes, and can be fitted inside static or moving, homogeneous or heterogeneous phantoms. Radiochromic film provides a means to measure both simple coronal dose planes in homogeneous media or liquid water (for beam delivery verification) and to measure dose planes in heterogeneous and humanoid phantoms (for dose calculation verification).

Radiochromic film dosimetry is not dependent on the use of black-box software, and users do not need to be reliant on commercial software documentation and support, unless it is desired [59]. Film analysis can be performed using standard image viewing and spread-sheeting tools, and independent film analysis codes can be written quickly and easily by individual users [90].

Prior to the widespread availability of radiochromic film, radiographic (silver halide) film was often used for quality assurance testing of modulated radiotherapy treatments [91] and modulated radiotherapy-delivery system performance [20,24,25,31,33]. However, contemporary radiographic film dosimetry relies on hardware that is declining in availability [41], including film stock, chemicals, and processing units. As a tool for modulated radiotherapy quality assurance, radiographic film has additional disadvantages that do not affect radiochromic film, such as the need for chemical processing (and chemicals and processors), the need to be kept in light-tight packaging and to be developed in a dark-room, the impracticality of cutting and using small pieces of film, the difficulty of marking the film (pinholes or folds in the packaging are often used), and a susceptibility to substantial energy dependence (see Butson, Cheung and Yu, for a comparison with radiochromic film [53]).

Despite the manifold advantages of radiochromic film, for modulated radiotherapy quality assurance, the expense and nonreusability of the film, combined with the need for time-consuming calibration irradiations as well as the (potentially day-long) pauses in the radiotherapy treatment workflow that are required to allow for film-development stabilization after irradiation, may limit the clinical utilization of radiochromic film to the commissioning of the modulated radiotherapy program (as described in Section 8.3) or to the investigation of especially challenging clinical cases that cannot be thoroughly investigated using any other dosimetry system (as described in Section 8.5).

8.4.3 USE OF RADIOCHROMIC FILM

To exercise the specific advantages of radiochromic film dosimetry, routine patient-specific quality assurance may involve making dose measurements with a high level of detail, using setup geometries that cannot be achieved with other types of dosimeter. Figure 8.5 shows several examples of measurement geometries that may be used for radiochromic film-based patient-specific quality assurance.

As modulated radiotherapy treatments involve the integration of numerous small-beam apertures, delivered using moving collimation systems, a modulated radiotherapy quality assurance program should involve routine verification of the accuracy of treatment delivery to heterogeneous anatomy (where dose distributions from small IMRT field segments are affected by increased non-equilibrium effects [92]) and to targets affected by respiratory motion (where the interplay of respiratory motion and MLC motion may compromise the delivery of dose to the target [93]). The ease with which sheets of radiochromic film can be cut to size and inserted into commercial radiotherapy phantoms (as shown in Figure 8.6) provides the opportunity to augment simpler quality-assurance measurements with routine film-based verification of the effects of density heterogeneities or respiratory motion.

The ability to use modulated radiotherapy to treat targets that abut, surround, are close to or are surrounded by critical OAR (see Section 8.1) means that a modulated radiotherapy treatment program is likely to involve treatments of numerous anatomical sites in which critical structures

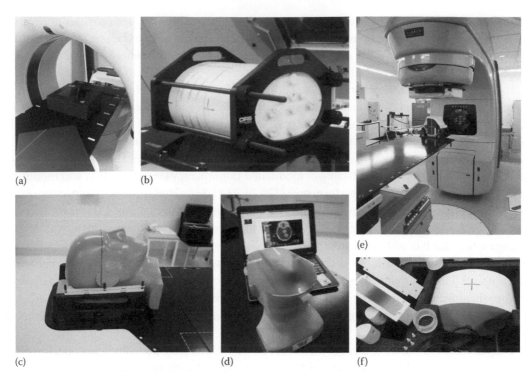

Figure 8.6 Examples of radiochromic film measurement geometries, showing film placed within homogeneous and heterogeneous phantoms, in the coronal (a), transverse (b-d), and sagittal planes (e) as well as in a moving and rotating insert for a respiratory phantom (f). (a: Courtesy of Emma L. Spelleken.)

lie directly anterior or posterior of the target. (Examples include the following: vertebra and spinal cord, prostate and rectum, left breast and heart, esophagus and trachea, cranial targets close to the brainstem or optic nerves, and abdominal nodes close to gastrointestinal organs.) In these cases, lower-resolution measurements in the coronal plane cannot provide detailed information about the dose gradient between the target and the abutting or nearby critical structure, which is required for clinical decision-making. In these cases, radiochromic film measurements in the transverse plane (as shown in Figures 8.6b-d) or in the sagittal plane (as shown in Figure 8.6e) are particularly valuable [94]. To provide similar information, measurements in the coronal plane would need to be repeated at multiple locations (as suggested for the *C-shape* test plan in the TG-119 test suite, in which a high-dose target wraps around a low-dose organ-at-risk [41]).

The near-tissue-equivalence and lack of a detectable energy-dependence, for film irradiated in the megavoltage range [53], means that film can be used to acquire measurements at various depths or orientations (including transverse and sagittal), in phantoms that may be heterogeneous and/or moving, using various beam angles (including arcs), with reference to calibration measurements acquired at one depth with one beam angle.

In addition, the ease with which radiochromic film can be marked to indicate the position of the film in relation to features in the phantom or with the radiation isocenter (as indicated by accurately aligned and verified room lasers), means that patient-specific quality-assurance measurements made using radiochromic film can be used to provide an indication of the treatment's geometric accuracy, as well as its dosimetric accuracy. If the film is clearly marked and aligned with the isocenter, and if these marks are used to register the film measurement with the planned dose plane, then any shifts in the film dose plane that are required to achieve dosimetric agreement will provide a valuable indication of the geometric uncertainty of the treatment delivery.

Ultimately, though it may often be preferable to examine dose profiles across specific regions of clinical interest [95], the two-dimensional dose plane provided by the film measurement is likely to be evaluated and compared with the dose plane produced by the treatment-planning system using a gamma evaluation [54]. The gamma evaluation is commonly used in radiotherapy dose-measurement analysis and in dose measurement analysis software [96,97] because it allows the dose within multidimensional, modulated dose distributions to be compared without being dominated by large dose differences in high-dose gradient regions.

When performing a gamma evaluation using a film measurement, the film-dose plane should be used as the reference data set to minimize the detrimental effects of noise on the gamma evaluation unless the planned dose distribution is unusually sparse (in which case the film dose plane may be smoothed and used as the evaluation dataset) [98]. The submillimeter spatial resolution of radiochromic film measurements and the ease of marking and aligning the film with the radiation isocenter mean that gamma evaluations with 1-mm distance-to-agreement criteria can provide meaningful and informative results (which is not always the case for lower resolution dosimeters [96]).

It should be understood that the high-resolution dosimetric information provided by radiochromic film measurements is not excessive or irrelevant data; measuring dose at the same (low—often 2 to 4 mm) resolution at which the treatment-planning system calculates dose should not be regarded as an adequate test of the accuracy or deliverability of a modulated radiotherapy-treatment plan. Rather, the additional spatial information provided by a radiochromic film measurement provides a valuable indication of the level of fine-scale variability in the dose that will actually be delivered to the patient. This information is needed by physicists to evaluate the accuracy with which the treatment plan can be delivered to the patient. This information should also be understood by prescribing oncologists, so that they can evaluate the overall suitability of the treatment plan for its intended clinical purpose.

8.5 USE OF FILM FOR INVESTIGATING CHALLENGING CLINICAL CASES

8.5.1 CHALLENGING CLINICAL CASES

In a busy radiation oncology department, it is not unusual for clinically challenging, unusual, one-off or first-time treatments to require dosimetric investigation. These investigations are undertaken, sometimes at short notice, to establish the safety and accuracy of treatments for specific patients.

The frequency with which these cases arise for investigation may depend on local factors including; patient workload, availability of treatment and QA equipment, physics and treatment planning staff availability, the commitment or enthusiasm of referrers and prescribing oncologists, the degree of compliance with local treatment protocols, and the degree to which those protocols rigidly exclude nonstandard cases from treatment.

Radiochromic film is an ideal medium for evaluating unusual and one-off clinical cases, in which a quantitative measurement might be useful but where a qualitative observation or demonstration might also be very valuable. The graphic nature of the film measurement result makes film especially valuable for demonstrating the degree of irradiation or sparing provided by particularly complex treatment plans.

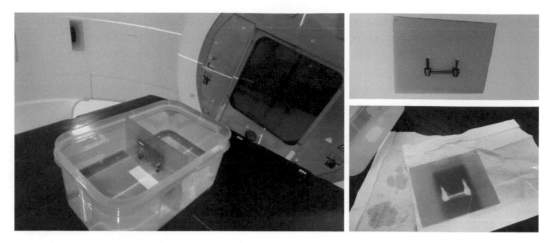

Figure 8.7 Example of the use of film to obtain qualitative information about the deliverability of a treatment plan, even without calibration and conversion to dose. Photographs show radiochromic film, with inlaid titanium vertebral reconstruction implant, immersed in a small water phantom and irradiated with a VMAT treatment plan. (The treatment used in this study was planned by Daniel Papworth.)

For example, Figure 8.7 shows a sheet of radiochromic film, cut to surround a titanium vertebral implant and irradiated while immersed in liquid water, as part of a one-off investigation into the use of VMAT to minimize the effects of the high-density vertebral implant on the dose to surrounding tissues. Even before conversion to dose, visual inspection of the film in Figure 8.7 provided a clear indication that the effects of the implant on the downstream dose are effectively avoided (there is no obvious low-dose shadow) by the use of a complex VMAT-treatment plan with carefully placed avoidance sectors.

A film measurement produces a compelling, tangible object that can be carried into meetings, displayed at chart rounds, or even used when discussing treatment options with the patient.

8.5.2 EXPERIMENTAL DESIGN

Challenging, one-off or first-time treatment investigations require experimental setups and procedures to be designed on a case-by-case basis and cannot, by definition, rely on established dosimetry processes. Specific attention must be paid to experimental design.

When undertaking a study of a clinically challenging case, it is important that the goal of the film measurement is known and that the contribution to be made to the study by the film measurement is understood. It is also advisable to have defined specific action criteria before making, analyzing, or reporting the measurement. (For example, the following questions should be considered: Will the test result be defined as a *pass* or a *fail*, or will the test result be used to provide warnings or advice to the prescribing oncologist? Are there any results that might lead to the proposed treatment being abandoned?)

Once the purpose of the measurement is understood, the experimental setup should be planned. This includes selection of the phantom, film orientation (coronal, sagittal, transverse, other?), film placement (including depth), and irradiation method (all fields on one piece of film, or each field delivered separately to different pieces of film). Care must be taken at this stage to ensure that the planned measurement will provide a result that is relevant to the purpose. For example, if the purpose of the measurement is to measure a junction dose, then setting up the film isocentrically may lead to the area of interest being missed or located close to the edge of the film. If the purpose of the measurement is to investigate the effect of a bolusing material on skin dose, then measurements at the outside surface of the bolus or at excessive depth in the phantom are irrelevant and measurements that do not use clinically realistic bolus thicknesses are likely to produce uninformative results. In particular, care should be taken when simplifying the measurement geometry, due to the many obvious differences between the complex and variable patient anatomy and the blocks of plastic generally used in radiotherapy dosimetry [99].

In addition, if modulated radiotherapy is to be used to treat the patient, then the use of simple, static test fields in the experimental scenario is inadvisable due to the important dosimetric characteristics of modulated beams (especially relating to small apertures, MLC leak age and transmission, and other features discussed in Section 8.3). It is important to plan and complete the measurement appropriately to avoid wasting time and film on measurements that do not provide conclusive results.

Similarly, the analysis of the film measurement results must be undertaken with care. Results should be evaluated against the original aim of the measurement and discussed with the prescribing oncologist as well as the staff involved in planning and delivering the treatment before any final recommendations are made. Recommendations should provide a clear indication of whether the study should be repeated in full, broadened, modified, or repeated in part, before any similar cases are treated, or whether this study provides sufficient information to release the proposed treatment technique for routine, clinical use.

8.5.3 SAMPLE INVESTIGATIONS

Due to the unique experimental design and analysis required, the process of investigating one-off clinical cases using radiochromic film dosimetry is best illustrated by example.

Figure 8.8 shows two examples of film measurements of complex, modulated radiotherapy treatment plans designed to treat cranial targets. One measurement provided inconclusive results as part of an evaluation of the use of VMAT to treat four discrete metastases in the brain (Figure 8.8a), and the other provided a compelling illustration of the deliverability of a VMAT treatment of the whole brain with hippocampal avoidance in a multiple brain metastases case (Figure 8.8b).

To produce the measurement of a VMAT treatment of four brain metastases shown in Figure 8.8a, radiochromic film was placed in the coronal plane, at the center of a 10-cm thick water-equivalent block of plastic. The verification treatment plan was produced, and the measurement was completed without consideration for the physical locations of the four high dose regions; the film was simply placed at the center of the phantom and irradiated isocentrically. Consequently, although the resulting measurement suggested that the dose was relatively accurately calculated and delivered at the measurement plane, the measurement was unable to provide a useful confirmation of the accuracy of the dose delivery to any one of the targets.

By contrast, the measurement shown in Figure 8.8b was carefully planned and set up to provide the specific information required by the prescribing oncologist; confirmation of the accurate delivery of the prescription dose to the brain tissue and a measurement of the degree of sparing provided to the hippocampus. In this case, the radiochromic film measurement was made in the transverse plane, using a humanoid phantom containing bone-like density heterogeneities. Ultimately, the qualitative appearance of the film provided compelling evidence of treatment plan deliverability that graphically supported the quantitative dose comparison.

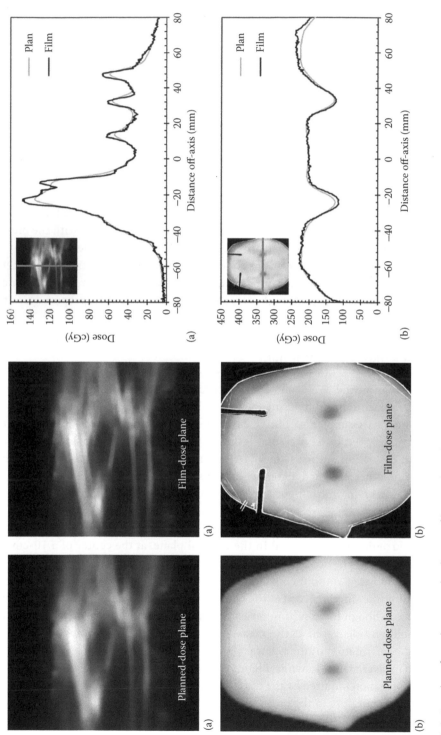

Figure 8.8 Results of irradiating radiochromic film irradiated using: (a) a multiple brain metastases treatment (in the coronal plane, in a homogeneous phantom) and (b) a whole brain radiotherapy with hippocampal avoidance treatment (in the transverse plane, in a heterogeneous phantom). (Gray lines in insets show locations of profiles.) (The treatment that produced result (a) was planned by Carly Albeiz and Charmaine Atkinson. The treatment that produced result (b) was planned by Joanne Mitchell and Elizabeth Jimmieson.)

8.6 USE OF FILM AS A RESEARCH AND AUDIT TOOL

Routine, internal, departmental auditing of treatment plan quality, and deliverability are important part of the process of continuous quality improvement in radiotherapy [35]. Regular external auditing of treatment plan dose calculation and delivery accuracy is also important for maintaining consistency of practice across different radiotherapy centers and thereby ensuring that reported outcomes are correlated with accurate appraisals of delivered dose.

The need for accurate dose reporting, and therefore the need for dosimetric intercomparison between radiotherapy departments, is especially apparent when clinical trials are considered. Here, systematic credentialing of participating centers necessarily includes treatment-plan evaluation and dosimetric measurement components, because the reliability of the data provided by the trial, and possibly the overall outcome of the trial, depends on all participating centers following the trial protocol, planning treatments similarly, and delivering them accurately [100].

The use of film dosimetry for the purpose of auditing is discussed in detail in Chapter 18. However, there are specific issues relating to the auditing of modulated treatments that warrant discussion in this section.

First, in a modulated radiotherapy treatment program, it is especially important to perform regular internal auditing to verify the consistency of treatment-plan deliverability over time. The performance of a group of radiotherapy treatment planner scan drift, over time. Generally, this drift should be in the direction of improved plan quality (target dose homogeneity and conformity, and organ-at-risk sparing) as the planners gain more experience. For inverse-planned treatments, however, there may also be a drift toward increased reliance on the optimizer (decreased time spent on manual arrangements of beams or collimator angles) that may lead to the production of increasingly complex, increasingly modulated treatment plans over time.

If low-resolution, array-based dosimeters are in clinical use for routine patient-specific quality assurance of modulated radiotherapy treatment plans, then regularly augmenting the quality assurance program with more detailed film evaluations of sample plans may identify subtle changes in treatment-planning procedure (especially, increased beam modulation), which may not be detectable using courser dosimeters.

In a department in which an electronic dosimeter is used for routine patient-specific quality assurance of modulated treatments, these frequent film-based audits of treatment plan deliverability also have the potential to identify any change or degradation in the performance of the electronic dosimeter.

Second, when film is used as a dosimetric intercomparison or clinical trial credentialing tool, there are specific issues that should be considered, when modulated radiotherapy treatment plans are to be examined. For modulated radiotherapy treatments, accurate measurements throughout dose gradients and low-dose regions may be as important as accurate measurements throughout high-dose regions. Consequently, if the film is to be scanned and evaluated at a central location by the auditor, then particular attention needs to be paid to calibration. It may be advisable to select test treatment plans and prescriptions to allow dose measurements to be performed in an approximately linear dose-response regime, which removes the need for additional calibration measurements or reduces the effect of calibration uncertainty.

Alternatively, intercomparison or trial planners may elect to allow the individual departments to provide their own analysis of their film measurements of test treatment plans, by using their own in-house method of calibrating the film and converting the result to dose. This option has the disadvantage of potentially confounding the analysis of questionable results, making it harder to differentiate between dose-calculation errors and treatment-delivery errors due to the additional possibility of calibration or analysis errors. However, this option also has the advantage of allowing the entire modulated radiotherapy treatment planning, delivery, and patient-specific quality assurance process to be evaluated or credentialed simultaneously.

8.7 SUMMARY

Modulated radiotherapy treatments (such as IMRT and VMAT or IMAT) are especially useful for treating convoluted or concave treatment targets that are close to or abutting one or more OAR. Modulated treatment techniques are being used with an increasing frequency to improve dose conformity and potentially reduce radiation side effects for a growing range of anatomical sites.

To ensure accurate modulated radiotherapy delivery, it is necessary to undertake thorough commissioning and ongoing testing of modulated treatment planning and delivery systems. Radiochromic film dosimetry techniques can be used to achieve these important clinical goals and thereby contribute to the ongoing quality improvement of a modulated radiotherapy program.

Radiochromic film is very well suited for completing the following recommended tests of Linacs delivering modulated radiotherapy treatments: MLC transmission and leakage, abutted leaf leakage, the MLC tongue-and-groove effect, MLC quadrant field test, MLC spoke shot, coincidence of the MLC radiation and light fields, picket fence test, and regular testing of sample treatment beams.

Results can be analyzed qualitatively (usual visual examination of film) or quantitatively (by scanning the film and converting the result to normalized net-optical density or by calibrating the film and converting the result to dose).

Radiochromic film dosimetry can also be used to measure and/or optimize the small number of specific factors that are used by the radiotherapy-treatment-planning system when generating modulated treatment plans (as opposed to the extensive set of profiles and factors that are measured during commissioning of the treatment-planning system for conventional, static-beam radiotherapy).

Investigation of the parameters required by the treatment-planning system for modulated radiotherapy fluence optimization (inverse-planning) and dose calculation may involve a process of iterative testing and modification, while seeking maximum agreement between the planned and measured dose distributions. End-to-end testing is a valuable tool in this process, and radiochromic film is the ideal dosimetry system for providing the necessary high-resolution measurements in heterogeneous phantoms.

Specific advantages of radiochromic film as a dosimetry system for use in modulated radiotherapy commissioning, routine quality assurance, and patient-specific quality assurance include the following:

- Submillimeter spatial resolution (limited by scanning resolution rather than detector spacing).
- Temperature and pressure independence.
- Angular independence.
- Approximate energy independence in the megavoltage range.

- Approximate tissue equivalence.
- Lightweight, flexible, and can be cut to size.
- Can easily be marked for reproducible alignment and registration with planned dose planes for use in geometric verification.
- Can be fitted inside static or moving, homogeneous or heterogeneous phantoms as well as in a liquid water phantom.
- Provides a means to measure dose in coronal, sagittal, and transverse planes.
- Can be analyzed without reliance on black-box codes.
- Is not subject to the disadvantages affecting radiographic film (including sensitivity to optical light, need for chemical processing and reliance on hardware that is declining in availability, including film stock, chemicals, and processing units).

To capitalize on the benefits of using radiochromic film for patient-specific quality assurance, measurements should use specific measurement depths and orientations that provide the most relevant clinical information about each treatment. Measurements in heterogeneous or moving phantoms may also be advisable, when verifying the doses from modulated radiotherapy treatments planned for anatomical sites that are especially heterogeneous or subject to respiratory motion.

Evaluation of results should extend beyond checking of the percentage pass rate from a gamma evaluation (which should itself use the film measurement as the reference dataset, to minimize the effects of noise, and may use unusually tight distance-to-agreement criteria), to include thorough examination of dose profiles and verification of dose points, especially in gradient regions between treatment targets and OAR. The film result should be understood as a valuable indication of the level of fine-scale variability in the dose that will be delivered to the patient.

If low-resolution, array-based dosimeters are in clinical use for routine patient-specific quality assurance of modulated radiotherapy treatment plans, then regularly augmenting the quality assurance program with more detailed film evaluations of treatment plans may help to identify any drift in treatment plan quality or complexity and also to identify any change or degradation in the performance of the electronic dosimeters.

Radiochromic film is also an ideal medium for evaluating unusually challenging, one-off modulated radiotherapy treatments. When undertaking such evaluations, it is important to plan and complete the measurement appropriately, to produce informative results that answer a specific clinical question, and provide guidance to the prescribing physician.

Lastly, radiochromic film may be used as a dosimetric intercomparison or clinical trial credentialing tool, provided that particular care is taken with the planning of test treatment doses and calibrations. This need arises due to the increased importance of dose accuracy throughout dose gradients and low-dose regions for modulated radiotherapy treatments, because targets may be treated around and through critical OAR.

ACKNOWLEDGMENTS

I wish to acknowledge the generous support and assistance provided by my clinical colleagues at Genesis Cancer Care Queensland, in the preparation of this chapter. Specifically: The treatment plans shown in Figures 8.1, 8.4, 8.5, 8.6(b) and 8.7 were planned by Daniel Papworth; the measurement shown in Figure 8.6(b) was completed by Somayeh Zolfaghari, Scott Crowe and Mark West; the commissioning calculations and measurements shown in Figure 8.4 were completed by Bess Sutherland; the data shown in Figure 8.3 was measured and analyzed by Bess Sutherland and Nigel D. Middlebrook; the photograph (a) in Figure 8.6 was provided by Emma L. Spelleken;

and the treatment plans shown in Figure 8.8 were planned by Carly Albeiz, Charmaine Atkinson, Joanne Mitchell, and Elizabeth Jimmieson.

REFERENCES

1. Boyer AL, Butler EB, Dipetrillo TA et al. Intensity-modulated radiotherapy: Current status and issues of interest. *Int J Radiat Oncol Biol Phys* 2001;51:880–914.
2. Crowe SB, Kairn T, Middlebrook N et al. Retrospective evaluation of dosimetric quality for prostate carcinomas treated with 3D conformal, intensity-modulated and volumetric-modulated arc radiotherapy. *J Med Radiat Sci* 2013;60:131–138.
3. Zelefsky MJ, Levin EJ, Hunt M et al. Incidence of late rectal and urinary toxicities after three-dimensional conformal radiotherapy and intensity-modulated radiotherapy for localized prostate cancer. *Int J Radiat Oncol Biol Phys* 2008;70:1124–1129.
4. Veldeman L, Madani I, Hulstaert F et al. Evidence behind use of intensity-modulated radiotherapy: A systematic review of comparative clinical studies. *Lancet Oncol* 2008;9:367–375.
5. Pfister DG, Ang K, Brockstein B et al. NCCN practice guidelines for head and neck cancers. *Oncology* 2000;14:163.
6. Van den Steen D, Hulstaert F, Camberlin C. Intensity-modulated radiotherapy. *KCE Reports 62* (Brussels, Belgium: Belgian Health Care Knowledge Centre) 2007.
7. Mackie TR. History of tomotherapy. *Phys Med Biol* 2006;51:R427–R453.
8. Yu CX, Tang G. Intensity-modulated arc therapy: Principles, technologies and clinical implementation. *Phys Med Biol* 2011;56:R31–R54.
9. Teshima T, Tsukamoto N, Terahara A et al. Japanese structure survey of radiation oncology in 2009 based on institutional stratification of the Patterns of Care Study. *J Radiat Res* 2012;53:710–721.
10. Bridge P, Dempsey S, Giles E et al. Practice patterns of radiation therapy technology in Australia: Results of a national audit. *J Med Radiat Sci* 2015;62:253–260.
11. Klein EE, Hanley J, Bayouth J et al. Task Group 142 report: Quality assurance of medical accelerators. *Med Phys* 2009;36:4197–4212.
12. Ezzell GA, Galvin JM, Low D et al. Guidance document on delivery, treatment planning, and clinical implementation of IMRT: Report of the IMRT subcommittee of the AAPM radiation therapy committee. *Med Phys* 2003;30:2089–2115.
13. James H, Beavis A, Budgell G et al. Guidance for the clinical implementation of intensity modulated radiation therapy. IPEM Report No 96. York, UK: Institute of Physics and Engineering in Medicine; 2008.
14. International Atomic Energy Agency. Transition from 2-D radiotherapy to 3-D conformal and intensity modulated radiotherapy. IEAE TECDOC No 1588. Vienna, Austria: International Atomic Energy Agency; 2008.
15. Bortfeld T, Boyer AL, Schlegel W et al. Realization and verification of three-dimensional conformal radiotherapy with modulated fields. *Int J Radiat Oncol Biol Phys* 1994;30:899–908.
16. Mackie TR, Holmes T, Swerdloff S et al. Tomotherapy: A new concept for the delivery of dynamic conformal radiotherapy. *Med Phys* 1993;20:1709–1719.
17. Yu CX. Intensity-modulated arc therapy with dynamic multileaf collimation: An alternative to tomotherapy. *Phys Med Biol* 1995;40:1435–1449.
18. Huq MS, Fraass BA, Dunscombe PB et al. The report of task group 100 of the AAPM: Application of risk analysis methods to radiation therapy quality management. *Med Phys* 2016;43:4209–4262.
19. Bayouth JE, Morrill SM. MLC dosimetric characteristics for small field and IMRT applications. *Med Phys* 2003;30:2545–2552.

20. Klein EE, Low DA. Interleaf leakage for 5 and 10 mm dynamic multileaf collimation systems incorporating patient motion. *Med Phys* 2001;28:1703–1710.

21. LoSasso T, Chui CS, Ling CC. Comprehensive quality assurance for the delivery of intensity modulated radiotherapy with a multileaf collimator used in the dynamic mode. *Med Phys* 2001;28:2209–2219.

22. Patel I, Glendinning AG, Kirby MC. Dosimetric characteristics of the Elekta Beam Modulator™. *Phys Med Biol* 2005;50:5479–5492.

23. Chui CS, LoSasso T, Spirou S. Dose calculation for photon beams with intensity modulation generated by dynamic jaw or multileaf collimations. *Med Phys* 1994;21:1237–1244.

24. Sykes JR, Williams PC. An experimental investigation of the tongue and groove effect for the Philips multileaf collimator. *Phys Med Biol* 1998;43:3157–3165.

25. Deng J, Pawlicki T, Chen Y et al. The MLC tongue-and-groove effect on IMRT dose distributions. *Phys Med Biol* 2001;46:1039–1060.

26. Boyer A, Biggs P, Galvin J et al. Basic applications of multileaf collimators. AAPM Report No 72. Madison, WI: American Association of Physicists in Medicine; 2001.

27. Kirby M, Ryde S, Hall C. Acceptance testing and commissioning of linear accelerators. IPEM Report No 94. York, UK: Institute of Physics and Engineering in Medicine; 2007.

28. Boyer AL, Li S. Geometric analysis of light-field position of a multileaf collimator with curved ends. *Med Phys* 1997;24:757–762.

29. Vial P, Oliver L, Greer PB et al. An experimental investigation into the radiation field offset of a dynamic multileaf collimator. *Phys Med Biol* 2006;51:5517–5538.

30. Kairn T, Asena A, Charles PH et al. Field size consistency of nominally matched Linacs. *Australas Phys Eng Sci Med* 2015;38:289–297.

31. Mubata CD, Childs P, Bidmead AM. A quality assurance procedure for the Varian multi-leaf collimator. *Phys Med Biol* 1997;42:423–431.

32. Mayles P, Nahum AE, Rosenwald JC. *Handbook of Radiotherapy Physics: Theory and Practice.* (Boca Raton, FL: Taylor and Francis Publishing Group); 2007.

33. Chui CS, Spirou S, LoSasso T. Testing of dynamic multileaf collimation. *Med Phys* 1996;23:635–641.

34. Crowe SB, Kairn T, Middlebrook N et al. Examination of the properties of IMRT and VMAT beams and evaluation against pre-treatment quality assurance results. *Phys Med Biol* 2015;60:2587–2601.

35. International Atomic Energy Agency. Setting up a radiotherapy programme: Clinical, medical physics, radiation protection and safety aspects. IEAE Publication No 1296. Vienna, Austria: International Atomic Energy Agency; 2008.

36. International Atomic Energy Agency. Commissioning of radiotherapy treatment planning systems: Testing for typical external beam treatment techniques. IEAE TECDOC No 1583 Vienna, Austria: International Atomic Energy Agency; 2008.

37. Das IJ, Cheng CW, Watts RJ et al. Accelerator beam data commissioning equipment and procedures: Report of the TG-106 of the Therapy Physics Committee of the AAPM. *Med Phys* 2008;35:4186–215.

38. LoSasso T, Chui CS, Ling CC. Physical and dosimetric aspects of a multileaf collimation system used in the dynamic mode for implementing intensity modulated radiotherapy. *Med Phys* 1998;25:1919–1927.

39. Mei X, Nygren I, Villarreal-Barajas JE. On the use of the MLC dosimetric leaf gap as a quality control tool for accurate dynamic IMRT delivery. *Med Phys* 2011;38:2246–2255.

40. Kielar KN, Mok E, Hsu A et al. Verification of dosimetric accuracy on the TrueBeam STx: Rounded leaf effect of the high definition MLC. *Med Phys* 2012;39:6360–6371.

41. Ezzell GA, Burmeister JW, Dogan N et al. IMRT commissioning: Multiple institution planning and dosimetry comparisons, a report from AAPM Task Group 119. *Med Phys* 2009;36:5359–5373.

42. International Atomic Energy Agency. Commissioning and quality assurance of computerized planning systems for radiation treatment of cancer. IEAE Technical Report Series No 430. Vienna, Austria: International Atomic Energy Agency; 2004.

43. Schneider U, Pedroni E, Lomax A. The calibration of CT Hounsfield units for radiotherapy treatment planning. *Phys Med Biol* 1996;41:111–124.

44. Morales J, Butson M, Alzaidi S et al. Development of an end-to-end audit process for stereotactic radiosurgery. *Australas Phys Eng Sci Med* 2016;39:271.

45. Kairn T, West M. Six isocenters and a piece of film: Comprehensive end-to-end testing of a cranial stereotactic radiosurgery system. *Australas Phys Eng Sci Med* 2016;39:341–342.

46. Seaby AW, Thomas DW, Ryde SJ et al. Design of a multiblock phantom for radiotherapy dosimetry applications. *Br J Radiol* 2002;75:56–58.

47. Ahnesjoe A, Aspradakis MM. Dose calculations for external photon beams in radiotherapy. *Phys Med Biol* 1999;44:R99–R155.

48. Ling CC, Zhang P, Archambault Y et al. Commissioning and quality assurance of RapidArc radiotherapy delivery system. *Int J Radiat Oncol Biol Phys* 2008;72:575–581.

49. Mancuso GM, Fontenot JD, Gibbons JP et al. Comparison of action levels for patient-specific quality assurance of intensity modulated radiation therapy and volumetric modulated arc therapy treatments. *Med Phys* 2012;39:4378–4385.

50. Olding T, Alexander KM, Jechel C et al. Delivery validation of VMAT stereotactic ablative body radiotherapy at commissioning. *J Phys Conf Ser* 2015;573:012019.

51. Kairn T, Papworth D, Crowe SB et al. Dosimetric quality, accuracy, and deliverability of modulated radiotherapy treatments for spinal metastases. *Med Dosim* 2016;41:258–266.

52. Kairn T, Hardcastle N, Kenny J et al. EBT2 radiochromic film for quality assurance of complex IMRT treatments of the prostate: Micro-collimated IMRT, RapidArc, and TomoTherapy. *Australas Phys Eng Sci Med* 2011;34:333–343.

53. Butson MJ, Cheung T, Peter KN. Weak energy dependence of EBT GAFchromic film dose response in the 50 kVp–10 MVp X-ray range. *Appl Radiat Isot* 2006;64:60–62.

54. Low DA, Harms WB, Mutic S et al. A technique for the quantitative evaluation of dose distributions. *Med Phys* 1998;25:656–661.

55. Richley L, John AC, Coomber H et al. Evaluation and optimization of the new EBT2 radiochromic film dosimetry system for patient dose verification in radiotherapy. *Phys Med Biol* 2010;55:2601–2617.

56. Aland TM, Kairn T, Kenny J. Evaluation of a GAFchromic EBT2 film dosimetry system for radiotherapy quality assurance. *Australas Phys Eng Sci Med* 2011;34:251–260.

57. Moylan R, Aland T, Kairn T. Dosimetric accuracy of GAFchromic EBT2 and EBT3 film for in vivo dosimetry. *Australas Phys Eng Sci Med* 2013;36:331–337.

58. Kairn T, Aland T, Kenny J. Local heterogeneities in early batches of EBT2 film: a suggested solution. *Phys Med Biol* 2010;55:L37–L42.

59. Micke A, Lewis DF, Yu X. Multichannel film dosimetry with nonuniformity correction. *Med Phys* 2011;38:2523–2534.

60. Zeidan O, Stephenson S, Meeks S et al. Characterization and use of EBT radiochromic film for IMRT dose verification. *Med Phys* 2006;33:4064–4072.

61. Charles PH, Crowe SB, Kairn T et al. The effect of very small air gaps on small field dosimetry. *Phys Med Biol* 2012;57:6947–6960.

62. Peet SC, Wilks R, Kairn T et al. Technical note: Calibrating radiochromic film in beams of uncertain quality. *Med Phys* 2016;43:5647–5652.

63. Low DA, Moran JM, Dempsey JF et al. Dosimetry tools and techniques for IMRT. *Med Phys* 2011;38:1313–1338.

64. Bedford JL, Lee YK, Wai P et al. Evaluation of the Delta4 phantom for IMRT and VMAT verification. *Phys Med Biol* 2009;54:N167–N176.

65. Korevaar EW, Wauben DJ, van der Hulst PC et al. Clinical introduction of a Linac head-mounted 2D detector array based quality assurance system in head and neck IMRT. *Radiother Oncol* 2011;100:446–452.

66. Kairn T, Ibrahim S, Inness E et al. Suitability of diodes for point dose measurements in IMRT/VMAT beams. *World Congress on Medical Physics and Biomedical Engineering*, June 7–12, Toronto, Canada. Cham, Switzerland: Springer International Publishing; 2015.

67. Kairn T, Crowe SB, Kenny J et al. Predicting the likelihood of QA failure using treatment plan accuracy metrics. *J Phys Conf Ser* 2014;489:012051.

68. Valdes G, Scheuermann R, Hung CY et al. A mathematical framework for virtual IMRT QA using machine learning. *Med Phys* 2016;43:4323–4334.

69. Lewis D, Micke A, Yu X et al. An efficient protocol for radiochromic film dosimetry combining calibration and measurement in a single scan. *Med Phys* 2012;39:6339–6350.

70. Devic S. Radiochromic film dosimetry: Past, present, and future. *Phys Med* 2011;27:122–134.

71. Devic S, Tomic N, Lewis D. Reference radiochromic film dosimetry: Review of technical aspects. *Phys Med* 2016;32:541–556.

72. Morales JE, Butson M, Crowe SB et al. An experimental extrapolation technique using the GAFchromic EBT3 film for relative output factor measurements in small x-ray fields. *Med Phys* 2016;43:4687–4692.

73. O'Connor P, Seshadri V, Charles P. Detecting MLC errors in stereotactic radiotherapy plans with a liquid filled ionization chamber array. *Australas Phys Eng Sci Med* 2016;39:247–252.

74. Jursinic PA, Sharma R, Reuter J. MapCHECK used for rotational IMRT measurements: Step-and-shoot, TomoTherapy, RapidArc. *Med Phys* 2010;37:2837–2846.

75. Dobler B, Streck N, Klein E et al. Hybrid plan verification for intensity-modulated radiation therapy (IMRT) using the 2D ionization chamber array I'mRT MatriXX—A feasibility study. *Phys Med Biol* 2010;55:N39–N55.

76. Li G, Zhang Y, Jiang X et al. Evaluation of the ArcCHECK QA system for IMRT and VMAT verification. *Phys Med* 2013;29:295–303.

77. van Elmpt W, McDermott L, Nijsten S et al. A literature review of electronic portal imaging for radiotherapy dosimetry. *Radiother Oncol* 2008;88:289–309.

78. Mans A, Wendling M, McDermott LN et al. Catching errors with in vivo EPID dosimetry. *Med Phys* 2010;37:2638–2644.

79. Nicolini G, Vanetti E, Clivio A et al. The GLAaS algorithm for portal dosimetry and quality assurance of RapidArc, an intensity modulated rotational therapy. *Radiat Oncol* 2007;3:24.

80. Nicolini G, Clivio A, Vanetti E et al. PO-0863: Dosimetric testing of the new aS1200 MV imager with FF and FFF beams. *Radiother Oncol* 2015;115:S439.

81. Spezi E, Angelini AL, Romani F et al. Characterization of a 2D ion chamber array for the verification of radiotherapy treatments. *Phys Med Biol* 2005;50:3361–3373.

82. Das IJ, Ding GX, Ahnesjö A. Small fields: Nonequilibrium radiation dosimetry. *Med Phys* 2008;35:206–215.

83. Scott AJ, Kumar S, Nahum AE et al. Characterizing the influence of detector density on dosimeter response in non-equilibrium small photon fields. *Phys Med Biol* 2012;57:4461–4476.

84. Kairn T, Charles PH, Cranmer-Sargison G et al. Clinical use of diodes and micro-chambers to obtain accurate small field output factor measurements. *Australas Phys Eng Sci Med* 2015;38:357–367.

85. Langen KM, Meeks SL, Poole DO et al. Evaluation of a diode array for QA measurements on a helical tomotherapy unit. *Med Phys* 2005;32:3424–3430.

86. Greer PB, Popescu CC. Dosimetric properties of an amorphous silicon electronic portal imaging device for verification of dynamic intensity modulated radiation therapy. *Med Phys* 2003;30:1618–1627.

87. Kirkby C, Sloboda R. Consequences of the spectral response of an a-Si EPID and implications for dosimetric calibration. *Med Phys* 2005;32:2649–2658.
88. Fuduli I, Porumb C, Espinoza AA et al. A comparative analysis of multichannel data acquisition systems for quality assurance in external beam radiation therapy. *J Instrum* 2014;9:T06003.
89. Rosenfeld AB. Electronic dosimetry in radiation therapy. *Radiat Meas* 2006;41:S134–S153.
90. Bennie N, Metcalfe P. Practical IMRT QA dosimetry using GAFchromic film: A quick start guide. *Australas Phys Eng Sci Med* 2016;39:533–545.
91. Pulliam KB, Followill D, Court L et al. A six-year review of more than 13,000 patient-specific IMRT QA results from 13 different treatment sites. *J App Clin Med Phys* 2014;15:196–206.
92. Jones AO, Das IJ, Jones Jr FL. A Monte Carlo study of IMRT beamlets in inhomogeneous media. *Med Phys* 2003;30:296–300.
93. Ceberg S, Ceberg C, Falk M et al. Evaluation of breathing interplay effects during VMAT by using 3D gel measurements. *J Phys Conf Ser* 2013;444:012098.
94. Kairn T, Harris S, Moutrie Z et al. Achievable dose gradients in spinal radiotherapy treatments delivered via Tomotherapy (abstract). *Australas Phys Eng Sci Med* 2016;39:325.
95. Kairn T, Middlebrook N, Hill B. Junction effects of RapidArc head and neck treatments (abstract). *Australas Phys Eng Sci Med* 2011;34:562.
96. Crowe SB, Sutherland B, Wilks R et al. Relationships between gamma criteria and action levels: Results of a multicenter audit of gamma agreement index results. *Med Phys* 2016;43:1501–1506.
97. Hussein M, Rowshanfarzad P, Ebert MA et al. A comparison of the gamma index analysis in various commercial IMRT/VMAT QA systems. *Radiother Oncol* 2013;109:370–376.
98. Low DA, Dempsey JF. Evaluation of the gamma dose distribution comparison method. *Med Phys* 2003;30:2455–2464.
99. Kairn T, Crowe SB, Kenny J et al. Dosimetric effects of a high-density spinal implant. *J Phys Conf Ser* 2013;444:012108.
100. Rischin D, Peters L, O'Sullivan B, et al. Tirapazamine, cisplatin, and radiation versus cisplatin and radiation for advanced squamous cell carcinoma of the head and neck (TROG 02.02, HeadSTART): A phase III trial of the Trans-Tasman Radiation Oncology Group. *J Clin Oncol* 2010, 28:2989–2995.

Film dosimetry for linear accelerator-based stereotactic radiosurgery and stereotactic body radiation therapy

MARIA CHAN AND YULIN SONG

9.1 HISTORY AND EVOLUTION OF LINAC-BASED STEREOTACTIC RADIOSURGERY AND STEREOTACTIC BODY RADIATION THERAPY

9.1.1 HISTORY OF LINAC-BASED STEREOTACTIC RADIOSURGERY

9.1.1.1 EARLY DEVELOPMENT AND COMBINATION OF STEREOTAXY AND IRRADIATION

A Swedish neurosurgeon and scientist, Lars Leksell (1907–1986) at the age of 42, designed and developed the world's first stereotactic frame based on three-dimensional Cartesian coordinate system [1,2]. Two years later in 1951, Leksell and a medical physicist named Borje Larsson developed the earliest methodology of radiosurgery by using X-ray beams generated from the Uppsala University cyclotron [3]. Their goal was to treat small intracranial tumors that were not accessible by conventional surgery noninvasively. The initial concept of radiosurgery was to deliver a series of slit beams of radiation from multiple directions and focus them on the target. In this way, the target can receive a lethal dose of radiation, whereas the nearby healthy tissue can be adequately spared.

In 1961, Leksell became the professor and the Chairman of Neurosurgery at Karolinska Institute in Sweden. He continued his research work on stereotactic radiosurgery (SRS). Even though the stereotactic proton beams had already replaced X-ray beams, Leksell persisted and kept working on a more compact, amenable, and clinically viable machine. In 1968, his persevering efforts eventually led to the birth of the world's first prototype Gamma Knife®, which was installed at the Sophiahemmet Hospital, Stockholm, Sweden [4]. The original design of Gamma Knife consisted of 179 ^{60}Co radioactive sources placed in a helmet-like spherical sector of 70° latitude and 160° longitude [5]. Today, the modern Gamma Knife units consist of 201 ^{60}Co radioactive sources. In the United States, the first Gamma Knife was installed at the University of Pittsburgh Medical Center in Pittsburgh in 1987 [6].

9.1.1.2 MEGAVOLTAGE ISOCENTRIC LINAC-BASED STEREOTACTIC RADIOSURGERY

The application of medical linear accelerator (Linac) to radiation oncology can be traced back to as early as the 1950s [7]. However, because of inadequate target localization accuracy at the time, Linac was not used as a radiosurgery device till 1982 [8], when the Argentine neurosurgeon Betti

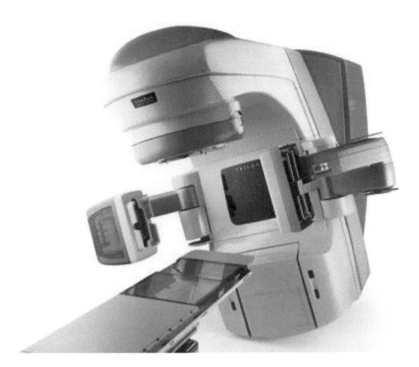

Figure 9.1 A Varian Trilogy linear accelerator commonly used in Linac-based SRS.

customized a Varian Linac and used it to treat the first Linac-based radiosurgery patient in Buenos Aires. Shortly after, the Argentine engineer Derechinsky developed a rotating chair, in which the patient rotated about the x-axis whereas the gantry rotated about the z-axis. Such a combination of orthogonal rotations provided multiple converging arcs, thus significantly improving the conformity of radiation-dose distribution around the target and sparing of the adjacent critical organs. From the late 1980s to middle 1990s, Linac-based SRS underwent a rapid development phase [9]. Dynamic conformal rotation was developed in 1987 and conical rotation was implemented in 1990 [10]. The world's first micro-multileaf collimator (MLC) was developed in 1992 [11]. The invention of MLC fundamentally changed the paradigm and practice of SRS in two most important aspects: treatment planning and treatment delivery (Figure 9.1).

9.1.1.3 FRAMELESS STEREOTACTIC RADIOSURGERY

Due to its special characteristics, SRS imposes stringent physical and clinical requirements on implementation. The two most important ones are stable patient immobilization and accurate target localization. Throughout the history of SRS evolution, the stable patient immobilization has been mainly achieved with rigid head frames that are affixed to patient's skull with screws [12–15]. The invasive nature of the head frames prohibits them from being repeatedly applied to the same patient. Consequently, the methodology of frame-based SRS cannot be directly extended to the hypofractionated SRS and stereotactic radiotherapy without major modifications. Such modifications have led to the developmental and clinical implementation of relocatable stereotactic frames [16]. One such popular system is the Gill–Thomas–Cosman frame [17,18], with which the patient immobilization is achieved with a custom-made bite block and is reinforced with an occipital plate and Velcro straps. However, bite blocks can only be used in patients who have relatively healthy teeth. They are generally not suitable for older patients and patients with dentures [19].

Recently, driven by strong interests in stereotactic radiotherapy and rapid advances in on-board imaging technologies, such as 2D kV X-ray imaging and 3D cone beam computed tomography [20,21], SRS has entered the frameless age. The goal of frameless SRS is to achieve the inherent accuracy and precision of the frame-based stereotactic techniques with the guidance of high-resolution on-board image systems. Various frameless stereotactic systems have been developed [22–24]. Regardless of their specific designs, the most common feature is a special stereotactic mask affixed to an indexed board for patient immobilization. Target localization is achieved through the on-board imaging system using either bony landmark or soft tissue-based image registration techniques. Alternatively, target localization is performed through surgically implanted fiducial markers using rigid body-based image registration techniques. Latest dedicated machines have the capability of performing real-time target tracking and six-degree of freedom (6D) online-positioning corrections using stereoscopic X-ray imaging [23,25,26].

9.1.2 History of Linac-based stereotactic body radiation therapy

9.1.2.1 EARLY DEVELOPMENT AND EXTENSION OF STEREOTACTIC RADIOSURGERY METHODOLOGY TO EXTRA-CRANIAL SITES

9.1.2.1.1 Initial concept in the early 1990s

Rapid advances in on-board imaging systems and the tremendous success of intracranial SRS encouraged investigators to extend the methodology of SRS to tumors in extra-cranial sites. The pioneering work was initiated at the Swedish Karolinska University Hospital in 1991 with tumors in the liver and lungs [27,28], followed by other investigational works by several groups in the United States. At the same time, a similar methodology was also explored in Japan. A decade-long research and investigation eventually led to the developmental and clinical implementation of stereotactic body radiation therapy (SBRT) in the mid-2000s [29,30]. Today, SBRT is being widely used to treat tumors in the lungs, liver, pancreas, head and neck, breast, prostate, and spine [31–34].

9.1.2.1.2 Frameless stereotactic body radiation therapy in the late 2000s

In the early developmental phase of SBRT, the clinical implementation of SBRT adopted the same SRS methodology for patient immobilization and target localization. Patients were immobilized with an indexed body frame. With rapid adoption of on-board imaging systems beginning in the late 2000s, the traditional frame-based methodology has been gradually abandoned. Today, just as SRS, SBRT has also entered the frameless era.

9.1.3 Commissioning of stereotactic radiosurgery and stereotactic body radiation therapy

9.1.3.1 BEAM DATA ACQUISITION FOR TREATMENT PLANNING SYSTEM

The process of commissioning a medical Linac for clinical SRS and SBRT treatment consists of systematic measurements of various dosimetric parameters for a series of small radiation fields. The specific beam data are dependent on the dose-calculation algorithm used in the treatment planning system (TPS). In general, the model-based approaches require much less beam data than the correction-based techniques. These beam data are needed for accurately modeling TPS for dose calculation. The commissioning process generally includes beam data acquisition, beam data entry into TPS, testing of beam modeling accuracy, development of specific clinical procedures, and training of relevant personnel. Prior to initiating the commissioning of Linac, the responsible medical physicist should be familiar with national or international recommendations on commissioning Linac, for example, the AAPM TG-106 Report [35]. As a rule of thumb, measured data should always be compared with published data on the same machine model or a national standard, such as Houston Imaging and Radiation Oncology Core (IROC). The following sections describe the minimum beam-data requirements for the commissioning of SRS and SBRT.

9.1.3.1.1 Nominal Linac output

The dose-calculation algorithm used in the TPS requires the nominal Linac output as one of the input parameters. The nominal Linac output describes the relation between monitor units (MU) and the absorbed dose to water under reference measurement conditions for a certain beam quality. Mathematically, the relation is defined as

$$\text{Nominal output} = \frac{D(F_{cal}, \ d_{cal}, \text{SSD}_{cal})}{\text{MU}} \qquad (9.1)$$

where:

Nominal dose is expressed in Gy/MU

D represents the measured dose in Gy

F_{cal} is the calibration field size in cm

d_{cal} is the calibration field depth in cm

SSD_{cal} is the source-to-surface distance used in calibration in cm

The nominal output can be measured by setting the ion chamber at the calibration depth d_{cal} (e.g., 10 cm), setting the MLC and jaws to F_{cal} (e.g., 10 × 10 cm²) and SSD_{cal} to 10 cm, delivering 100 MU (Figure 9.2). The measured electrometer reading in (nC) can be converted into dose in Gy by

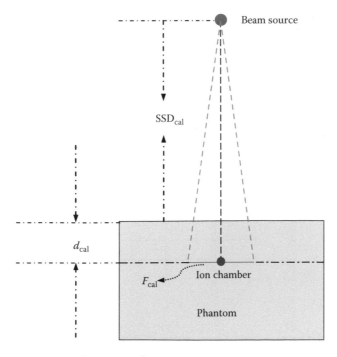

Figure 9.2 Measurement setup for nominal Linac output determination.

applying all necessary conversion and correction factors. Measurement accuracy can be improved by averaging the repeated measurements.

9.1.3.1.2 Central axis percentage depth doses

For small-field measurements, it is strongly recommended to use a high-definition detector. The measurement setup for the percentage depth doses (PDD) is similar to that for the nominal Linac output measurement. For intracranial SRS applications, the field sizes for PDD measurements should be from 10×10 mm^2 up to 150×150 mm^2. The measurement depth should be up to 25 cm. However, for extracranial SBRT treatments, the measurement depth should be up to 40 cm or even more.

9.1.3.1.3 Output factors

The output factor (scatter factor) in a medium is defined as the ratio of the output in that medium for a given field size at a given depth to that for a reference field size (e.g., 10×10 cm) and a reference depth (e.g., 10 cm). The output factors should be measured for a matrix of combinations of square MLC apertures and square jaw settings. Figure 9.3 gives a sample matrix of recommended output factor measurements for a Varian Trilogy machine. Only the green cells in the table should be measured.

9.1.3.1.4 Transverse beam profiles

Off-axis dose profiles are required for 3D dose calculation. The number of transverse profiles measured at the different depths depends on the specific TPS. The most efficient way of measuring transverse profiles is using a water phantom and a high-resolution detector. It is important to position the detector away from the interleaf gap and ensure that the detector is at least 2 cm away from the interleaf end junction line. To improve signal-to-noise ratio of the measured data, it is imperative to use a finer step size (e.g., 0.5 mm) at the penumbral region. For a Varian Trilogy machine, the transverse profiles should be measured at two depths: 10 and 20 cm with jaw setting being 15×15 cm and the MLC parked.

9.1.3.1.5 Diagonal radial profiles

Diagonal radial profiles are used to correct for off-axis beam profile fluctuation for the open field. The measurement setup is similar to that for the transverse profile measurements. The MLC should

	Jaw (mm)								
MLC (mm)	12×12	22×22	32×32	42×42	60×60	80×80	100×100	120×120	150×150
10×10									
20×20									
30×30									
40×40									
60×60									
80×80									
100×100									
120×120									
150×150									

Figure 9.3 A sample matrix for output factor measurements.

be parked, and jaws should be set to 15 × 15 cm². The measurement should be performed from one corner of the field to the other. The diagonal radial profiles should be measured for the following depths: 0.5, 1.4, 2.5, 5, 10, 20, and 35 cm.

9.1.3.1.6 Dosimetric characteristics of micro-multileaf collimator

Generally, the MLC is characterized by its radiation leakage through the device. There are three types of radiation leakage through the MLC. The first type is called the intraleaf leakage. It is the radiation leakage transmitted through the MLC leaves. It is the smallest one among the three in terms of magnitude. The second type is called the interleaf leakage. It is the radiation leakage through the gaps formed between the adjacent MLC leaf sides. The third type is called the interleaf end leakage. It is the radiation leakage through the gaps formed between the rounded leaf ends when they are in a parked position or in motion. This is the largest leakage among the three. These three types of leakage can be determined with various detectors; however, film provides the best suited for such purposes. Regardless of the detector used, the methodology involves measuring the radiation leakage with the MLC leaves completely closed and comparing the result with that of an open reference field (e.g., 10 × 10 cm²). As unused MLC leaves in treatment plans are generally parked under the jaws, the mean of intraleaf leakage and interleaf leakage is normally used in the TPS as one of the MLC-characteristic parameters. Depending on the TPS, other MLC-characteristic parameters may also be required for accurate beam modeling. For example, BrainLab also requires an MLC parameter called the dynamic leaf shift (DLS) for IMRT implementation. The concept of DLS is similar to that of dynamic leaf gap. It represents an effective leaf shift or gap caused by the rounded leaf end design. The DLS or dynamic leaf gap phenomenon is apparent even if the leaf pairs are completely closed. The DLS can be experimentally determined by measuring isocenter doses for a series of dynamic leaf gaps of different widths sweeping across the field. The measured doses and leakages are then fitted to the following empirical formula:

$$D - D_{leak} = k(LG + 2 \cdot DLS) = k \cdot LG + c \tag{9.2}$$

$$c = 2k \cdot DLS \tag{9.3}$$

where:
   LG represents the dynamic leaf gap width (mm) used for the dose measurement (e.g., 1, 5,..., 100 mm)
   D_{leak} is the measured MLC leakage dose (nC or Gy) with MLC leaves completely closed
   k and c are two constants
   DLS is the effective leaf shift per leaf (mm)

The factor 2 indicates that each leaf pair has two leaves. The physical meaning of k is linear dose rate in Gy/mm. k and c can be easily determined by fitting the measured data to the above-mentioned linear equation. The DLS is then calculated by

$$DLS = \frac{c}{2k} \tag{9.4}$$

9.1.3.2 ORGANIZE THE BEAM DATA IN THE FORMAT REQUIRED BY TREATMENT PLANNING SYSTEM

After all required beam data have been acquired and processed, they need to be reformatted according to the data format recognized by the TPS. To facilitate the data conversion, most vendors provide a data Editor. For example, BrainLab uses Beam Profile Editor for data entries and refor-matting. The formatted data are then imported into the TPS for initial beam modeling and testing.

9.1.3.3 DEVELOP TEST PLANS

The purposes of the test plans are to verify the accuracy of modeled beam data and dose calcula-tion. The test plans should consist of various field sizes and depths that are commonly used in clinical treatments. They can be created on a flat solid water phantom or a commercial SRS head phantom. The test plans can also be created on a commercial validation SRS head phantom, such as an IROC SRS head phantom. Beam profiles, depth–dose distributions, and penumbra accuracy should be verified for various regular and irregular field sizes, and different beam configurations. Measured results should be compared with those measured with different techniques. In addition, test plans for verifying interleaf and intraleaf leakages should also be developed. Furthermore, the TPS calculated MU should be verified by a third-party software.

9.1.3.4 VERIFY THE DOSIMETRIC ACCURACY OF THE TEST PLANS

Most modern Linacs have a quality assurance (QA) mode so that these test plans can be delivered using this mode for repeated measurements. The accuracy of the test plans can be measured by using an ion chamber positioned at the isocenter for a point dose measurement or using radiochro-mic film, RCF (EBT3 or EBT-XD) that has been described in previous chapters. RCF is used for dose distribution and profile measurement. For test plans with a prescription dose less than 10 Gy, EBT3 film should be adequate. However, for those test plans with a prescription dose above 10 Gy, the latest EBT-XD film is strongly recommended in order to avoid reduced response sensitivity as observed on EBT3 film. Alternatively, 2D-beam profilers, such as MapCHECK2, can also be used for the dose distribution and profile measurement.

9.2 END-TO-END TEST FOR STEREOTACTIC RADIOSURGERY WITH EBT3/EBT-XD FILMS

9.2.1 SIMULATION

9.2.1.1 SELECTION OF A SUITABLE DOSIMETRY PHANTOM

There are various types of phantom for SRS testing. IROC uses its own phantom for validating the clinical data. In either situation, RCF is used for dosimetry.

The SRS end-to-end test requires the use of an anthropomorphic phantom. There are many commercial anthropomorphic phantoms that are specially designed for this purpose. The suitable ones should consist of at least one film insert for film dosimetry and one ion chamber insert for point dose measurement. The same film insert should be easily positioned in different planes. The inserts should fit in the phantom tightly without significant air gaps. They should contain bony structures that emulate a human head for the purpose of volume delineation. They should be light opaque so that 3D optical surface imaging systems can be used on them. They should also be MRI compatible.

9.2.1.1.1 Film inserts

Figure 9.4 shows a Kesler head phantom (CIRS, Norfolk Virginia) specially designed for SRS appli-cation. The head phantom consists of two detachable parts. The film insert, consisting of four iden-tical components, is located in the superior part. Each individual component is clearly marked so

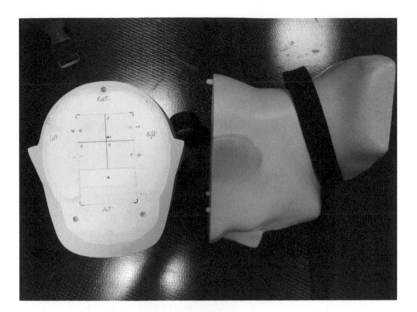

Figure 9.4 The Kesler head phantom can be separated into two parts: the superior part (left) and the inferior part (right). The superior part contains a film insert for dosimetry measurement. The two parts can be put together by aligning the pins in the inferior part with the holes in the superior part.

that misplacement can be avoided. When inserting the film insert loaded with the film into the insert slot, the user should not apply excessive pressure on it. Otherwise, it may be very difficult to remove the film insert after exposure.

9.2.1.1.2 Film cutting, marking, and positioning

Film cutting is crucial in film dosimetry. Poorly cut film strips will be difficult to be inserted into the film insert. In addition, inaccurately cut film strips will affect the target localization accuracy measurement. There are commercial film and photo cutters available in the supermarkets. Once the properly cut film strips are ready, their orientation with respect to the original film should be clearly labeled. These labels will ensure that correct film orientation can be restored during film scanning. Either orientation could be used, but the key is consistency, with all calibration, background, and measurement films to be scanned in the same orientation. Film data are sensitive to the film orientation relative to the laser-light polarization direction; hence, it is critical to assure the consistent film orientation. Figure 9.5 shows three pieces of a precisely cut EBT3 film, each with correct orientation labels and a width of 63 mm. The larger piece of the film strips will be placed in the sagittal plane, and the two smaller pieces will be placed in the coronal plane in the middle of the sagittal film, with one piece on each side. However, such an arrangement of the film strips will produce a long and narrow air gap between the two coronal film strips, resulting in some level of dosimetric perturbation in the film junction area and making the target localization accuracy measurement very difficult.

9.2.1.2 PHANTOM IMMOBILIZATION

Figure 9.6 shows an end-to-end test simulation performed on a CT scanner. In this case, the Kesler head phantom was placed on a simulation CDR board (CDR Systems, AB, Canada) in a neutral position. Due to the shape of the head phantom, a Size F headrest was used for head phantom positioning.

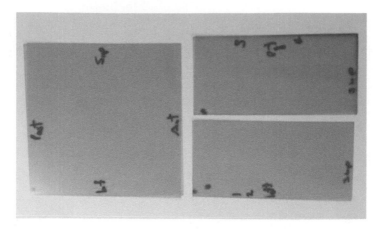

Figure 9.5 Three pieces of a precisely-cut EBT3 film with annotations.

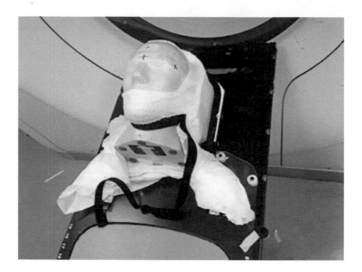

Figure 9.6 The Kesler head phantom setup prior to CT scan.

The forehead and chin were aligned and parallel to the CT scanner table. Efforts have been made to ensure that the sagittal film was perpendicular to the CT scanner table, and the coronal films were parallel to it. A mold and open face mask were then fabricated for the purpose of phantom immobilization. During the mold expanding and hardening process, attention was focused on the head phantom so that it would not deviate from the desired position. The lateral side of the mold was at the middle line of the head phantom in the sagittal view. The open face mask was symmetric about the nose. Two three-inch mask strips were also prepared and applied to the chin and forehead, respectively, to increase the face mask strength and rigidity. This was based on our clinical experience with this type of mask material. Three small triangulation markers were placed on the face mask in the superior part of the head phantom.

9.2.1.3 CT DATA ACQUISITION

The head phantom should be scanned with an SRS protocol. The CT scan should have a slice thickness of 1.0 mm and an in-plane matrix size of 512 × 512 for accurate target delineation.

The acquired CT images should be reviewed slice-by-slice for completeness and quality check before being transferred to the TPS.

9.2.2 TREATMENT PLANNING

9.2.2.1 CT DATA IMPORT INTO TREATMENT PLANNING SYSTEM

Once the CT scan, RTPLAN, and RTSTRUCT are imported to the TPS, the medical physicist should verify whether there are any missing slices, points, and contours. If the end-to-end test also requires an MRI scan for the purpose of testing image fusion function, it can be imported from the PACS.

9.2.2.2 TARGET AND ORGANS AT RISK DELINEATION

After necessary preplanning preparations, such as imaging fusion, reference point selection, and surface segmentation, are completed, relevant organs at risk , including the brainstem, cord, and optic structures, will have to be delineated manually on the CT scan. The autosegmentation tool does not work on any of the head phantoms. For the best results, the ideal shape of the target is spherical. This ensures that both the center of the sphere and isocenter are precisely placed at the geometric center of the film insert. In this way, all three pieces of the film pass through the isocenter. Normally, two spherical planning target volumes (PTV) with different diameters, such as 12 and 20 mm, may be sufficient for the purpose of the end-to-end test. As a reference, IROC recommends the diameter of the PTV to be ≥14 mm for the best measurement results. Figure 9.7 shows

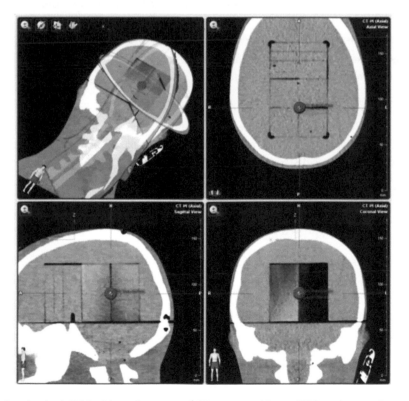

Figure 9.7 A spherical PTV with a diameter of 12 mm used in an SRS end-to-end test in a Kesler head phantom.

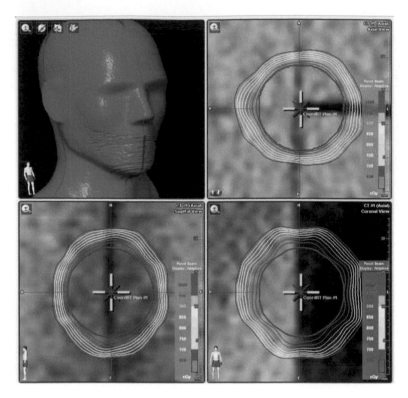

Figure 9.8 The isodose distribution of a test plan displayed in zoom-in mode. The inner red solid circle represents the PTV.

an example of a spherical PTV with a diameter of 12 mm. The isocenter is placed at the geometric center of the PTV, as indicated by the green cross.

9.2.2.3 OPTIMIZATION OF TEST PLANS

The test plans should emulate a real patient treatment in terms of prescription dose, the number of beams or arcs, the number of couch rotations, MLC margin, and plan normalization technique. Figure 9.8 shows the isodose distribution of a test plan computed with 10 non-coplanar beams.

9.2.2.4 PLAN APPROVAL

The final test plans should be approved by a medical physicist and a radiation oncologist. Certain TPSs require plans to be approved by two people.

9.2.2.5 DOSE PLANE EXPORT FOR MEASUREMENT AND COMPARISON

All TPSs provide a utility for computing 2D dose across a plane and exporting the dose file for measurement and comparison. Film provides unique opportunity to validate TPS-generated plan by measurements. In Dose Export utility, the planes through the films should be selected for dose calculation and export. As the sizes of the films are generally very small, a suitable region of interest (ROI) about the size of the film should be defined for dose calculation. In addition, as the film dosimetry has a continuous dose distribution and the geometric match between the film measurement and the planned dose requires a submillimeter accuracy, the dose planes should be calculated with a higher grid resolution, for example, ≤ 0.5 mm. This is to avoid excessive interpolation by the film analysis software.

9.3 COMMON DOSIMETRY TOOLS USED IN STEREOTACTIC RADIOSURGERY AND STEREOTACTIC BODY RADIATION THERAPY END-TO-END TEST

SRS and SBRT involve small-field dosimetry. A small field can be defined as a field with dimensions smaller than the lateral range of charged particles. Consequently, there are two challenging issues associated with small-field dosimetry. One is related to the beam itself. Small fields lack lateral charged particle equilibrium. Chapter 13 provides details of film dosimetry in small field. The other issue is related to the detector used. The size of the detector is generally large compared with the beam dimensions. Therefore, measured dose more represents the mean dose over a volume than a point. Regardless of the types of the detectors used, they all need some sort of correction as discussed in TRS-483 [36].

RCF has been used in SRS/SBRT intensively due to its high spatial resolution, energy independence over therapeutic X-ray, and a reasonable sensitivity that is important in a situation of dose-rate independence and near tissue-equivalence. Figure 9.9 shows the films used in the end-to-end test with the Kesler phantom scanned with a 600 dpi on an Epson 10,000XL scanner.

For verification of the geometrical accuracy, RCF has been used in Winston–Lutz test [37] for verification of computed tomography image guidance of patient positioning for Linac-based SRS with localization frame [38] and frameless technique [39], and in submillimeter end-to-end alignment test for SRS commissioning [40]. The use of RCF has also been reported for verifying the dosimetric accuracy of the treatment planning calculations and patient-specific QA [41–53]. Good agreement between the planned-dose and measured-dose was found within the target region.

As small photon fields are utilized in SRS, accurate dosimetry is especially challenging. The small field dosimetry in megavoltage beams will be extensively reviewed in Chapter 13 of this book. Gonzales-Lopez et al. [54] compared the PinPoint ion chamber with EBT2 film in measurements of field sizes of 0.5×0.5 cm^2, 0.7×0.7 cm^2, 1×1 cm^2, 2×2 cm^2, 3×3 cm^2, 6×6 cm^2, and 10×10 cm^2 for both 6 and 15 MV photon beams. They found the PinPoint chamber and EBT2 in agreements for the field sizes of 2×2 cm^2 and above; Monte Carlo and EBT2 in agreements for 1×1 cm^2 field size; for field sizes smaller than 1×1 cm^2. For the 0.7×0.7 cm^2, in respect to Monte Carlo calculation, the differences with the EBT2 measurements are 19% and 35% for the 6-MV beams and 17% and 33% for the 15-MV, respectively. In the case of the 0.5×0.5 cm^2 field, the differences with the RCF measurements are 38% and 52% for the 6 MV beams and 32% and 50% for the 15 MV beams, respectively. EBT film dosimetry was also used of comparison with Monte Carlo simulation in TPS and good agreements were observed [43,51,55]. Bouchard et al. 2015 provided a theoretical

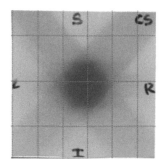
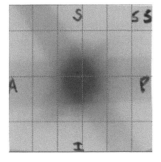

Figure 9.9 EBT-XD films after irradiated in the Kesler SRS phantom (left: coronal cut; right: sagittal cut.)

explanation for significant quality correction factors in megavoltage small photon fields and the underlying concepts relevant to dosimetry under such conditions [56].

In addition, for the verifying geometrical accuracy of SBRT, RCF is used to evaluate the alignment accuracy of a CT on rail system with a detachable micro-multileaf collimator unit and Linac isocenter [57]. The RCF has also been used in treatment planning dose verification for conformal field SBRT [58,59]. Gallo reported an RCF study of end-to-end test to verify the dose calculations by various TPS and also dose delivery system of SBRT [60]. Hardcastle reported the use of RCF in patient specific QA of SBRT [61]. Followill reported the design and implementation of pelvis and thorax phantom containing RCF for QA of SBRT [62].

9.4 EBT3/EBT-XD FILM DOSIMETRY ANALYSIS FOR STEREOTACTIC RADIOSURGERY/STEREOTACTIC BODY RADIATION THERAPY

9.4.1 CHARACTERISTICS OF EBT3 AND EBT-XD FILM

To improve the photon energy dependence of the newer RCFs, minor changes in their atomic composition have been made in EBT2 and EBT3 since their introduction [63]. In early 2015, a new film, EBT-XD, similar in composition and construction to EBT3, was added. The reduced crystal size of the active component makes the EBT-XD film less sensitive, and the increased slope of the dose–response curve at doses above 10 Gy is more suitable for measurements >10 Gy, which is the dose range of SRS and SBRT [64–66]. Detailed discussion on the type of film is presented in Chapter 2.

The dose–response curves for EBT3 film (6-MV, 15-MV, and 50-kV photons) and EBT-XD (6 and 18 MV photons) analyzed on an Epson 10,000XL scanner have been reported [64,67]. Though the EBT2 and EBT3 family are useful for most clinical radiation therapy applications, EBT-XD model, with a useful dose range of 0.04–40 Gy, is designed for single fraction SRS, SBRT applications. If a choice of film models exists for the dose range of interest considers that for a particular RCF film dose uncertainty depends on its sensitivity. For equal doses, the film with the lesser sensitivity will have higher contrast resulting in lower dose uncertainty. Thus for doses >10 Gy, EBT-XD would usually be preferred over EBT2 or EBT3 [66].

9.4.2 EQUICOLOR CALIBRATION DOSES IN TRIPLE-CHANNEL DOSIMETRY

Several papers have reported on improvements in RCF dosimetry using triple-channel methods [68–71]. Dosimetry of SRS is critical in which RCF could provide satisfactory dose. However, RCF dosimetry requires correction. A protocol with a far fewer dose points (e.g., seven-point) required for generating a calibration curve has been proposed [72] using the triple-channel method. The fewer number of dose points comes from the adoption of rational data-fitting functions (Equation 9.5) having natural behavior similar to that of RCF.

$$X(D) = \frac{a+b}{(D-c)} \tag{9.5}$$

in which:

$X(D)$ represents the response at dose D

a, b, and c are constants

The goal of selecting calibration doses is to equilibrate error in high- and low-dose areas. The equicolor dose scheme ensures the high- and low-dose regions are equally well approximated for EBT3 film. At higher dose such as SRS/SBRT cases, the higher dose needs more support as the sensitivity toward errors increases quickly, and the equicolor values cannot be recommended. Above 10 Gy, one should change to EBT-XD film. Individual irradiation of calibration film strips is recommended in a water-equivalent phantom at the depth of at least 5 cm to avoid the uniformity of the beam profiles.

The range of doses used to expose the calibration films should be sufficient to encompass the range of all anticipated measurements. Extrapolation of calibration curves beyond measured points is not recommended. The number of dose points necessary generally depends on the calibration fitting function used. For polynomial fits, several points covering an even logarithmic spacing spanning at least two decades of dose are recommended [73]. Users employing rational calibration functions may require fewer points [72].

Clustering data points near the low-dose part of the curve in which the calibration changes rapidly with dose is advisable, whereas fewer points are generally required at higher dose levels in which the calibration curve changes more gradually. Acquiring and averaging multiple calibration films at the same dose level have the advantage of improving dose statistics and averaging out anomalous behavior within individual films at the expense of more materials and time required for exposure and analysis.

SRS fields usually do not have the size restriction imposed by the lateral response artifact; however, the placement of films for scanning and under some circumstances (e.g., off-center film placement on the scanner, doses >5 Gy, single-channel dosimetry) could introduce dose uncertainty >10%. The effect is mitigated through the use of multichannel dosimetry and can be completely eliminated through a one-time procedure to characterize the effect of a given film model on a particular scanner. The resulting correction coefficients can be applied to remove the artifact from any subsequent image obtained for that film model on the scanner [74]. From our experience, the region of interest used in the film calibration is better with a circular shape rather than a square or rectangular for SRS dosimetry as shown in Figure 9.10. The calibration was based on an output factor of 0.8183 cGy/MU for Trilogy 6-MV under the conditions of 100 cm source to axis distance (SAD), depth of 5 cm, and 3×3 cm^2 MLC/jaw setting. The calibration dose levels were 2880, 2138, 1611, 1178, 814, 504, 235, and 0 cGy, respectively, following the equicolor scheme for triple-channel dosimetry. For EBT-XD and EBT3 with equal exposure, the effect of lateral response artifact is markedly less for the former, making it preferred to EBT3 for doses >10 Gy [66].

9.4.3 Other factors affecting the accuracy of stereotactic radiosurgery film dosimetry

Sheets or smaller pieces of most RCF models are unlikely to be perfectly flat, and this can lead to substantial uncertainty in the measured response values if the film is not coplanar with the glass window in a flatbed scanner. Lewis [75] and Palmer [65] reported that RCF responses vary with the distance between the film and light source. To address this issue, a simple expedient is to use a clear 3–4 mm thick and clear glass plate over the film on flatbed scanners. The glass plate should be free of visual defects and cover the entire glass window of the scanner including the calibration area at the *scan start* end of the window. The weight of the glass plate flattens the film ensuring that it is in a fixed and reproducible position coplanar with the surface of the scanner glass. For EBT3 and EBT-XD films, plain plate glass is sufficient as the polyester substrate has a surface treatment

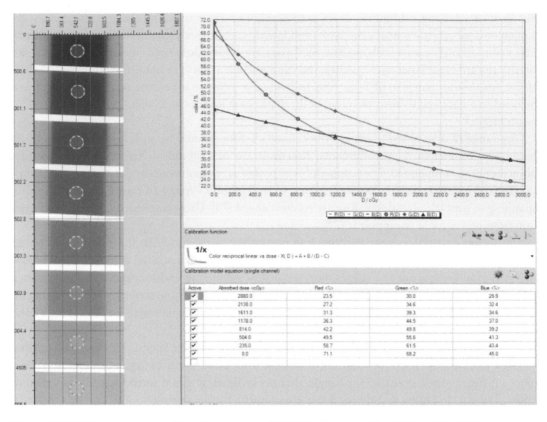

Figure 9.10 Triple-channel calibration curves of EBT-XD films used for SRS with 0–28 Gy.

preventing the formation of Newton's Rings interference patterns. Other crucial factors affecting the accuracy of SRS film dosimetry include keeping/storing films in the dark that can minimize exposure to light when necessary for use and not to expose them to temperatures above 60°C. At last, it is strongly recommended to precut the film for either calibration or measurements several days in advance, enabling the films to equilibrate with the environment. Not doing so could result in large dose differences as well as nonuniform response in the plane of the film [76].

9.5 SUMMARY

In the current chapter, we present the detailed procedure of commissioning SRS and SBRT with RCF and SRS end-to-end test. RCF, especially, EBT2 and EBT3 have been the predominant types of GAFchromic™ films used in commissioning SRS and SBRT in radiation oncology for the past few years. In spite of increased dynamic dose range, studies, including our own investigations, have found that they are not optimal for applications beyond 10 Gy, such as SRS and SBRT, due to diminished response sensitivity. This leads to increased measurement uncertainty at the high dose region. For this very reason, many clinical medical physicists prefer to use a dose much lower than the actual clinical dose for their commissioning and validation tests. However, this practice significantly deviates from the fundamental essence of the SRS end-to-end tests. For nonintensity-modulated modalities, such as 3D-CRT and 3D-conformal arc, this dose reduction approach in end-to-end tests may be acceptable. This is due to the fact that beam apertures are static and the gantry angular speed is not a function of dose. Therefore, if a machine has an acceptable dose-MU linearity, then the measurement error introduced by dose reduction will be in the order of dose-MU linearity uncertainty. However, for intensity-modulated modalities, such as IMRT and VMAT, MLC leaf speed, the gantry angular speed, and the angular dose rate are all a function of dose per control point. Consequently, their mechanical performance is greatly affected by dose magnitude. As IMRT- and VMAT-based SRS has a long delivery time, this could potentially lead to a big cumulative dosimetric error, which is underestimated or even completely missed in dose-reduction approach. Owing to this, the dose-reduction technique should not be used in end-to-end tests for IMRT and VMAT-based SRS and SBRT. In addition, both EBT2 and EBT3 films have also been shown to exhibit pronounced lateral scan effect. The films scanned in the middle of the scanning window yield different dosimetric results from the films scanned at the lateral positions of the window. In our recent SRS commissioning and end-to-end tests for Varian Trilogy and TrueBeam machines at our institution, we used the latest EBT-XD GAFchromic film for dosimetry measurement and target localization accuracy assessment. EBT-XD film was released to the market for clinical evaluation and applications in 2015. Our initial clinical tests and evaluation with EBT-XD film showed that it has much higher response sensitivity at the dose range used in SRS and SBRT treatment (18–24 Gy) than both EBT2 and EBT3 films and less lateral scan effect. Thus, it is ideal for SRS and SBRT commissioning and end-to-end tests.

REFERENCES

1. Leksell L. A stereotaxic apparatus for intracerebral surgery. *Acta Chir Scand* 1949;99:229.
2. Steiner L. (Ed.). *Radiosurgery: Baseline and Trends*. New York: Raven Press; 1992.
3. Leksell L. The stereotaxic method and radiosurgery of the brain. *Acta Chir Scand* 1951;102:316–319.
4. Leksell L. Stereotactic radiosurgery. *J Neurol Neurosurg Psychiatry* 1983;46:797–803.
5. Larsson B, Lidén K, Sarby B. Irradiation of small structures through the intact skull. *Acta Radiol* 1974;13:512–534.
6. Lunsford LD, Maitz A, Lindner G. First United States 201 source cobalt-60 gamma unit for radiosurgery. *Appl Neurophysiol* 1987;50:253–256.
7. Ginzton EL, Mallory KB, Kaplan HS. The Stanford medical linear accelerator: I. Design and development. *Stanford Med Bull* 1957;15(3):123–140.
8. Betti OO, Derechinsky YE. Irradiations stereotaxiques multifaisceaux. *Neurochirurgie* 1982;28:55–56.
9. Das IJ, Downes MB, Corn BW et al. Characteristics of a dedicated linear accelerator-based stereotactic radiosurgery-radiotherapy unit. *Radiother Oncol* 1996;38:61–68.
10. Loeffler JS, Alexander E, Siddon RL et al. Stereotactic radiosurgery for intracranial arteriovenous malformations using a standard linear accelerator: Rationale and technique. *Int J Radiat Oncol Biol Phys* 1989;17:1327–1335.
11. Schlegel W, Pastry O, Bortfeld T et al. Computer systems and mechanical tools for stereotactically guided conformation therapy with linear accelerators. *Int J Radiat Oncol Biol Phys* 1992;24:781–787.
12. Brown RA, Roberts TS. *Stereotactic Cerebral Irradiation: Proceedings of the INSERM Symposium on Stereotactic Irradiation*. Amsterdam, the Netherlands: Elsevier/North-Holland Biomedical Press; 1979. Vol. 13, p. 25–27.
13. Brown RA, Roberts TS, Osborn AE. Stereotaxic frame and computer software for CT-directed neurosurgical localization. *Invest Radiol* 1980;15:308–312.
14. Dubois PJ, Nashold BS, Perry J et al. CT-guided stereotaxis using a modified conventional stereotaxic frame. *Am J Neuroradiol* 1982;3:345–351.

15. Leksell L, Leksell D, Schwebel J. Stereotaxis and nuclear magnetic resonance. *J Neurol Neurosurg Psychiatry* 1985;48:14–18.

16. Burton KE, Thomas SJ, Whitney D et al. Accuracy of a relocatable stereotactic radiotherapy head frame evaluated by use of a depth helmet. *Clin Oncol (R Coll Radiol)* 2002;31–39.

17. Kooy HM, Dunbar SF, Tarbell NJ et al. Adaptation and verification of the relocatable Gill-Thomas-Cosman frame in stereotactic radiotherapy. *Int J Radiat Oncol Biol Phys* 1994; 30:685–691.

18. Kassaee A, Das IJ, Tochner Z et al. Modification of Gill-Thomas-Cosman frame for extracranial head-and-neck stereotactic radiotherapy. *Int J Radiat Oncol Biol Phys* 2003;57:1192–1195.

19. Fu L, Perera H, Ying X et al. Importance of CBCT setup verification for optical-guided frameless radiosurgery. *J Appl Clin Med Phys* 2014;15:1–9.

20. Chang J, Yenice KM, Narayana A et al. Accuracy and feasibility of cone-beam computed tomography for stereotactic radiosurgery setup. *Med Phys* 2007;34:2077–2084.

21. Masi L, Casamassima F, Polli C et al. Cone beam CT image guidance for intracranial stereotactic treatments: Comparison with a frame guided set-up. *Int J Radiat Oncol Biol Phys* 2008;71:926–933.

22. Ackerly T, Lancaster CM, Geso M et al. Clinical accuracy of ExacTrac intracranial frameless stereotactic system. *Med Phys* 2011;38:5040–5048.

23. Gevaert T, Verellen D, Engels B et al. Clinical evaluation of a robotic 6-degree of freedom treatment couch for frameless radiosurgery. *Int J Radiat Oncol Biol Phys* 2012;83:467–474.

24. Li G, Ballangrud A, Kuo LC et al. Motion monitoring for cranial frameless stereotactic radiosurgery using video-based three-dimensional optical surface imaging. *Med Phys* 2011;38:3981–3994.

25. Yang G, Wang Y, Wang Y et al. CyberKnife therapy of 24 multiple brain metastases from lung cancer: A case report. *Oncol Lett* 2013;6:534–536.

26. Tamari K, Suzuki O, Hashimoto N et al. Treatment outcomes using CyberKnife for brain metastases from lung cancer. *J Radiat Res* 2015;56:151–158.

27. Lax I, Blomgren H, Näslund I et al. Stereotactic radiotherapy of malignancies in the abdomen. Methodological aspects. *Acta Oncol* 1994;33:677–683.

28. Blomgren H, Lax I, Näslund I et al. Stereotactic high dose fraction radiation therapy of extracranial tumors using an accelerator. Clinical experience of the first thirty-one patients. *Acta Oncol* 1995;34:861–870.

29. Timmerman R, Papiez L, McGarry R et al. Extracranial stereotactic radioablation: Results of a phase I study in medically inoperable stage I non-small cell lung cancer. *Chest* 2003;124:1946–1955.

30. McGarry RC, Papiez L, Williams M et al. Stereotactic body radiation therapy of early-stage non-small-cell lung carcinoma: Phase I study. *Int J Radiat Oncol Biol Phys* 2005;63:1010–1015.

31. Hazelaar C, Dahele M, Mostafavi H et al. Subsecond and submillimeter resolution positional verification for stereotactic irradiation of spinal lesions. *Int J Radiat Oncol Biol Phys* 2016;94:1154–1162.

32. Scorsetti M, Franceschini D, De Rose F et al. Stereotactic body radiation therapy: A promising chance for oligometastatic breast cancer. *Breast* 2016;26:11–17.

33. Strom T, Wishka C, Caudell JJ. Stereotactic body radiotherapy for recurrent unresectable head and neck cancers. *Cancer Control* 2016;23:6–11.

34. Myrehaug S, Sahgal A, Russo SM et al. Stereotactic body radiotherapy for pancreatic cancer: Recent progress and future directions. *Expert Rev. Anticancer Ther* 2016;16:523–530.

35. Das IJ, Cheng CW, Watts RJ et al. Accelerator beam data commissioning equipment and procedures: Report of the TG-106 of the Therapy Physics Committee of the AAPM. *Med Phys* 2008;35:4185–4215.

36. IAEA TRS 483, Dosimetry of small static fields used in external beam radiotherapy: An IAEA-AAPM international code of practice for reference and relative dose determination, technical report series No. 483. 2017; Vienna, Austria.

37. Lutz W, Winston KR, Maleki N. A system for stereotactic radiosurgery with a linear accelerator. *Int J Radiat Oncol Biol Phys* 1988;14:373–381.

38. Fukuda A. Pretreatment setup verification by cone beam CT in stereotactic radiosurgery: Phantom study. *J Appl Clin Med Phys* 2010;11:122–129.

39. Chang Z, Wang Z, Ma J et al. Six degree-of-freedom image guidance for frameless intracranial stereotactic radiosurgery with kilo-voltage Cone-Beam CT. *J Nucl Med Radiat Ther* 2010;1:1000101.

40. Grimm J, Grimm SYL, Das IJ et al. A quality assurance method with submillimeter accuracy for stereotactic linear accelerators. *J Appl Clin Med Phys* 2011;12:182–198.

41. Ramani R, Lighthouse AW, Mason DLD et al. The use of radiochromic film in treatment verification of dynamic stereotactic radiosurgery. *Med Phys* 1994;21:389–392.

42. Paskalev KA, Seuntjens JP, Patrocinio HJ et al. Physical aspects of dynamic stereotactic radiosurgery with very small photon beams 1.5 and 3 mm in diameter. *Med Phys* 2003;30:111–118.

43. Keller BM, Beachey DJ, Pignol JP. Experimental measurement of radiological penumbra associated with intermediate energy X-rays (1 MV) and small radiosurgery field sizes. *Med Phys* 2007;34:3996–4002.

44. Sturtewagen E, Fuss M, Paelinck L et al. Multi-dimensional dosimetric verification of stereotactic radiotherapy for uveal melanoma using radiochromic EBT film. *Med Phys* 2008;18:27–36.

45. Hoffmann L. Implementation and experimental validation of the high dose rate stereotactic treatment mode at Varian accelerators. *Acta Oncol* 2009;48:201–208.

46. Coscia G, Vaccara E, Corvisiero R et al. Fractionated stereotactic radiotherapy: A method to evaluate geometric and dosimetric uncertainties using radiochromic films. *Med Phys* 2009;36:2870–2880.

47. Kairn T, Kenny J, Crowe SB et al. Technical Note: Modeling a complex micro-multileaf collimator using the standard BEAMnrc distribution. *Med Phys* 2010;37:1761–1767.

48. Kairn T, Crowe S, Kenny J et al. Investigation of stereotactic radiotherapy dose using dosimetry film and Monte Carlo simulations. *Radiat Meas* 2011;46:1985–1988.

49. Wong JHD, Knittel T, Downes S et al. The use of a silicon strip detector dose magnifying glass in stereotactic radiotherapy QA and dosimetry. *Med Phys* 2011;3:1226–1238.

50. Hardcastle N, Basavatia A, Bayliss A et al. High dose per fraction dosimetry of small fields with gafchromic EBT2 film. *Med Phys* 2011;38:4081–4085.

51. Chan MF, Zhang Q, Li J et al. The verification of iPlan commissioning by radiochromic EBT2 films. *Int J Med Phys Clin Eng Radiat Oncol* 2012;1:1–7.

52. Barbosa N, da Rosa LAR, Batista DVS et al. Development of a phantom for dose distribution verification in stereotactic radiosurgery. *Phys Med* 2013;29:461–469.

53. Esparza-Moreno KP, Garcia-Garduno OA, Ballesters-Zebadua P et al. Comparison of trigeminal neuralgia radiosurgery plans using two film detectors for the commissioning of small photon beams. *J Appl Clin Med Phys* 2013;14:1–9.

54. Gonzalez-Lopez A, Vera-Sanchez JA, Lago-Martin JD. Small fields measurements with radiochromic films. *J Med Phys* 2015;40:61–67.

55. Garcia-Garduno OA, Larraga-Gutierrez JM, Rodriguez-Villafuerte M et al. Small photon beam measurements using radiochromic film and Monte Carlo simulations in a water phantom. *Radiother Oncol* 2010;96:250–253.

56. Bouchard H, Seuntjens J, Simon D et al. Detector dose response in megavoltage small photon beams. I. Theorectical concepts. *Med Phys* 2015;42:6033–6047.

57. Oita M, Takegawa Y, Yagi H et al. Quality control (QC) of CT on rail system (FOCAL Unit) with a micro-multi leaf collimator (mMLC) using new GafChromic film for stereotactic radiotherapy. *Nihon Hoshasen Gijutsu Gakkai Zasshi* 2006;62:711–713.

58. Dobler B, Walter C, Knopf A et al. Optimization of extracranial stereotactic radiation therapy of small lung lesions using accurate dose calculation algorithms. *Radiat Oncol* 2006;1:45.

59. Cho GA, Ralston A, Tin MM, et al. *In vivo* and phantom measurements versus Eclipse TPS prediction of near surface dose for SBRT treatments. *J Phys Conf Series* 2014;489:012008.

60. Gallo JJ, Kaufman I, Powell R et al. Single-fraction spine SBRT end-to-end testing on TomoTherapy, Vero, TrueBeam, and CyberKnife treatment platforms using a novel anthropomorphic phantom. *J Appl Clin Med Phys* 2015;16:170–182.

61. Hardcastle N, Clements N, Chesson B et al. Results of patient specific quality assurance for patients undergoing stereotactic ablative radiotherapy for lung lesions. *Australas Phys Eng Sci Med* 2014;37:45–52.

62. Followill DS, Evans DR, Cherry C et al. Design, development, and implementation of the radiological physics center's pelvis and thorax anthropomorphic quality assurance phantoms. *Med Phys* 2007;34:2070–2076.

63. Bekerat H, Devic S, DeBlois F et al. Improving the energy response of external beam therapy (EBT) GafChromic™ dosimetry films at low energies ($\leq$ 100 keV). *Med Phys* 2014;41:022101.

64. Grams M, Gustafson JM, Long KM et al. Technical Note: Initial characterization of the new EBT-XD Gafchromic film. *Med Phys* 2015;42:5782–5786.

65. Palmer AL, Dimitriadis A, Nisbet A et al. Evaluation of Gafchromic EBT-XD film, with comparison to EBT3 film, and application in high dose radiotherapy verification. *Phys Med Biol* 2015;60:8741–8752.

66. Lewis DF, Chan MF. Technical Note: On GafChromic EBT-XD film and the lateral response artifact. *Med Phys* 2016;43:643–649.

67. Massillon-JL G, Chiu-Tsao ST, Domingo-Murioz I et al. Energy dependence of the new Gafchromic EBT3 film—Dose response curves for 50 kV, 6 and 15 MV X-ray beams. *Int J Med Phys Clin Eng Radiat Oncol* 2012;1:60–65.

68. Micke A, Lewis DF, Yu X. Multichannel film dosimetry with nonuniformity correction. *Med Phys* 2011;38:2523–2534.

69. van Hoof SJ, Granton PV, Landry G et al. Evaluation of a novel triple-channel radiochromic film analysis procedure using EBT2. *Phys Med Biol* 2012;57:4353–4368.

70. Hayashi N, Watanabe Y, Malmin R et al. Evaluation of triple channel correction acquisition method for radiochromic film dosimetry. *J Radiat Res* 2012;53:930–935.

71. Chan MF, Lewis DF, Yu X. Is it possible to publish a calibration function for radiochromic film? *Int J Med Phys Clin Eng Radiat Oncol* 2014;3:25–30.

72. Lewis DF, Micke A, Yu X et al. An efficient protocol for radiochromic film dosimetry combining calibration and measurement in a single scan. *Med Phys* 2012;39:6339–6350.

73. Bouchard H, Lacroix F, Beaudoin G et al. On the characterization and uncertainty analysis of radiochromic film dosimetry. *Med Phys* 2009;36:1931–1946.

74. Lewis DF, Chan MF. Correcting lateral response artifacts from flatbed scanners for radiochromic film dosimetry. *Med Phys* 2015;42:416–429.

75. Lewis D, Devic S. Correcting scan-to-scan response variability for a radiochromic film-based reference dosimetry system. *Med Phys* 2015;42:5692–5701.

76. Niroomand-Rad A, Blackwell CR, Coursey BM et al. Radiochromic film dosimetry: Recommendations of AAPM radiation therapy committee task group 55. *Med Phys* 1998;25:2093–2115.

Gamma Knife®

STEVEN J. GOETSCH AND ANDY (YUANGUANG) XU

10.1 REVIEW OF GAMMA KNIFE® FOR RADIOSURGERY

10.1.1 HISTORY OF GAMMA KNIFE

The creation of the concept of minimally invasive surgery can be traced to a paper published in 1951 by Lars Leksell, a Swedish neurosurgeon [1]. Leksell had previously developed a human stereotactic head frame and used it successfully to guide a needle and selectively destroy abnormal brain tissue with radiofrequency rhizotomy. Beginning with his 1951 paper, he described the substitution of a precisely directed, concentrated beam of X-rays without needing to open the skull and risk infection. He coined the term *radiosurgery* to describe this process using kilovoltage X-rays. The low-energy X-rays available at that time were inadequate, so Leksell began using high-energy protons from a cyclotron at Uppsala University just west of Stockholm. Together with physicist Borge Larson in 1967, he invented a hospital-based device known as the Leksell Gamma Unit (Figure 10.1) that contained a large amount of radioactive cobalt-60 sources, each precisely directed to the Unit Center Point, later referred to as the isocenter [2].

Leksell's device, later popularly known as the Leksell Gamma Knife®, was restricted to intracranial use only and required placement of the Leksell Model G stereotactic frame, at first with screws penetrating the skull and later with sharp pins. The only form of neuroimaging available at that time was pneumoencephalography, which required performing a spinal tap and displacing most of the cerebrospinal fluid with air. This was difficult and excruciatingly painful. The air-filled ventricles could then be imaged by anterior–posterior and lateral X-ray films, and the anterior commissure–posterior commissure line could be determined. Previously available brain maps [3] could then be used to determine the stereotactic coordinates (X: lateral, Y: anterior–posterior, and Z: inferior–superior) to direct the beams of gamma radiation to the anatomical site of interest. Thus, only functional disease such as Parkinson's disease and trigeminal neuralgia could be treated for many years.

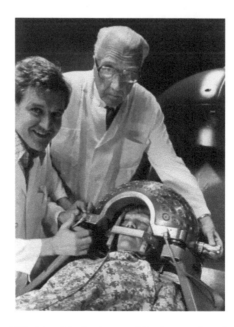

Figure 10.1 Lars Leksell and Christer Lindquist positioning patient in the original Leksell Gamma Unit, about 1968 in Stockholm, Sweden.

Leksell's sons Dan and Lawrence persuaded their father to create the Elekta Corporation in 1972 to market the Gamma Knife and Leksell's other surgical inventions. In 1985, Leksell permitted his blueprints to be used to construct two other Gamma Units in Buenos Aires, Argentina [4] and in Sheffield, England [5]. Shortly after that, neurosurgeon Robert Rand was given permission to move Leksell's first Gamma Unit (which had been in storage) to University of California Los Angeles (UCLA) Medical Center, where it was used to treat a number of patients [6].

The world's first commercial Leksell Gamma Unit (which later became popularly known as the Model U) was manufactured in Sweden by Elekta and installed at the University of Pittsburgh Medical Center in 1987 [7]. The unit was quite successful, and many more units were installed in the United States and many other countries, especially in Europe, Japan, Korea, and China. At the time of this writing, more than 300 units are installed worldwide, with over 1 million radiosurgery treatments reported through the end of calendar year 2015 as shown in Figure 10.2.

The Leksell Gamma Unit Model U came with a primitive computer system, which allowed the clinical users to see lines of equal radiation dose (isodose lines) portrayed on a single computed tomography (CT) axial slice for a single isocenter (or shot). This process took about 15 min for a single slice. Over the decades since that time, the Leksell GammaPlan® treatment planning system has become more sophisticated with thousands times faster. Now users can see axial, coronal, and sagittal images (as well as angiographic images) in many slices at the same time. This has increased the sophistication of planning and made the process extremely rapid. In addition, there is no limit on the number of isocenters that can be employed.

The original 201 source Gamma Unit Model U required the stereotactic coordinates (X, Y and Z) to be manually set by the clinical staff. It featured 4 mm, 8 mm, 14 mm, and 18 mm nearly spherical volumes of irradiation. The unit required a very challenging manual process to set the lateral (X) coordinate with trunnions, attached to an external helmet and the Y–Z sliders affixed to the Leksell coordinate frame. Each isocenter required the patient to be inserted into the machine (placed within a large concrete bunker) and then removed for the setting of the coordinates of the next point. This was a time-consuming process and could lead to errors, either in the accuracy of the coordinates or due to the inadvertent interchange of digital parameters.

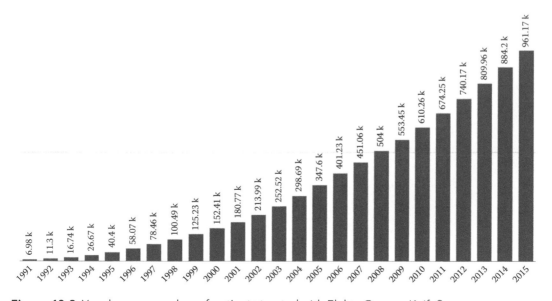

Figure 10.2 Year-by-year number of patients treated with Elekta Gamma Knife®.

The next generation Gamma Knife (Models, B, C, and 4C) introduced in about 1988 and 1999, respectively, were far more sophisticated. The change out of secondary collimator helmets was much simpler, and a computer monitored and supervised the treatment of the patient, eliminating the possibility of wrong coordinates, or wrong helmets [8]. The Model 4C featured the automatic positioning system that could move the patient to multiple coordinate locations without the staff having to reenter the treatment room (as long as the shots were all within a 2-cm radius). This speeded up the delivery of treatment. The construction of the new Models B, C, and 4C also made replacement of depleted Cobalt-60 sources (typically performed every 5 to 6 years) much quicker and less difficult.

The Gamma Knife Perfexion was introduced in 2006 as shown in Figure 10.3. It was a completely revolutionary design in which the external helmets were eliminated. A total of 192 sources were placed on 8 actuator arms bearing 24 sources each, which could be positioned over 4 mm, 8 mm, or 16 mm diameter collimators. The initial clinical report suggested that even with nine fewer sources, the totally automated patient positioning system (using a high precision couch) reduced the average patient treatment time by 20% or more [9].

The latest innovation in the Leksell Gamma Knife is the Perfexion ICON system, which includes a swing-arm cone beam CT scanner, an infrared patient motion monitoring system, and reinforced thermoplastic face mask for imaging and fractionated treatment. This system makes it possible for Gamma Knife users to perform either single fraction radiosurgery with the Leksell Model G frame or fractionated treatments with daily patient position verification. It was introduced in Europe in 2015 and in the United States in 2016.

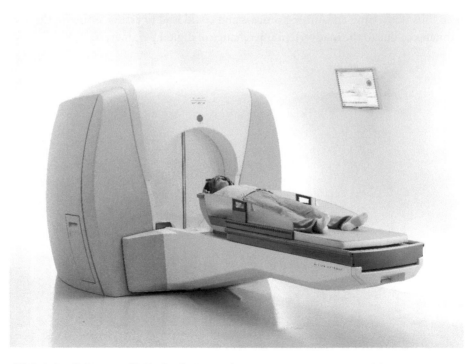

Figure 10.3 Leksell Gamma Knife Perfexion with patient in treatment position.

10.1.2 GAMMA KNIFE APPLICATIONS IN STEREOTACTIC RADIOSURGERY

The original Leksell Gamma Unit was invented to ablate small intracranial targets to treat chronic pain (such as trigeminal neuralgia) and movement disorder (such as Parkinson's disease). There is a case report of an acoustic neuroma with bilateral tumors treated with Gamma Knife radiosurgery in 1969 [10].

Clinical indications for Gamma Knife radiosurgery were greatly expanded in 1970 when Leksell's neurosurgical colleague Ladislau Steiner used the newly developed technique of cerebral angiography to extend Gamma Knife radiosurgery to arteriovenous malformations [11]. These hereditary genetic anomalies caused bleeding in the brain, which could result in seizures, coma, or death. Steiner performed Gamma Knife radiosurgery on these lesions with a high rate of success and brought the technique to a new Gamma Knife Center at the University of Virginia in 1989. The next advance in Gamma Knife radiosurgery came in the late 1970s when the revolutionary new CT imaging technique was first integrated into stereotactic radiosurgery [12]. This new imaging technique suddenly made intracranial tumors visible, and they became viable targets for Gamma Knife radiosurgery [13].

Metastatic brain tumors became a rising indication for Gamma Knife radiosurgery in the 1990s. Once detection and localization of metastatic tumors in the brain became routinely possible with CT and magnetic resonance imaging scans, clinicians rapidly attacked the problem [14]. Localized control of metastatic brain tumors became so successful that the existing paradigm of recommending whole brain radiation therapy for all such patients began to be challenged. Other malignant brain tumors, especially gliomas, were also treated with Gamma Knife radiosurgery [15]. Patients with benign disease such as arteriovenous malformations, acoustic neuromas, meningiomas, and pituitary adenoma, began to select Gamma Knife radiosurgery over conventional craniotomy in increasing numbers [16–19]. The traditional role of Gamma Knife radiosurgery in the treatment of functional disease has not been neglected. A recent tally conducted by the Leksell Society indicates that more than 61,000 patients have been treated worldwide through calendar year 2015 for trigeminal neuralgia [20].

10.1.3 A BRIEF SURVEY OF GAMMA KNIFE STEREOTACTIC RADIOSURGERY DEVICES

The conspicuous success of the Leksell Gamma Knife device in the 1980s and 1990s lead to the design and manufacture of similar devices, especially in China. The OUR rotating gamma unit was introduced in China in the 1990s and consisted of 30 high intensity cobalt-60 radioactive sealed sources mounted on a large rotating disk [21]. The absorbed dose rate, radiation falloff, and mechanical accuracy were comparable with the static Leksell gamma unit. At least three such units were imported into the United States beginning in 2001 [22]. At this writing, one other manufacturer has obtained a license to distribute a similar rotating gamma stereotactic device within the United States.

10.1.4 GAMMA KNIFE HELMETS AND COLLIMATORS

The Leksell Gamma Knife Perfexion Model has largely replaced the previous Leksell gamma units. Leksell Gamma units prior to this model had very large 400-km external helmets that had to be mechanically hoisted into place on the patient positioning couch above the patient's head.

The prototype Gamma Unit in Stockholm had helmets that gave projected nominal beam diameters of 4 mm, 8 mm, and 14 mm. The commercial Leksell Model U added an 18-mm helmet. The new Gamma Knife Perfexion is designed with a very large, precisely machined tungsten collimator internal to the unit that allows nominal beam sizes of 4 mm, 8 mm, and 16 mm diameter.

10.2 PHYSICAL CHARACTERISTICS OF GAMMA KNIFE BEAMS

10.2.1 BEAM AND ENERGY FLUENCE OF GAMMA KNIFE BEAMS

The hallmark characteristics of the Leksell Gamma Knife are the mechanical pointing accuracy (alignment of collimators) and the rapid dose falloff [23]. The Leksell Gamma Knife Models U, B, 4C and Perfexion have all been warranted by the manufacturer to deliver an output at the isocenter of the unit of at least 3 Gy/min (using the largest collimator) at time of acceptance testing. Clearly the absorbed dose rate decreases with the 5.26-year half-life of cobalt-60, approximately 1% per month.

10.2.2 ABSOLUTE DOSIMETRY AND TOTAL SCATTER FACTORS

Gamma stereotactic radiosurgery device manufacturers rely on extremely precise machining and manufacturing techniques and quality assurance to make each device virtually identical to other devices of the same model, differing only in the unique radioactive source loading for an individual serial number device. Therefore, manufacturers typically specify collimator (helmet) output factors and dose falloff tables, which are embedded in the customized radiation therapy treatment planning system. Historically the clinical medical physicist had a limited ability to verify these factors, but recently, techniques have been developed for verification in which radiochromic film (RCF) plays an important role.

The Elekta corporation revised upward by 7% their recommended helmet factor for the 4-mm helmet (smallest size) in about 2000 [24]. Many Gamma Knife centers utilized newly developed dosimetry technology (micro-thermoluminescent dosimeters (TLDs), RCF, diamond detectors, liquid filled ionization chambers) to verify this recommended change [25]. Ultimately, most investigators converged on an output factor for this remarkably small collimator that was in close agreement with the recommended value.

10.2.3 RELATIVE DOSIMETRY: PDDs AND OFF-AXIS RATIOS

Gamma stereotactic radiosurgery units are limited to treatment at the isocenter, so percent depth dose (PDD) is only important in dose calculation for each individual treatment plan. The geometry of the Gamma Knife unit makes it physically impossible to measure percent depth dose, and the concept would be meaningless due to the superposition of all 192 gamma ray beams (formerly 201 beams). It is only possible to measure a valid PDD in the laboratory of the manufacturer with a single source loaded into a collimator.

Off-axis ratios (beam profiles) are supplied by the manufacturer and embedded in the respective treatment planning algorithm. Individual users may verify these data with RCF. Some investigators have modeled the Leksell Gamma Knife collimation system as well as rotating gamma units with Monte Carlo calculations [26,27].

10.2.4 COMMONLY DETECTORS USED FOR GAMMA KNIFE DOSIMETRY

The ground truth calibration of Gamma Knife and rotating gamma unit radiosurgery systems is performed with calibrated ionization chambers. This poses several problems. The most commonly used Farmer-type cylindrical ionization chambers have a volume of approximately 0.6 cm^3 and dimensions too large for the 16- to 18-mm diameter maximum field sizes of these units. A new generation of microionization chambers (with volumes of approximately 0.1 cm^3) as defined in TG-106 [28] has been manufactured for small-field dosimetry. However, these extremely small ionization chambers are not as robust as somewhat larger ionization chambers [29].

Annual quality assurance requirements for Gamma Knife dosimetry vary from state to state but commonly require verification of output factors, isocenter verification, and beam profiles. This can be done using RCF [30]. A special device, called the pinprick film holder, places a pinprick hole in a piece of film precisely at the isocenter. The film is exposed in the Gamma Knife. Upon processing and scanning, the X and Y distance from the center of the measured dose distribution to the pinprick itself is measured and checked to see if it is within 0.5-mm distance.

10.3 RADIOCHROMIC FILM FOR GAMMA KNIFE RADIOSURGERY DOSIMETRY

10.3.1 MATERIALS AND EQUIPMENT

10.3.1.1 RADIOCHROMIC FILM MODELS: HS, MD, AND EXTERNAL BEAM THERAPY

Historical views of RCF can be found in Chapter 2 with evolution of HD-810/DM-1260 [31,32] and MD-55 [33] to measure beam profile parameters for Gamma Knife models U and B. A more sensitive model, the MD-55-2 [34], was produced by adding two MD-55 layers together.

Although the sensitivity of the MD-V2-55 model GAFchromic™ film is relatively high, its dose response was reported to be not uniform enough (8%–15% variation) for certain clinical applications. Consequently, the HS GAFchromic film model was introduced with a single and slightly thicker (40 μm) sensitive layer leading to higher sensitivity and better uniformity than the MD-V2-55. With the improvement in the uniformity and reproducibility of the films, the relative output factors from the film measurements for different collimators of the Gamma Knife Perfexion and C models agreed better with those from the Monte Carlo calculations [35,36]. Further development in RCF production led to the EBT (which stands for External Beam Therapy) GAFchromic film model, initially designed to replace the radiographic films for intensity modulated radiation therapy (IMRT) quality assurance procedures [37,38]. Several versions of the EBT films have been produced over time, typically with improved dose sensitivity, lower energy dependence, and wider dose range. The current model of the EBT films (EBT-3) works in the dose range from 1 to 40 Gy with a specified dose uniformity of better than ±3% and is a suitable tool for routine quality assurance in Gamma Knife radiosurgery.

10.3.1.2 SCANNERS AND DENSITOMETERS

Chapter 4 provides details of scanners and densitometer that can be looked in the context of RCF. The point dose measurement tools such as densitometers and spectrophotometers have been used for Gamma Knife quality assurance only when 2D detectors were not available. It is in general difficult to get a stable attenuation reading from an irradiated region of a film using a point dose detector because of the nonuniformity and the noises in the film. Attempts were also made during the early stage of the clinical film dosimetry to acquire beam profiles and 2D dose distributions for Gamma Knife radiation using scanning densitometers [32,39].

The acquired image files from the films scanners can be processed using different imaging processing software for dosimetric analysis. Commonly used imaging processing software packages for physics quality assurance checks in Gamma Knife radiosurgery include FilmQA, ImageJ, RIT, dose lab, and so on.

10.3.1.3 PHANTOMS

Various head phantoms have been designed and fabricated for research and patient care purpose in Gamma Knife radiosurgery. Elekta has produced two types of phantoms for routine physics dosimetric measurements, one made from acrylonitrile butadiene styrene and the other from water equivalent plastic (Figure 10.4). Both phantoms are 160 mm in diameter and can be used for dose rate measurements with ion chambers and for output factor and beam profile measurements with films. The acrylonitrile butadiene styrene phantom is made of polystyrene and has an opening in the central region for ion chamber inserts or film insert. It can be used to irradiate only one film at a time. The solid-water phantom is composed of certified therapy grade solid water. It was introduced along with the Perfexion Gamma Knife unit. It can be used to measure 2D dose distributions on multiple planes.

10.3.2 RADIOCHROMIC FILM CHARACTERISTICS IN CO-60 BEAMS

The dose response of a batch of RCFs is usually characterized with a dose response curve or a calibration curve. A dose response curve for a batch of film can be obtained by irradiating a series of films with graded doses. The irradiation is usually done on Gamma Knife machines using the largest collimator with the exact experimental setup for all films, that is, same phantom, same size film pieces from the same batch, same film positioning, and so on. The irradiated film pieces should be placed exactly at the same scanning position and scanned with the same scanner setting to exclude any variation in scanner response over scan field. Particular attention is also needed for the orientation of the GAFchromic film samples in the scanner. Klassen et al. [40] showed that reading of the GAFchromic film may change substantially due to the polarization of the light source. This effect is attributed to the preferred orientation of the polymer chains, which absorb light in the microcrystals of the radiosensitive layers. Detailed and general precautions have been provided in Chapter 3.

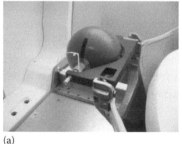
(a)

(b)

(c)

Figure 10.4 Dosimetric phantoms for routine quality assurance in Gamma Knife radiosurgery: (a) acrylonitrile butadiene styrene (ABS) phantom, (b) film insert for the ABS phantom, and (c) solid-water phantom and film inserts.

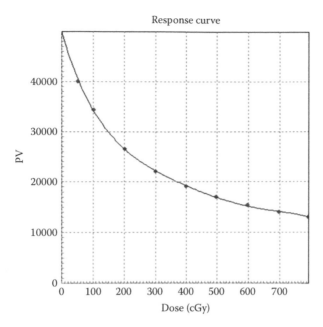

Figure 10.5 Dose response curve for the GAFchromic™ EBT film; the solid line represents the fifth degree polynomial fit to the 10 film points after background corrections; PV stands for pixel value. (From Novotny, J. et al., *Med. Phys.*, 36, 1768–1774, 2009.)

Theoretically, a dose response curve needs to be obtained for every dosimetric measurement with films. The response of the films may change with time, as may the temperature/humidity environment. A dose response curve usually gives a one-to-one correspondence between the delivered dose and a predefined film response parameter (i.e., optical density, image number, etc.) and is desired to cover the entire dose range for the study. Figure 10.5 gives a sample dose response curve for the EBT film generated using the Film QA software [41]. The dose range for this curve is 0–8 Gy.

10.3.3 MEASUREMENT OF OUTPUT FACTORS

The radiation fluence of a Gamma Knife unit is usually characterized by the dose rate measured at the center of a 16-cm spherical phantom using an ionization chamber for the largest collimator (16 mm for Perfexion; 18 mm for models U, B, C, and 4C). The dose rates for smaller collimators cannot be measured using ionization chambers because the radiation fields from these collimators are not large enough to provide electronic equilibrium and due to partial volume effects as described by Das et al. [42] and IAEA TRS-483 [43]. Therefore, a relative output factor is defined for a given collimator as the quotient of the output measured with this collimator to that measured with the largest size collimator [44]. Historically, the relative output factors for different Gamma Knife machines have been obtained from Monte Carlo simulations and also from film measurements. A set of output factors is usually suggested by the manufacturer for all Gamma Knife units of the same model, even though the individual user can use their own output factors in their planning system based on in-house measurements but in comparison with manufacturer data.

The output factors for a Gamma Knife unit can be measured using radiochromic film pieces placed at the center of the 16-cm diameter spherical polystyrene or solid water phantom. Two irradiations with a preselected irradiation time are needed to determine the output factor for a given collimator: An irradiation using this collimator and an irradiation using the largest collimator,

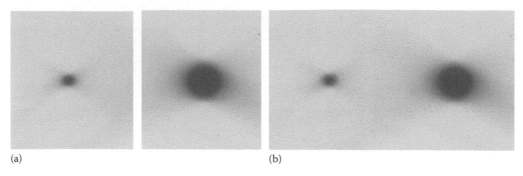

Figure 10.6 Illustration of exposed film samples for the 4-mm collimator output factor measurement using the following: (a) single-shot exposure method and (b) double-shot exposure method. (From Ma, L. et al., *Phys. Med. Biol.*, 54, 3897–3907, 2009.)

both done with films placed in the XY plane. The irradiated films are then scanned and converted into 2D dose maps using the calibration curve for this film batch. An averaged dose reading is then taken from the central region of each radiation field, and the output factor for a given collimator is calculated as the ratio of the two dose readings.

Two methods have been developed for output measurement using films (Figure 10.6). The first approach [41] is to cut RCFs into square pieces (i.e., 5 × 5 cm) and then place the center of the film pieces on the radiation focus point during irradiation (100 mm, 100 mm, and 100 mm in the Leksell stereotactic coordinate system). The underlying assumption for this approach is that the dose responses from all film pieces are exactly the same. So special care is needed with the handling of the film pieces (i.e., use film pieces from the same batch of film, irradiate and scan the films in the same position etc.). The other approach [45] for estimation of relative helmet output factors is to cut film pieces into rectangular pieces (5 cm by 13 cm) and irradiate the film with two shot-positions (70, 100, and 100 mm) and (130, 100, and 100 mm), and so on. The underlying assumptions for this approach are as follows: (1) The two fields are far enough that the scattering effect is negligible, and (2) the signal is uniform from the scanner across the scan field as the two radiation fields cannot be placed on the same position in the scanner.

10.3.4 MEASUREMENT OF BEAM PROFILE PARAMETERS AND OFF-AXIS RATIOS

The off-axis ratio table for dose calculation in Gamma Knife radiosurgery is usually generated from Monte Carlo simulation (and verified by film measurements). The beam profile information for a given collimator can be obtained from films placed in the center of the 16-mm phantom and irradiated in the XY or the XZ planes. Figure 10.7 gives the beam profile analysis results for the 4-mm collimator of a Perfexion Gamma Knife unit from the FilmQA software. The horizontal and the vertical directions are the X and Y direction of the stereotactic coordinate system, respectively. The calibrated 2D dose distribution map is shown on the upper right corner of the pictures. The full width at half maximum (FWHM) of the beam profile is the length of the horizontal line around the 50% isodose. The FWHM obtained for the X and the Y directions are 6.0 mm and 6.4 mm, respectively.

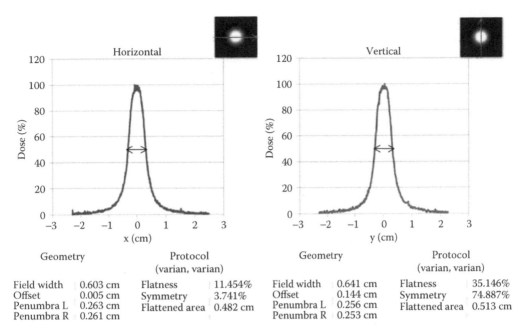

Geometry		Protocol (varian, varian)		Geometry		Protocol (varian, varian)	
Field width	0.603 cm	Flatness	11.454%	Field width	0.641 cm	Flatness	35.146%
Offset	0.005 cm	Symmetry	3.741%	Offset	0.144 cm	Symmetry	74.887%
Penumbra L	0.263 cm	Flattened area	0.482 cm	Penumbra L	0.256 cm	Flattened area	0.513 cm
Penumbra R	0.261 cm			Penumbra R	0.253 cm		

Figure 10.7 Horizontal and vertical profiles from the FilmQA software for a film irradiated in the *XY* plane using the 4-mm collimator of a Perfexion Gamma Knife and an ABS phantom.

10.4 RADIOCHROMIC FILM FOR QUALITY ASSURANCE IN GAMMA KNIFE

A comprehensive quality assurance program is essential for the delivery of accurate and reliable Gamma Knife radiosurgery treatments. The components of the quality assurance program for Gamma Knife radiosurgery have been changed over time in accordance with the introduction of new Gamma Knife machines and new treatment techniques. The licensing guide for the medical use of byproduct material from the United States Nuclear Regulatory Commission [46] requires that the following basic tests and measurements for Gamma Knife radiosurgery:

1. Measurement of the absolute dose rate. This is usually done on a monthly basis for the largest collimator of a Gamma Knife unit (18-mm collimator for the models B and C, 16-mm collimator for models Perfexion) using a calibrated ionization chamber and a 16-cm head phantom.
2. Measurement of the relative output factors (ROFs) for smaller collimators.
3. Agreement of measured beam profiles with Leksell GammaPlan calculations for all collimator sizes in the *XY*, *YZ*, and *XZ* planes.
4. Quality assurance for the radiation-focal point. For models B and C, a special tool is used to check the mechanical isocenter of each collimator. For the Perfexion model, coincidence of the mechanical isocenter of the patient-positioning system with the radiation-focal point can be checked using diode test tool and film tests.
5. For the cone beam CT capability on the ICON machine, regular check for the cone beam CT (CBCT) imaging accuracy and the CBCT image dose.

As a major tool for dosimetric measurement in radiation oncology physics, film dosimetry is an integral part of the routine clinic practice for Gamma Knife radiosurgery practice. The use of radiochromic films for the periodical quality assurance for Gamma Knife radiosurgery is summarized in the following sections.

10.4.1 OUTPUT FACTORS AND BEAM PROFILES

The output factor and beam profile measurements are required by the Nuclear Regulatory Commission (NRC) guideline to be done during initial commissioning of a Gamma Knife unit and during the following annual quality assurance process. Typical numbers of films that may be needed for output factor and beam profile measurements during an annual QA include the following:

1. One piece of film, irradiated in the *XY* plane for each collimator with a predefined time setting (i.e., 1 min etc.) for output measurement.
2. Two pieces of films for each collimator irradiated in the *XY* and the *XZ* planes with medium dose (i.e., 4 Gy for the EBT-3 films) for beam profile measurement.
3. A series of calibration films, irradiated with graded doses (up to a maximum dose significantly bigger than the dose on other films) using the largest collimator for the calibration curve.

In general, the output factors from the annual QA film measurements should agree with the values from the planning system and the commissioning report within a few percent (3%–5%). The FWHM should agree with those from the expected values within 1 mm. Table 10.1 gives the relative output factors for a 4C Gamma Knife unit from an annual QA report. Table 10.2 gives the FWHM for the 4, 8, 14, and 18 mm collimators.

Table 10.1 Relative output factor for Perfexion Gamma Knife® unit from 2015 annual calibration

	Dose for 1.4-min exposure (cGy)	Measured ROF	Expected ROF	Deviation (%)
4 mm	301.1	0.848	0.881	−3.8
8 mm	334.3	0.941	0.956	−1.6
14 mm	349.1	0.983	0.985	−0.2
18 mm	355.2	1.000	1.000	0

Table 10.2 Full width at half maximum for Perfexion Gamma Knife unit from 2015 annual calibration

	Measured (mm)			Expected (mm)			Deviation (mm)		
	X	Y	Z	X	Y	Z	X	Y	Z
4 mm	6.0	6.4	4.8	6.1	6.1	4.8	−0.1	0.3	0
8 mm	11.4	11.7	9.3	11.3	11.3	9.1	0.1	0.4	0.2
14 mm	19.3	19.8	15.8	19.4	19.4	15.7	−0.1	0.4	0.1
18 mm	24.4	24.9	20	24.4	24.4	20	0	0.5	0.1

10.4.2 FOCUS PRECISION TESTS

The coincidence of patient positioning system and the radiation-focal point for the Gamma Knife B and C models can be checked by a pinhole test tool (Figure 10.8a). Elekta introduced a scanning diode tool and a new film test tool (Figure 10.8b) for focus precision test on the Perfexion and ICON Gamma Knife units. The pin-prick test tool has a small needle in the center and is made with narrow geometric tolerances. When the tool is aligned in the Gamma Knife unit the tip of a sharp needle, located in the tool, exactly points toward the mechanical isocenter. Just prior to the exposure, a small piece of RCF, also located in the tool, is pierced by the tip of the needle. The film is then scanned, and the orthogonal beam profiles are analyzed. By measuring the asymmetry of the position of the hole in relation to the density distribution at approximately FWHM, the coincidence of the radiation focus point (radiation isocenter) and the patient positioning system (PPS) calibration center point (mechanical isocenter) is determined as shown in Figure 10.8c [47].

Two films are usually irradiated in perpendicular planes in the film holder in order to evaluate the coincidence in the X, Y and Z directions (ΔX, ΔY, ΔZ). A total radial deviation can then be calculated as root mean square of displacement, $\sqrt{\Delta X^2 + \Delta Y^2 + \Delta Z^2}$. As the collimator settings [4,8,16] on the Perfexion and the ICON units are independent, the check for the coincidence of the PPS and the radiation focus point (RFP) must be done for each collimator. Elekta's specification for the 4-mm collimator is that is that ΔX, ΔY, and ΔZ are all ≤ 0.3 mm and that the total radial deviation must be less than 0.4 mm (typically about 0.15 mm).

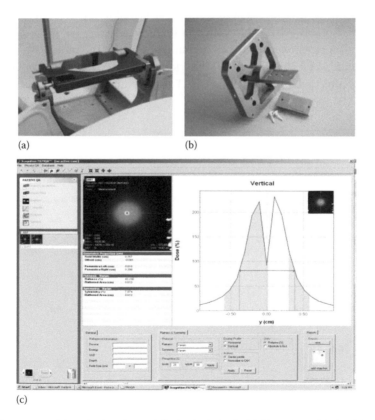

(a) (b)

(c)

Figure 10.8 Focus precision test using pinhole test tool: (a) the pin-prick test tool for Gamma Knife B, C models, (b) the new film test tool for Perfexion, and (c) focus precision test results from the FilmQA software.

10.4.3 END-TO-END TEST

10.4.3.1 EXTENDED SYSTEM BASED FRACTIONATED TREATMENT

A study was conducted on the verification of Gamma Knife Extend system for fractionated treatment planning using EBT2 film as shown in Figure 10.9 [48]. A humanoid head–shaped phantom with a dental impression positioning device was fabricated and scanned using a CT scanner. A treatment plan with two 8-mm collimator shots and three sectors blocking for each shot was created. A dose of 4 Gy at 100% was planned for two fractions. The RCFs were cut and placed inside the film insert of the phantom for treatment dose delivery. The films were irradiated twice within the insert simulating two fractionated treatments. Gamma index analysis between the film measurement and the treatment planning system showed a high pass rates >90% for a 1% and 1 mm criteria. The measured and the computed dose distributions on the sagittal and the coronal planes were in close agreement.

10.4.3.2 GEOMETRIC ACCURACY IN CBCT-BASED PATIENT POSITIONING

A special phantom derived from the pinhole focus precision test tool was developed to test the accuracy in patient positioning during fractionated Gamma Knife radiosurgery with the ICON. The phantom has multiple chambers for film pieces. A tiny hole was made on each film piece before the phantom was mounted on the treatment couch. A reference image set of the phantom were acquired by the CBCT, and a 4-mm shot was planned on each film at the exact same position at which the film was pierced. The phantom was undocked and then redocked to the patient positioning system. A treatment CBCT was acquired and coregistered to the first reference CBCT, and the planned treatment shot was delivered to each film (Figure 10.10).

The irradiated film pieces were analyzed the same way as for the films with a pin-hole test. An overall mean deviation of 0.25 mm was observed for the X direction, and a value of 0.27 mm was observed for the Z direction, indicating that CBCT-based patient positioning is a reliable process.

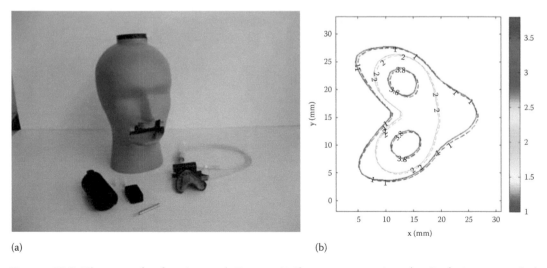

(a) (b)

Figure 10.9 Film tests for fractionated Gamma Knife treatment using the Perfexion extended system: (a) human head phantom and film insert and (b) isodose dose distribution in the XY plane. (From Natanasabapathi, G. and Bisht, R.K., *Med. Phys.*, 40, 122104-1–122104-9, 2013.)

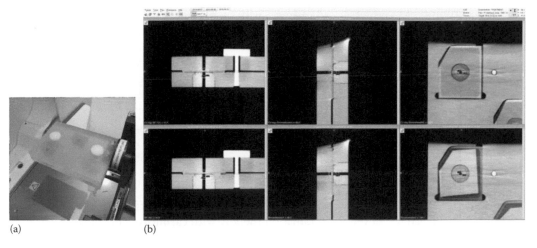

Figure 10.10 End-to-end test tool for the CBCT positioning accuracy with Gamma Knife ICON: (a) phantom with 3 film inserts and (b) CBCT images on the axial, coronal, and sagittal planes for the reference CBCT (upper row) and the treatment CBCT (lower row).

10.5 SUMMARY

The Elekta Gamma Knife (and other similar units) has been in use clinically since 1967. The highly collimated gamma rays from cobalt-60 create a challenging environment for definitive radiological dosimetry. Conventional ionization chambers are too large to measure all but the largest collimated beams from these units. Radiosensitive film, especially radiochromic film, is invaluable in assessing and verifying the manufacturer's suggested output (helmet) factors and off-axis profiles. The radiochromic film can be also used to perform the focus precision tests for different collimators and to evaluate the accuracy of the CBCT imaging–based patient positioning. Most Gamma Knife medical physicists find radiochromic film indispensable for commissioning and annual quality assurances tests.

REFERENCES

1. Leksell L. The stereotactic method and radiosurgery of the brain. *Acta Chir Scand* 1951;102:316–319.
2. Schaltenbrand G, Bailey P. *Introduction to Stereotaxis with an Atlas of the Human Brain.* Stuttgart, Germany: Georg Thieme Verlag; 1959.
3. Leksell L. *Stereotaxis and Radiosurgery: An Operative System.* Springfield, IL: Thomas; 1971.
4. Bunge HJ, Guevara JA, Chinela AB. Stereotactic brain radiosurgery with gamma unit III RBS 5000. *Proceedings of the 8th European Congress of Neurosurgery*, Barcelona, Spain; 1987.
5. Walton LC, Bomford CK, Ramsden D. The Sheffield stereotactic radiosurgery unit: Physical characteristics and principles of operation. *Br J Radiol* 1987;60:897–906.
6. Rand RE, Khonsary A, Brown WJ. Leksell stereotactic radiosurgery in the treatment of eye melanoma. *Neurol Res* 1987;9:142–146.
7. Wu A, Lindner, G, Maitz AH et al. Physics of Gamma Knife approach on convergent beams in stereotactic radiosurgery. *Int J Radiat Oncol Biol Phys* 1990;18:941–949.
8. Goetsch SJ. Risk analysis of Leksell Gamma Knife model C with automatic positioning system. *Int J Radiat Oncol Biol Phys* 2002;52:869–877.

9. Regis J, Tamura M, Guillot C. Radiosurgery with the world's first roboticized Leksell Gamma Knife Perfexion in clinical use: A 200-patient prospective, randomized, controlled comparison with the Gamma Knife 4C. *Neurosurgery* 2009;64:346–356.

10. Leksell L. A note on the treatment of acoustic tumors. *Acta Chir Scand* 1971;137(8):763–765.

11. Steiner L, Lindquist C, Adler JR et al. Clinical outcome of radiosurgery for cerebral arteriovenous malformations. *J Neurosurg* 1992;77:1–8.

12. Bergström M, Greitz T. Stereotaxic computed tomography. *Am J Roentgenol* 1976;127:167–170.

13. Lindquist C. Gamma knife surgery for recurrent solitary metastasis of a cerebral hypernephroma: Case report. *Neurosurgery* 1989;25:802–804.

14. Shehata MK, Young B, Reid B et al. Stereotactic radiosurgery of 468 brain metastases ≤2 cm: Implications for SRS dose and whole brain radiation therapy. *Int J Radiat Oncol Biol Phys* 2004;59:87–93.

15. Souhami L, Seiferheld W, Brachman D et al. Randomized comparison of stereotactic radiosurgery followed by conventional radiation therapy with carmustine to conventional radiotherapy with carmustine. *Int J Radiat Oncol Biol Phys* 2004;609:853–860.

16. Singh M, Aggarwal D, Kale SS. Gamma knife stereotactic radiosurgery in the management of large cerebral arteriovenous malformations. *Neurosurgery* 2016;61(Supp):203.

17. Hasegawa T, Kida Y, Kato T et al. Long term safety and efficacy for vestibular schwannomas: Evaluation of 440 patients more than 10 years after treatment with Gamma Knife surgery. *J Neurosurg* 2013;118:557–565.

18. Cohen-Inbar O, Lee CC, Sheehan JP. The contemporary role of stereotactic radiosurgery in the treatment of meningiomas. *Neurosurg Clin N Am* 2016;27:215–228.

19. Lee CC, Sheehan JP. Advances in Gamma Knife radiosurgery for pituitary tumors. *Curr Opin Endocrinol Diabetes Obes* 2016;23:331–338.

20. Regis J, Tuleasca C. Fifteen years of Gamma Knife surgery for trigeminal neuralgia in the Journal of Neurosurgery: History of a revolution in functional neurosurgery. *J Neurosurg* 2011;115: 2–7.

21. Goetsch SJ, Murphy BD, Schmidt R et al. Physics of rotating gamma systems for stereotactic radiosurgery. *Int J Radiat Oncol Biol Phys* 1999;43:689–696.

22. Kubo, HD, Araki F. Dosimetry and mechanical accuracy of the first rotating gamma system installed in North America. *Med Phys* 2002;29:2497–2505.

23. Novotny J, Bhatnagar JP, Niranjan A et al. Dosimetric comparison of the Leksell Gamma Knife Perfexion and 4C. *J Neurosurg* 2008;108:8–14.

24. Goetsch SJ. 4 mm Gamma Knife Helmet Factor (Letter). *Int J Radiat Oncol Biol Phys* 2002;54:300.

25. Mack A, Scheib SG, Major J et al. Precision dosimetry for narrow photon beams used in radiosurgery—Determination of Gamma Knife output factors. *Med Phys* 2002;29:2080–2089.

26. Al-Dweri FMO, Lallena AM, Vilches M. A simplified model of the source channel of the Leksell Gamma Knife tested with PENELOPE. *Phys Med Biol* 2004;49:2687–2703.

27. Cheung JYC, Yu KN. Rotating and static sources for gamma knife radiosurgery systems: Monte Carlo studies. *Med Phys* 2006;33:2500–2505.

28. Das IJ, Cheng CW, Watts RJ et al. Accelerator beam data commissioning equipment and procedures: Report of the TG-106 of the Therapy Physics Committee of the AAPM. *Med Phys* 2008;35:4186–4215.

29. Drzymala RE, Alvarez PE, Bednarz G et al. A round-robin gamma stereotactic radiosurgery dosimetry inter-institution comparison of calibration protocols. *Med Phys* 2015;41:6745–6756.

30. Maitz AH, Wu A, Lunsford LD et al. Quality assurance for Gamma Knife stereotactic radiosurgery. *Int J Radiat Oncol Biol Phys* 1995;32:1465–1471.

31. Sonders M, Sayeg J, Coffey C et al. Beam profile analysis using GafChromic films. *Stereotact Funct Neurosurg* 1993;61(suppl 1):124–129. Proceedings of the 1992 Meeting of the Leksell Gamma Knife Society, Buenos Aires, Argentina.

32. McLaughlin WL, Soares CG, Sayeg JA et al. The use of a radiochromic detector for the determination of stereotactic radiosurgery dose characteristics. *Med Phys* 1994;21:379–388.

33. Butson MJ, Mathur JN, Metcalfe PE. Radiochromic film as a radiotherapy surface-dose detector. *Phys Med Biol* 1996;41:1073–1078.

34. Meigooni AS, Sanders MF, Ibbott GS, Szeglin SR. Dosimetric characteristics of an improved radiochromic film. *Med Phys* 1996;23:1883–1888.

35. Cheung YC, Yu KN, Ho RTK, Yu CP. Monte Carlo calculations and GafChromic film measurements for plugged helmets of Leksell Gamma Knife Unit. *Med Phys* 1999;26:1252–1256.

36. Tsai J, Rivard M, Engler MJ et al. Determination of the 4 mm Gamma Knife helmet relative output factor using a variety of detectors. *Med Phys* 2003;30:986–992.

37. Lynch B, Ranade M, Li J et al. Characteristics of a new very high sensitivity radiochromic film. *Med Phys* 2004;31:1873.

38. Sankar A, Ayyangar KM, Nehru RM et al. Comparison of Kodak EDR2 and GafChromic EBT film for intensity-modulated radiation therapy dose distribution verification. *Med Dosim* 2006;31:273–282.

39. Dempsey JF, Low DA, Kirov AS, Jeffrey FW. Quantitative optical densitometry with scanning-laser film digitizers. *Med Phys* 1999;26:1721–1731.

40. Klassen N, Zwan L, Cygler J. GafChromic MD-55: Investigated as a precision dosimeter. *Med Phys* 1997;24:1924–1934.

41. Novotny J, Bhatnagar JP, Quader MA et al. Measurement of relative output factors for the 8 and 4 mm collimators of Leksell Gamma Knife Perfexion by film dosimetry. *Med Phys* 2009;36:1768–1774.

42. Das IJ, Ding X, Ahnesjo A. Small fields: Non-equilibrium radiation dosimetry. *Med Phys* 2008;35:206–215.

43. International Atomic Energy Agency, Technical Report Series No. 483. Dosimetry of small static fields used in external beam radiotherapy: An IAEA-AAPM International Code of Practice for reference and relative dose determination. IAEA TRS 483, 2017.

44. Perks J, Gao M, Skubic S, Goetsch SJ. Glass rod detectors for small field, stereotactic radiosurgery dosimetry audit. *Med Phys* 2005;32:726–732.

45. Ma L, Kjall P, Novotny J et al. A simple and effective method for validation and measurement of collimator output factors for Leksell Gamma Knife Perfexion. *Phys Med Biol* 2009;54:3897–3907.

46. U.S. Nuclear Regulatory Commission. *Medical Use of Byproduct Material.* 10 CFR Part 35: 67 FR 20370. Washington, DC: U.S. Nuclear Regulatory Commission; 2002.

47. Novotny JJ, Bhatnagar JP, Xu, Y, Huq MS. Long term stability of the Leksell Gamma Knife Perfexion patient positioning system. *Med Phys* 2014;41:031711.

48. Natanasabapathi G, Bisht RK. Verification of Gamma Knife extend system based fractionated treatment planning using EBT2 film. *Med Phys* 2013;40:122104-1–122104-9.

Use of radiochromic films in commissioning and quality assurance of CyberKnife®

EVAGGELOS PANTELIS AND AZAM NIROOMAND-RAD

11.1 INTRODUCTION TO CYBERKNIFE® SYSTEM

The CyberKnife® (CK) Robotic Radiosurgery System as shown in Figure 11.1 is a stereotactic radio-surgery (SRS) device manufactured by Accuray, Inc. (Sunnyvale, CA, and USA) [1–3] that combines advanced robotic, image guidance, and linear accelerator (Linac) technologies to deliver high doses of ionizing radiation to lesions anywhere in the body [4–8]. It has a 6-MV Linac mounted on a robotic arm capable of pointing the radiation beam at the target from an increased number of noncoplanar directions. The robotic arm is spatially calibrated to an image-guidance system consisting of 2-kV X-ray generators and two digital detectors fixed on the ceiling and the floor of the treatment room, respectively. During treatment, the image guidance system takes images of the target region and compares them with corresponding digital reconstructed radiographs (DRRs) [9] obtained from the planning computed tomography (CT) scan. The calculated deviations of the position and orientation of the target are corrected automatically by adjusting the position and direction of each treatment beam [3].

It should be noted that the position of the Linac on a robotic arm distinguishes CK from other radiosurgery/radiotherapy modalities and provides flexibility to treat in addition to intracranial lesions, other lesions of the body that are challenging to treat with high doses of radiation, for example, lesions inside the spinal canal, or lesions of the lung or liver that move with respiration. Moreover, the use of an increased number of noncoplanar and nonisocentric small beams that are precisely registered to the treated lesions minimize the doses to healthy surrounding tissues. Therefore, the CK system is reasonably characterized as one of the most advanced and effective SRS equipment.

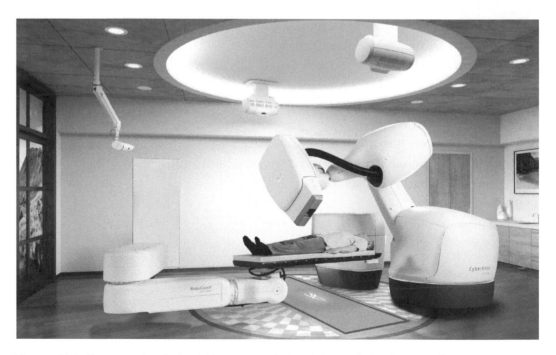

Figure 11.1 Photograph of the M6 version of the CyberKnife® robotic radiosurgery system. (Courtesy of Accuray Incorporated—©2016 Accuray Incorporated, Sunnyvale, CA.)

11.1.1 TECHNICAL DESCRIPTION OF THE CYBERKNIFE SYSTEM

The current CK system, as presented in Figure 11.1, consists of the following:

- A compact lightweight Linac that uses an X-band cavity magnetron and a standing wave, side-coupled accelerating waveguide, to produce a 6-MV X-ray treatment beam with a dose rate up to 1000 cGy/min at 800-mm source-to-axis distance (SAD). Bending magnet and flattening filter are not used. Due to the absence of flattening filter, the energy spectrum of the emitted X-rays contains an increased component of lower energy photons [10]. Secondary collimation of the treatment beam is provided using 12 fixed circular collimators with diameters ranging from 5 to 60 mm defined at 800-mm SAD. These collimators can be fitted manually or automatically using the Xchange® Robotic Collimator Changer. Alternatively, the Iris® Variable Aperture Collimator allows the same set of 12 circular field sizes to be achieved with a single-variable aperture and therefore provides the flexibility to apply any field size at any beam position without the need to swap collimators during treatment [3,11]. In addition to the circular radiation fields, arbitrary field shapes can be produced using the InCise® micro-multileaf collimator (MLC) [12]. This micro-MLC consists of 26 leaf pairs, with a thickness of 90 mm and a width of 3.85 at 800 mm SAD that can create a maximum field size of 115 mm in the direction of leaf motion and 100 mm in vertical direction at 800 mm SAD.
- A robotic manipulator (Kuka Roboter GmbH, Augsburg, Germany) with six joints, allowing for a wide range of dynamic motion with 6 degrees of freedom (DOF). The manipulator facilitates the position of the X-ray target of the Linac at any point within a spherical shell of 600 to 1,000 mm from the treated lesion with a precision better than 0.12 mm [3]. The robotic manipulator allows each treatment beam to be directed at a unique point in space (i.e., there is no isocenter) and also removes any coplanar constraint on the beam geometry. This flexibility in geometry means that the vault in which the system is installed requires a larger primary barrier than is typical for a gantry-mounted Linac as the beam directions are much less constrained. A version with reduced irradiation points can also be provided for CK system to fit in existing vaults.
- A robotic treatment couch (RoboCouch®, Accuray) with 6-DOF and six joints is able to remotely position the patient to the desired location with fine maneuvers. A 5-DOF standard treatment couch capable of performing all six motions except jaw is also commercially available.
- A kV-imaging target-locating subsystem (TLS) is used for patient alignment and target tracking during treatment delivery. The TLS consists of two pairs of diagnostic X-ray tubes and amorphous silicon digital flat-panel detectors. The X-ray tubes are installed on the ceiling of the treatment room at 45° angles with beams orthogonal to each other. At the point at which the central axes of these beams intersect, the X-ray field size is approximately 150×150 mm. The flat panel X-ray detectors, which are mounted horizontal in the floor, consist of cesium–iodide scintillator deposited directly on amorphous silicon photodiodes and generate high-resolution digital images ($1,024 \times 1,024$ pixels of 0.4 mm in size with 16-bit resolution).
- A stereo-camera system mounted on a boom arm is attached to the ceiling. The camera is used to continuously capture the position of optical markers, attached to the patient during treatment. There are three charged coupled device (CCD) cameras within this system. In combination with the X-ray imaging system, this enables the robotic manipulator to track tumors that move with respiration using the Synchrony® Respiratory Tracking System.

11.1.2 OVERVIEW OF TREATMENT PROCEDURE

11.1.2.1 TREATMENT PLANNING

Treatment planning begins with obtaining one or more three-dimensional (3D) images that allow the target volume and nearby organs at risk to be visualized. Once acquired, the 3D images are transferred to the MultiPlan™ treatment-planning system (TPS) via a dedicated database server. The minimum requirement is for a volumetric CT study of the patient in treatment position. This CT study is used to generate a 3D patient model, which is correlated with the image-guided subsystem and robotic arm 3D geometries during the *align* step and the generation of DRRs of the treated region. This correlation enables the position of treatment beams to the patient model. Each treatment beam is described by a vector linking the source point that coincides with the focal spot of the Linac and a direction point that lays inside the target volume. Source points can be directed to the target points from the same robot position that is called *node* in CK terminology. The complete set of nodes comprises a treatment path set. Different path sets are constructed to provide a range of noncoplanar beam directions for intra- and extracranial treatment sites. The appropriate path set for each patient is selected manually at the start of the treatment-planning process. The number of nodes in the different path sets ranges around 20 up to more than 100 [3].

For circular collimator treatments, both isocentric and nonisocentric treatment delivery techniques are available with the system. In the isocentric mode, the user is allowed to position one or more pseudoisocenters within the patient model resulting in one beam from each node to each pseudoisocenter. The nonisocentric mode takes advantage of the ability of the robotic manipulator to direct each beam at a unique point within the patient, without any need to reposition the patient between beams. For the circular collimators, this works by generating a large number of direction points (2,000–6,000) semirandomly on the surface of the target volume and distributing these uniformly among the nodes to form the candidate beam set. The user selects one or more treatment beam diameters, between 5 and 60 mm, to be considered during treatment optimization. In general, the fields with dimensions larger than 20 mm are used for extracranial treatments, whereas the fields with dimensions less than 20 mm are used for intracranial applications. The isocentric mode produces dose distributions comprising approximately spherical dose clouds around each pseudoisocenter similar to those in other radiosurgery systems using circular collimators. The nonisocentric mode represents a very different treatment geometry that is more similar to those achieved using multiple pencil beams. From a single node, a modulated fluence pattern can be delivered using multiple beams directed at unique points within the target volume, each of which has an independent field size and beam weight. The optimal set of relative weighting factors for the candidate beam set (i.e., the dose delivered per beam) is obtained by inverse-planning methods [3]. For MLC treatment planning, one or more treatment beams are defined at each node, depending on number and size of target volume(s). Moreover, multiple possible MLC apertures are generated for each beam depending on options selected by the user and the target of interest and organ at risk volumes. After optimization, the 3D patient model includes the position and orientation of each treatment beam in a stereotactic coordinate system defined by the target anatomy itself or an internal surrogate fixed relative to it (i.e., in target space), together with the field size or shape and monitor unit (MU) setting of each beam. This information is stored as part of a treatment plan and is transferred to the treatment-delivery system via the database server. Compared with circular fields, treatment plans with the InCise MLC collimator are characterized by reduced total MUs and treatment time [13,14].

11.1.2.2 TREATMENT DELIVERY TECHNIQUES

Beam alignment at the time of treatment is based on automatic registration of DRRs generated from the 3D patient model, with live images acquired using the kV X-ray imaging system in the treatment room. At the start of every treatment, the TLS is used to align the patient at treatment position using the adjustable treatment table. The purpose of this initial alignment is to reduce the corrections that will be required from the robotic manipulator below maximum limits, which are ±10 or ±25 mm in each direction and ±1 to ±5 degrees about each axis depending on the tracking mode, treatment path set, and couch design. After the patient is aligned within these limits, TLS determines the additional translational and rotational corrections needed to precisely align each treatment beam. These corrections are relayed to the robotic manipulator and used to automatically compensate for small-target movements by repositioning the Linac. During treatment, the robot moves in sequence through the nodes selected during treatment planning. An optimized path traversal algorithm allows the manipulator to travel only between nodes at which one or more treatment beams are to be delivered, or through the minimum number of additional zero-dose nodes required to prevent the robot trajectory intersecting fixed room obstacles or a *safety zone* surrounding the couch and patient. At each node, the manipulator is used to reorient the Linac such that each beam originating at the node can be delivered. Image acquisition, target localization, and alignment corrections are repeated continually during treatment delivery, typically between 60 and 120 s; the imaging interval can be adjusted during treatment based on the stability of the target position. The robotic manipulator compensates for small translations and rotations based on the corrections obtained from the most recently acquired image pair; large translations and rotations automatically pause the treatment and prompt the operator to reposition the patient before proceeding. Dose-placement accuracy is assured by imaging and correcting beam aim frequently throughout each treatment fraction. For targets that move with respiration, an additional tracking system enables beams to move in real time to follow the target while the patient breathes freely [6].

11.2 USE OF RADIOCHROMIC FILMS FOR CYBERKNIFE COMMISSIONING AND QUALITY ASSURANCE

Use of image guidance along with robotic manipulator allows for the delivery of conformal 3D dose distributions created by an increased number of small pencil beams. These dose distributions are characterized by steep spatial dose gradients, which, combined with the increased complexity of the system, render the quality assurance (QA) of the CyberKnife a challenging task to perform.

Radiochromic films (RCFs) are able to provide a visual information of the dose distribution on a plane and therefore they are used for quick verification of various QA procedures needed in CK. Moreover, having the favorable characteristics of fine spatial resolution, near tissue equivalence and self-developing, the RCFs have become the detector of choice for measuring the geometric accuracy component owing to each subsystem (e.g., the precision and accuracy of manipulator movements and target locating system), comprising the total targeting error of the CK [7,8,15–19]. In fact, as will be discussed in the following sections of this chapter, Accuray, Inc. provides RCF kits with precut films for these tests with special phantoms and cassettes such as the Ball-Cube-II™ product [19]. In addition to the geometrical related parameters, RCFs are used for characterizing the dosimetric properties of the CK small fields, for the validation of the implemented dose calculation algorithms and for patient-specific dosimetry [20–28]. In the following sections, the use of RCFs for the commissioning and QA of the CK system are presented in detail.

11.3 DOSIMETRY OF SMALL FIELDS FOR CYBERKNIFE SYSTEM

Treatment-planning dose calculation for the CK system are based on output factors (OFs), off-axis ratios (OARs), and depth dose profiles of the used beams [7,8,15]. Measurement of these dosimetric quantities is performed during commissioning or periodic QA of the system using mainly diode detectors or microchambers [7,8,15]. It should be noted that, although the methodology and main characteristics of the appropriate detectors for determining the above-mentioned dosimetric quantities for large photon fields are well described [29], corresponding measurements for the small fields (of less than 15-mm diameter) are characterized by increased uncertainty [7,21–26,30–34]. The increased uncertainty in small-field dosimetry measurements is attributed to the following: (a) loss of lateral charge particle equilibrium on the beam axis, (b) partial occlusion of the primary photon source by the collimating devices on the beam axis, and (c) perturbation of the electron fluence caused by the presence of nonwater equivalent dosimeters, such as the diode and microchamber detectors [35–41]. The detector setup positional uncertainties and volume averaging effects are also noticeable in small-field dosimetry measurements [34,41–44]. For small-field dosimetry, the ideal dosimeter should have small sensitive volume, water equivalence, energy independence, high sensitivity and stability, linear or known dose response, and increased accuracy and precision [22–25,34,41,44]. As there is no dosimeter fulfilling all of the above-mentioned characteristics, it is advisable and considered good practice to repeat measurements with at least two different detectors and apply corrections to the readings according to the methodology suggested by Technical Report Series No. 483 (TRS-483) [34] to calculate their average readings [44].

In Figure 11.2, values of OFs for the Iris 5-mm circular field measured using a large series of detectors including EBT RCFs are presented. Large differences reaching up to 36% between measured OFs with different dosimeters can be observed in accordance with corresponding findings

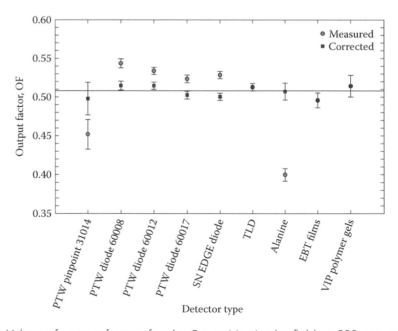

Figure 11.2 Values of output factors for the 5-mm Iris circular field at 800-mm source-to-axis distance measured using a variety of dosimeters. The weighted average of the EBT films, thermoluminescent dosimeter, alanine, and N-Vinylpyrrolidone polymer gel output factor results is presented with black solid line.

in the literature [21,22,24,25,30–34]. As RCFs are tissue equivalent, they do not have the partial volume effects that ion chambers could have. Measurement setup uncertainties are also minimized as the symmetry of the dose delivered to the film can be used to calculate the central-beam axis. As a result, RCFs have been used either alone or along with other water equivalent dosimeters, as reference detectors to experimentally determine the $k_{Q_{clin}, Q_{msr}}^{f_{clin}, f_{msr}}$ correction factors described by Alfonso et al. [33] and TRS-483 [34] for diode detectors and microchambers [22–25,45,46]. An example of this methodology is given in Figure 11.2 in which the weighted average of the measured OFs using EBT films, thermoluminescent dosimeter, Alanine (after correcting the measured value for volume average effects), and N-Vinylpyrrolidone based polymer gels was considered as reference to calculate the $k_{Q_{clin}, Q_{msr}}^{f_{clin}, f_{msr}}$ correction factors of the used diode and PinPoint 31014 (PTW, Freiburg, Germany) ion chamber. Applying these corrections to the diode and ion chamber measured OFs results in an improved agreement between the presented OF values.

In addition to OF measurements, RCFs have also been used to determine tissue maximum ratios (TMR), percent depth doses (PDD), and OAR values for the CK small fields [20,47]. There are also monthly consistency checks for dose output, energy constancy, beam shape, and beam symmetry that mostly use the largest, 60-mm diameter field [7]. Note that the beam shape is peaked as there is no flattening filter and hence no flatness measurement [7]. Instead, constancy of the beam profile is checked relative to the TPS at various radii from the beam center for a 1% agreement over time. For off-center ratio, tissue phantom ratio, and output checks, diodes are mentioned because of their ease of use, but instead of ion chambers, RCFs are preferred to benchmark diodes, especially for fields <20 mm in diameter [7]. Actually, some investigators suggest *not* to use diodes at all, especially for scatter factor measurements [48]. Moreover, because of the small fields, accurate heterogeneity testing is important [27,49] and therefore anthropomorphic phantoms are considered for the TG-135 protocol [7] for the automated QA (AQA) test and the *end-to-end* (E2E) tests [50].

Another significant dosimetry issue for small fields of CK system is that a 100 × 100 mm homogeneous reference field cannot be established as the largest field size is 60 mm in diameter at 800 mm and there is no flattening filter [22,33,51]. It has been shown that this field can satisfy electronic equilibrium on the central axis under this condition [33]. The small fields inherent to the CK raise the question of whether ratios in detector readings equate to ratios of absorbed dose. In addition, it has been shown that small collimator sizes can affect the photon spectrum [52]; however, the averaged stopping power ratios are insensitive to spectral changes. Some detectors have concerns with high doses characteristic of SRS. The RCFs have good dynamic range for use of high SRS doses, but the green channel for EBT2 has been shown to be more responsive above 8–10 Gy than the red channel [53,54]. Detailed information on the small-field dosimetry issues as well as on the use of RCFs for performing such type of measurements can be found in Chapter 13. For more details on the small-field advantages and issues in general with RCF dosimetry, see Chapter 13 of this book.

11.4 USE OF RADIOCHROMIC FILM FOR THE QUALITY ASSURANCE OF CYBERKNIFE LINAC

11.4.1 ALIGNMENT OF LASER AND RADIATION-BEAM AXIS

The CK system does not have a light field to represent the radiation beam. Instead, a laser beam is used to denote the Linac beam axis and radiation pointing direction [7,8]. The laser beam is used to calibrate the robotic operating system to determine the radiation beam pathway from each node of the available treatment paths to a precise point in room space called isocenter.

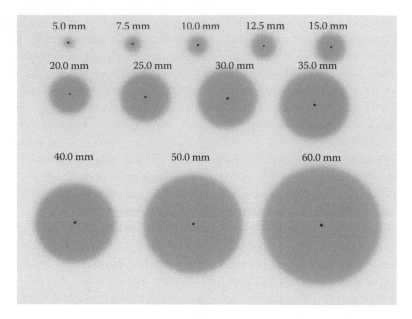

Figure 11.3 A 16-bit image of EBT film irradiated using fixed collimators at 800-mm source-to-axis distance. The laser beam is marked using a radiopaque marker.

The isocenter is physically defined by a small diode crystal fixed at the end of a metal rod, called isopost, that is mounted at the center of a metal frame which is positioned on the treatment room floor to hold the flat panels [7]. Path calibration is performed by instructing the robotic operating system to scan with the laser beam and find the position and direction of the Linac that corresponds to the maximum signal of the diode.

As the laser beam is used to calibrate the Linac, the alignment of the radiation beam and laser beam should be verified. For this test, the RCFs are positioned horizontal on few water-equivalent plastic slabs, and the Linac is aligned vertical to the RCF plane. Axes coincidence is checked by irradiating the RCFs with at least two different field sizes of each collimation system and at three different SADs, for example, 650 mm, 800 mm, and 1000 mm. A buildup slab is not required but sufficient MUs (e.g., 200 MUs) should be delivered to each film to obtain a measurable net optical density (OD). Prior to irradiation, the laser denoting the radiation beam axis is marked on the film using a radiopaque marker as shown in Figure 11.3.

The coincidence of the laser and radiation-beam axis can be tested qualitatively by visual inspection of the film. A quantitative evaluation of the laser and radiation beam coincidence requires use of a flatbed optical scanner in transmission mode with a resolution higher than 150 dpi. The film image can be analyzed using standard image-analysis methods to calculate the coordinates of the radiopaque dot and radiation-beam centroids. The Euclidean difference of the radiopaque dot and radiation beam centroids is calculated and used as a measure of the laser and radiation-beam axis coincidence that should be less than 1 mm [7,8].

11.4.2 CHARACTERISTICS OF SECONDARY COLLIMATION SYSTEMS

RCFs are routinely used to measure the geometrical characteristics of the Iris and the InCise secondary-collimation systems. The EBT2 and EBT3 films are used to measure the precision and accuracy of the dimensions of each Iris field [55]. This test is performed using a dedicated film-based

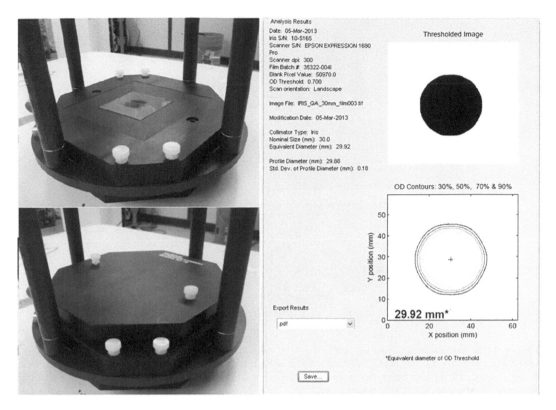

Figure 11.4 The Iris QA tool mounted on birdcage assembly with film (upper left) and buildup plate on film (lower left). A print screen of the Iris QA software used to analyze the exposed films and calculate the field size is shown on the right. (Adapted from Accuray Physics Essentials Guide, *CyberKnife System Manual v.10*, Accuray, Sunnyvale, CA. With permission.)

tool (Iris QA tool), which is provided by the vendor and assures measurement setup reproducibility. As shown in Figure 11.4, the Iris QA tool consists of a base plate of 50-mm thickness with a 15-mm thick buildup plate that fits into the birdcage assembly. The birdcage assembly is mounted on the Linac head assuring that the film is positioned at 800-mm SAD during measurement. During commissioning of the Iris fields, a series of measurements are performed by irradiating RCFs with 600 MUs. The films are scanned in a flatbed optical-transparency scanner using a resolution of 300 dpi. The film images are analyzed with the aid of the Iris QA software as shown in Figure 11.4, and the diameter of each field is measured. These results serve as a baseline dataset during periodic QA, or after an upgrade, recalibration, or major service of the Iris collimator.

RCFs are used for two tests that are required to check leaf calibration and the performance of the InCise micro-MLC. The TG-50 pattern also referred to as a *Picket Fence* pattern is used for a quick qualitative daily QA check of leaf-positioning accuracy due to its increased sensitivity [56]. The Bayouth or *Garden Fence* test is used for a quantitative measurement of leaf-positioning accuracy [57]. Both methods use multiple fields so the leaf positions can be measured at multiple positions across the field. The Garden Fence test is used for leaf calibration by a service engineer and also for measuring the leaf-positioning accuracy during acceptance and periodic QA.

A film-based tool (MLC QA tool) that mounts directly to the accessory mounting points is provided by the vendor for performing the Garden Fence and Picket Fence tests as shown in Figure 11.5. This tool holds a RCF and two radio-opaque markers that allow for very precise film

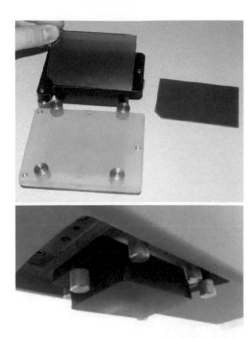

Figure 11.5 Experimental film setup for the MLC picket fence and garden fence quality assurance test patterns (upper). The MLC QA tool loaded with film mounted just below the MLC on the Linac head (lower). (Adapted from Accuray Physics Essentials Guide, *CyberKnife System Manual v.10*, Accuray, Sunnyvale, CA. With permission.)

marking to determine the beam center on the film and rotation of the film relative to the X/Y MLC axes. A film template is provided to cut the test film to match the insert area in the MLC QA tool as close as possible. The sides ($X1$, $X2$) of the two banks of the MLC are also marked near the short edges of the film. The test patterns can be run using automated software tools of the treatment-delivery system.

The irradiated film is scanned using an optical-transparency scanner with a resolution higher than 150 dpi. Some software is used to automatically determine the center and rotate the film based on the radiopaque marks. The uniformity of the measured dose (or OD) on the film is used to evaluate the alignment of leaf positions and perform Picket Fence and Garden Fence tests.

In Figure 11.6, Picket Fence test patterns created using EBT2 films are presented. The uniform exposure of the film shown on the left indicates a correctly calibrated MLC. Areas of underexposure

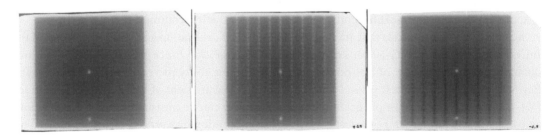

Figure 11.6 Picket Fence test patterns created using EBT2 film with a correctly calibrated InCise MLC (left), with one bank underextending shown by the underexposed areas (middle), and one bank overextending shown by the overexposed areas (right). (Adapted from Accuray Physics Essentials Guide, *CyberKnife System Manual v.10*, Accuray, Sunnyvale, CA. With permission.)

on the central film image of Figure 11.6 show that one bank is underextending (by 0.5 mm). Overextending of one bank would be manifested as areas of overexposure such as those observed on the right film image of Figure 11.6. Tilt errors can also be detected by Picket Fence tests by the variation in the thickness of under- and overexposed film areas as seen in Figure 11.6.

Transmission and leakage of the MLC leafs are also measured using RCFs. The film is irradiated at 800-mm SAD at a range of robot positions with the MLC leafs closed. The MLC leakage and transmission should be less than 0.5% with a mean value of less than 0.3%.

11.4.3 RADIATION-BEAM CHARACTERISTICS

During commissioning of the CK system, the dosimetric parameters of the used radiation fields are measured [7,8,15]. These parameters include beam-quality specification, beam profiles, TMR, OFs, output calibration, reproducibility, linearity, constancy at various Linac orientation, collimator transmission, leakage radiation, and end effect. Furthermore, small-field dosimetry issues such as lack of lateral electronic equilibrium, perturbation of the electron fluence by the detector materials, and finite dimensions of the detector, as compared with the radiation field, must be taken into account for detector choice and reliable measurements especially for collimator sizes less than 15 mm [7,21–26,30–44].

The favorable characteristics of water equivalence, minimal field perturbation, and high spatial resolution of RCFs make them suitable for measuring the dosimetric parameters of the small radiation fields. In practice however, RCFs are used to support beam-data measurements obtained with other dosimeters such as diode detectors [20,22–25,45,46]. This is mainly due to the passive nature of measurement, the relatively increased dosimetric uncertainty and handling difficulties of RCFs as discussed in Chapter 3. Application of RCFs in CK usually includes their irradiation within a water-equivalent slab phantom or the stereotactic dose verification phantom (Standard Imaging Inc., WI, USA). The films are scanned using a flatbed transparency scanner and conversion of the measured signal to dose using appropriate protocols [20,23–25,49]. Results from multiple films are averaged to reduce the uncertainty of the obtained dosimetry results.

11.5 MANIPULATOR AND IMAGE-GUIDANCE SYSTEM QUALITY ASSURANCE—THE AUTOMATED QUALITY ASSURANCE TEST

The AQA test is a daily test to check target reproducibility, the mastering of the treatment manipulator, and the stability of the image-guidance system. The AQA test is analogous to the Winston–Lutz gantry Linac stereotactic QA technique of placing a radiopaque ball at the treatment isocenter and observing the concentricity of the beam and shadow of the ball [58]. For a nonisocentric image-guided system such as the CK, the technique has been modified, that is, the target ball is not mechanically placed precisely at the room isocenter, but instead it is inserted into a specifically designed phantom as shown in Figure 11.7. The fiducials that are inserted in the phantom allows the target-locating system to direct the Linac radiation beam at the ball. As the ball is spherically symmetric and is located at the center of the phantom, this technique can determine the translational targeting error, which is a combination of tracking and delivery error.

For this test, two films are inserted in the coronal and sagittal planes of the AQA phantom and irradiated by a two beam plan (one anterior and one lateral) as shown in Figure 11.7. The shadow of the radiopaque ball is exposed on the films. Films are scanned in a flatbed transparency scanner

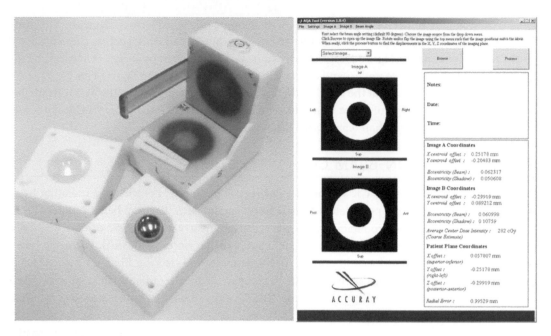

Figure 11.7 An AQA phantom (left) and vendor provided software to analyze concentricity of radiation beam and shadow caused by metal ball (right). (Adapted from Accuray Physics Essentials Guide, *CyberKnife System Manual v.10*, Accuray, Sunnyvale, CA. With permission.)

and analyzed for concentricity of the beam and shadow of the ball using a dedicated software tool provided by the vendor. Targeting errors should not deviate ≥1 mm from the baseline value set at time of calibration [7,8].

11.6 MEASUREMENT OF THE CYBERKNIFE TOTAL SYSTEM ERROR

11.6.1 END-TO-END TEST

RCFs are used to assess the total system error (TSE) of the CK system [3,8,19]. For this test, a hidden test object is used (called *ball cube*) containing two RCFs positioned orthogonally within a spherical target structure that is placed within an anthropomorphic phantom as shown in Figure 11.8 [7,8]. The test, also called E2E, integrates all components of the therapeutic procedure including CT scan, treatment planning (CT data import, contouring, and dose calculation), software generating DRRs, and treatment delivery using the robot, the target locating method, the linear accelerator, and the patient safety components. The E2E test is performed during commissioning to establish the accuracy of the system for each treatment-delivery modality as well as during periodic QA.

After scanning the phantom, the images are imported in the TPS, and the spherical structure situated at the center of the ball cube is delineated. The tested target-locating method is chosen (i.e., the 6D skull, Fiducial, or Xsight™ spine) and a single beam isocentric conformal treatment plan is developed. Treatment plan is created so that the 70% isodose line conforms the spherical contour (Figure 11.8). The EBT-2, -3 RCFs are used for this test [19]. Nominally, a dose of 420 cGy at 70% isodose line is prescribed, and the plan is saved as deliverable to the system database. After plan delivery, the films are scanned in a transparency scanner using 48 bit and 300 dpi resolution

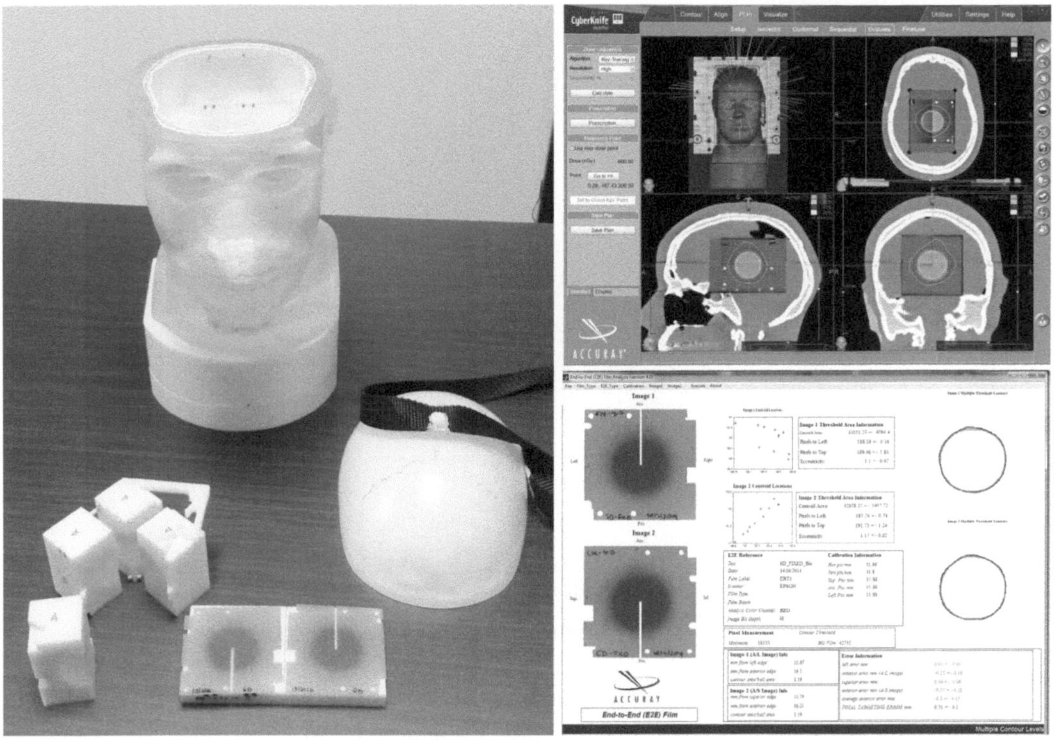

Figure 11.8 An anthropomorphic phantom (left) for placement of ball cube and film and vendor provided software (right) to analyze irradiated film for end-to-end test and assessment of total system error.

without applying any color corrections. Analysis of the exposed films is performed using a film analysis software provided by the vendor and includes alignment of the films, calculation of the ODs, and corresponding relative dosimetry values as shown in Figure 11.8. The Euclidean difference between the centroid of the delivered dose distribution on the exposed films and the geometrical center of the ball inside the cube is used to measure the TSE of the system.

Figure 11.9 shows TSE results of a G4 CK system, using RCFs for each target locating method with Whisker boxplots. It should be noted that these values correspond to TSE measurements that were performed in the clinic for a period of time of 10 years and include different ball cube versions, Linac designs, and collimation systems (i.e., Fixed and Iris) [8,19]. Figure 11.9 displays that the TSE of the CK is less than 1 mm for all static target-locating methods. The average and standard deviations of the presented TSE results are equal to (0.41 ± 0.19) mm, (0.38 ± 0.16) mm, and (0.56 ± 0.18) mm, for the 6D skull, Fiducial, and Xsight™ spine target locating methods, respectively.

11.6.2 MOTION-TRACKING QUALITY ASSURANCE

For targets that move with respiration the Synchrony-treatment delivery technique is used [3]. The TSE for this delivery technique is measured using a methodology similar to the one used for static target-locating methods as described in Section 11.6.1. The Synchrony delivery technique can be combined with the fiducial or lung-optimized target-locating methods depending on the clinical application. Different phantoms are used to measure the TSE of each target-locating method.

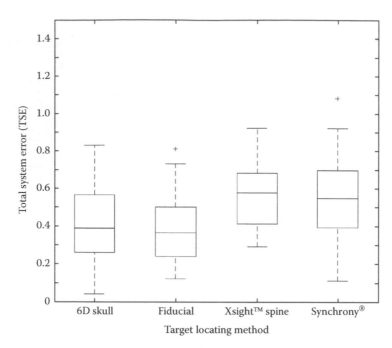

Figure 11.9 Whisker plot of Total System Error (TSE) for each target locating method at Iatropolis Clinic in Athens, Greece. Outliers are depicted with red cross and are defined as TSE results that deviate more than 1.5 times the inter-quartile range from the box top or bottom.

For measuring the Synchrony-TSE, a ball cube loaded with two orthogonal situated RCFs is inserted inside a dome and positioned on a motion table provided by the vendor as shown in Figure 11.10. With a so-called *motion table* option, target can be moved linearly in the superior–inferior (SI) direction with variable phase shifts between the moving target and the moving chest in the anterior–posterior direction. Two LED markers are used to track the chest motion. The Synchrony delivery technique is enabled to synchronize the movement of each beam with the target movement. The exposed films are scanned with a flatbed transparency scanner and analyzed with the E2E-analysis software for targeting accuracy evaluation as shown in Figure 11.8.

Figure 11.10 Synchrony quality assurance tool setup with film loaded ball cube and dome (left) and loaded cube inside dome on treatment couch (right). (Adapted from Accuray Physics Essentials Guide, *CyberKnife System Manual v.10*, Accuray, Sunnyvale, CA. With permission.)

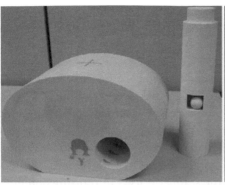

Figure 11.11 Thorax phantom (left) with orthogonal films (right) for assessment of total system error using synchrony treatment delivery system. (Adapted from Accuray Physics Essentials Guide, *CyberKnife System Manual v.10*, Accuray, Sunnyvale, CA. With permission.)

When the Synchrony delivery technique is combined with the lung-optimized target (LOT) locating method, a thorax mimicking phantom is used (Figure 11.11). A lung-equivalent rod containing a small tissue-equivalent cube is used to simulate lesion movement during treatment delivery. A denser spherical structure hidden inside the small ball cube is used for tracking with the LOT locating method. A pair of orthogonal RCFs are positioned inside the small ball cube (Figure 11.11), and an isocentric treatment plan dose distribution is delivered to the spherical structure. During treatment delivery, the lesion is moved along the SI direction simulating the largest lung-lesion movements. The Synchrony-delivery technique is enabled to synchronize the movement of each beam with the target movement. The TSE of this combination is measured by comparing the centroid of the delivered dose distribution with the geometrical center of the spherical structure inside the ball cube.

In addition to shifting the dose centroid, respiratory motion may blur the dose distribution (i.e., reduce the steepness of the dose gradient around the target). It is suggested to study this effect during commissioning and periodic QA by comparing the dose distributions delivered with the RCFs. For this test, the distance between the 20% and 80% isodose lines is measured in the SI direction (the axis of greatest motion) at the edges of the target. Motion-induced blurring is quantified by the change in the 20%–80% isodose distance for treatments with and without motion.

The TSE results for the Synchrony treatment-delivery technique combined with fiducial and LOT locating methods are also included in Figure 11.9. The average TSE of the CK for moving targets was found equal to 0.61 ± 0.30 mm. This TSE data also suggest that the CK system is able to deliver the planned dose distribution to moving with respiration targets with an error of less than 1.0 mm [8,19].

11.7 VERIFICATION OF TREATMENT-PLANNING DOSIMETRY CALCULATIONS

During CK commissioning, a complete set of TMR values, OARs, and OFs are measured for all clinically available treatment fields. These data are imported into the MultiPlan TPS and interpolated to a fine grid to enable fast searching when dose calculations to the patient model geometry are required. For PDD measurements or off-axis distance measurements outside the corresponding measured ranges, extrapolation is performed using exponential or least square functions. The accuracy of the performed dose calculations is suggested to be verified prior their clinical use [7].

RCFs are commonly used for testing the accuracy of the dose values calculated by the TPS dose-calculation algorithms [20,27]. This is due to their ability of recording the dose distribution to a plane with high spatial resolution. RCFs are not absolute dosimeters and therefore their dose response function must be determined prior to any experimental measurements [59]. This calibration procedure is performed by positioning film pieces of the same batch inside a solid water or polymethylmethacrylate (PMMA)-based solid phantom and irradiating them to known radiation doses. The irradiated films are then scanned (usually after 24 h to allow for postirradiation OD growth) using a flatbed optical scanner [60]. The scanner is used in transmission mode and films are scanned in 48-bit red-green-blue mode with a resolution ranging from 72 dpi (pixel size 0.35 mm) to 300 dpi (pixel size 0.085 mm). The film images are analyzed to define regions of interest, calculate the mean pixel values, and corresponding ODs [59]. The obtained mean OD values are then correlated with the corresponding absorbed doses to obtain the calibration dose response curve of the used film batch (Figure 11.12). It should be noted that although originally, the response of the film in the red channel was used for dosimetric measurements, but at present, with newly established protocols, all three channels (i.e., R, G, B) can be used in an effort to remove systematic uncertainties related with film thickness and scanner lamp intensity non-uniformities [61,62].

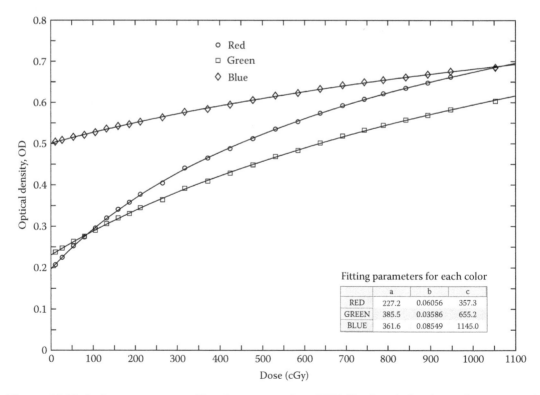

Figure 11.12 A dose response calibration curve of an EBT2 film batch for the red, green, and blue channels. In the inset the fitting parameters obtained by fitting a function of the form of $OD = -\log\big((a + b * Dose)/(c + Dose)\big)$ on the presented calibration data is shown.

For experimental measurements with the CK system, one or more RCFs can be positioned inside water-equivalent slab phantoms and be irradiated with their plane either situated perpendicular or aligned with beam axis. Fiducial markers are generally embedded inside the phantom to enable image guidance during irradiation. To perform TPS dose calculations, a volumetric CT study of the phantom in the experimental position is acquired using the same imaging protocol used for clinical applications, and the obtained images are imported in the CK data server. A single beam treatment plan can be developed using the ray tracing and/or Monte Carlo TPS dose calculation algorithms. The phantom loaded with films is then positioned inside treatment room, and the plan is delivered using the image-guidance system. After delivery, the films are scanned following the same steps as in the calibration procedure. The dose-response function of the film batch shown in Figure 11.12 is used to convert the measured ODs to absorbed doses [59,61,62]. Validation of the TPS dose calculations is performed by comparing the measured film-dose results with corresponding values extracted by the TPS in the form of RTDOSE digital imaging and communications in medicine files. To test the accuracy of the dose-calculation algorithms in the presence of tissue inhomogeneities, corresponding slabs can be inserted inside the slab phantom, and measurements can be repeated [27].

Wilcox et al. [27] used EBT films to verify the dose calculations performed by the ray tracing and Monte Carlo dose calculation algorithms of the MultiPlan TPS for a slab phantom containing a bone or lung-equivalent inhomogeneity. Results of their work are presented in Figure 11.13 and in which PDD curves along the central-beam axis of the 7.5 and 20 mm collimators are shown. For comparison, the corresponding TPS data using the effective path length-based ray tracing and Monte Carlo dose calculation algorithms are also presented in the same figure. Comparison of EBT results with corresponding TPS calculated data shows that the ray-tracing algorithm succeeds in calculating the dose before and after the bone or lung-tissue heterogeneities [27]. However, within and at close distances before and after the heterogeneities, substantial differences between the RCF and ray-tracing TPS dose results can be observed. Besides tissue heterogeneity, these differences are also field-size depended [27]. The differences between ray tracing and RCF inside the bone heterogeneity can be attributed mostly due to the fact that RT is estimating dose to variable density water, as compared with dose to water in medium that are measured using films [63].

It should be pointed out that the fast Monte Carlo dose calculations performed by the TPS are in excellent agreement with corresponding RCF results at all measurement points except for those lying inside the bone heterogeneity. These dose differences, however, are attributed to the fact that the water dose in bone is measured with the RCFs, whereas the medium dose in bone is calculated by the Monte Carlo TPS. When the Monte Carlo dosimetry results were scaled with a factor of 1.117 for the 6 MV photon energies [64], a better agreement between the experimental and the TPS dose values was observed. This agreement is another example of the important role that RCF films play in CK commissioning.

These findings are indicative of the dose-calculation errors performed by the corresponding calculation algorithms of the TPSs of modern radiotherapy systems that use small and/or non-standard fields. As shown by Jones et al. [65], for field sizes greater than 30 mm, the available dose calculation algorithms, such as ray tracing or convolution superposition, can accurately predict the dose in and beyond inhomogeneous regions. For field sizes smaller than 30 mm, equivalent path length or correction-based dose-calculation algorithms, such as the above-mentioned ray-tracing algorithm, failed to predict the dose close and inside low density in homogeneities. Corresponding dosimetry results obtained using a convolution superposition dose-calculation algorithm were found in close agreement with corresponding Monte Carlo dose data for the majority of calculated points.

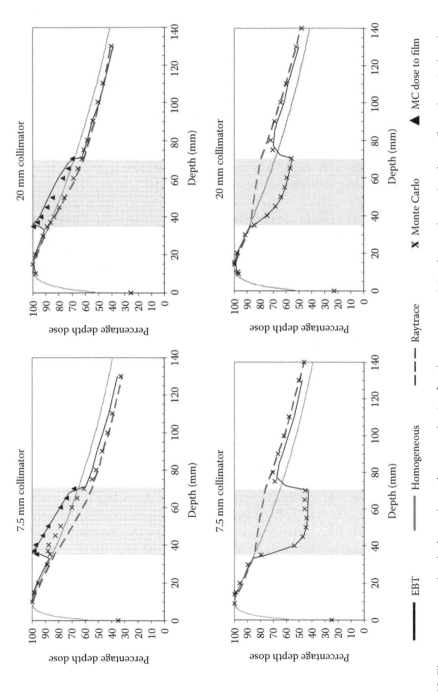

Figure 11.13 The percentage depth dose along the central axis of a phantom containing bone- (upper) or lung- (lower) equivalent heterogeneity at a depth of 35 to 70 mm (shown in shaded area). Irradiation was done using the 7.5 mm or the 20 mm collimators. The EBT film measurements are compared to calculations using the ray-tracing and Monte Carlo algorithm. The percentage depth dose in homogeneous water is shown for comparison. (Adapted from Wilcox, E.E., and Daskalov, G.M., *Med. Phys.*, 35, 2259, 2008. With permission.)

11.8 TREATMENT-DELIVERY QUALITY ASSURANCE

The E2E test allows the user to quantify the TSE of the CK system for each target-locating method. This test however is based on isocentric beam delivery, and it does not give any information about dose-delivery accuracy to complex targets that are treated using nonisocentric treatment plans that are the majority of plans delivered with the CK system. Currently, the nonisocentric targeting accuracy is assumed to be correct if the isocentric targeting is within specification. Estimation of the TSE of nonisocentric beam delivery can be performed by comparing E2E results with the phantom situated at the center and off center [8]. However, this is an indirect test, and it is suggested to perform patient specific dosimetry measurements using RCFs or other detectors with equally high resolution on a phantom [7].

Dosimetry QA (DQA) tests should be performed during machine commissioning, treatment-planning updates, and monthly QA. DQA results should be analyzed using the gamma index or other similar indices [66–68]. The acceptance criteria of these tests should be a 90% pass rate of distance-to-agreement of 2%/2 mm for the tumor, critical structures, and in the high-dose region down to the 50% isodose line. For Synchrony, the 90% pass rate for the distance-to-agreement should be within 3%/3 mm for a region encompassing the 50% isodose line [7]. The recommended phantom for DQA in an inhomogeneous environment, such as the lung, should contain a low-density region enclosing a higher density lung *tumor*.

A plan QA procedure is included in the MultiPlan TPS for performing patient-specific dosimetry measurements. The general steps for performing film-based DQA verification tests within the CK are shown in Figure 11.14. As shown in this diagram, the procedure can be separated in three major steps. In the first step, a patient treatment plan and a phantom CT image study are selected. The treatment plan is then registered to the phantom image study. This allows all nonzero beams, robot data, path set, and anatomy settings of the patient plan to be registered to the phantom plan. After confirming the registration, the TPS recalculates the dose distribution to take into account the geometric characteristics and materials of the phantom. At the end of this step, the user can rescale the total number of MUs of the plan so that the doses delivered to the film remain well within the calibration dose range.

In the second step, the phantom is loaded with one or more RCFs and aligned on the treatment couch using the image-guidance system of the CK. It is noted that image guidance is usually performed with the aid of fiducial markers that have been embedded in the phantom prior CT scan. After treatment delivery, the phantom is disassembled, and the film is kept in a dark and cool space to allow for growth of post-irradiation OD. In the third step, the film is scanned in a flatbed optical-transparency scanner using the same protocol followed during the calibration procedure but with a resolution equal or higher than 150 dpi. The calibration curve of the film batch is used to convert the measured OD values to corresponding doses on a pixel-by-pixel basis. The measured dose values are then registered with the TPS calculation-coordinate system and compared with corresponding TPS dose calculations using the gamma tool [66–68].

An example of a DQA test is presented in Figure 11.13 for a patient treated for a vestibular schwannoma of 3.1 cm^3 in Iatropolis (Athens, Greece) Clinic. The treatment plan consisted of 204 beams with the 5 and 10 mm fixed diameter collimators. The prescribed dose of 21 Gy at the 75% isodose covering the 98% of the target volume was delivered in three fractions. The ray-tracing dose calculation algorithm was used. Prior to treatment, the treatment plan was overlaid on the CT images of a spherical water-equivalent phantom containing an EBT2 RCF using the corresponding TPS tools. A dose of 6 Gy at 75% was prescribed to the film, and the DQA plan was saved as deliverable.

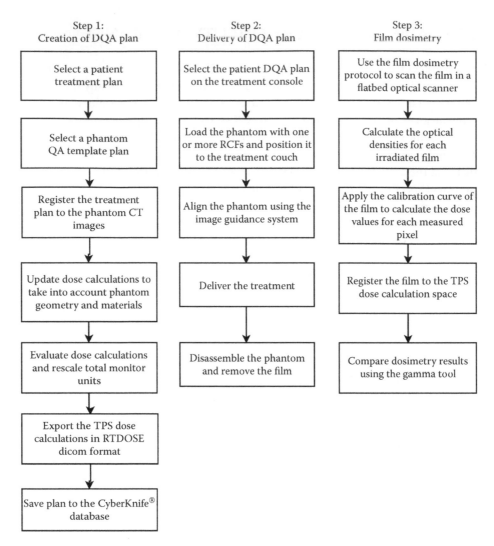

Figure 11.14 Diagram of the film-based dosimetry protocol for performing patient specific treatment plan dose verification tests.

The phantom contained metal pins to create small holes on the film surface and aid the spatial registration of film results with corresponding TPS dose calculations exported in RTDOSE digital imaging and communications in medicine (DICOM) format. The same fiducials were also used for the alignment of the phantom on the treatment couch using the fiducial tracking algorithm of the image-guidance system. One day post verification plan delivery, the film was scanned with an Epson Expression 1680Pro flatbed optical scanner using 150 dpi and 48 bit red–green–blue color mode. A triple channel-based protocol was used to convert measured pixel values to absorbed dose [61,62]. The results for doses greater than 2 Gy are presented in Figure 11.15. As shown, a good agreement between the film and the TPS data can be observed. The presented dose distributions were also compared using the Gamma analysis tool with the TG-135 recommended pass criterion of 2% dose difference and 2 mm distance to agreement [7]. The proportion of pixels meeting this criterion was 95%.

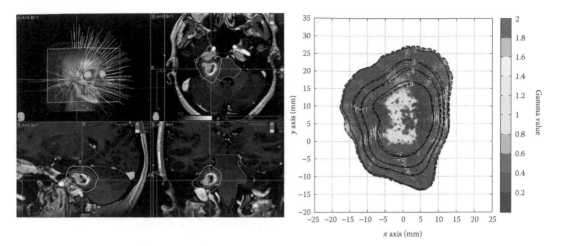

Figure 11.15 A treatment plan for a vestibular schwannoma case (left). EBT3 film dosimetry verification results (broken lines) and corresponding treatment planning system calculated dose values (solid lines) for the schwannoma plan (right). The corresponding gamma map calculated using 2%/2 mm acceptance criteria is also superimposed.

11.9 SUMMARY

The CK system is a frameless SRS device capable of delivering high doses of radiation to lesion of the whole body with stereotactic accuracy. Having the favorable characteristics of fine spatial resolution, near tissue equivalence, and no need for development, RCFs have become the detector of choice for measuring the geometric accuracy components of CK subsystem such as the precision and accuracy of manipulator movements and target locating system, including the total targeting error of the CK. The RCFs are also used for characterizing the dosimetric properties of the small fields of CK Linac as well as for the validation of the implemented dose calculation algorithms and for patient-specific dosimetry measurements. For new CK machines, it is recommended to perform Monte Carlo calculations on sample lung cases simulating anthropomorphic phantoms in which small pieces of RCFs are placed [7]. RCFs should also be used to check the QA of nonisocentric delivery techniques of CK Linac.

ACKNOWLEDGMENT

We would like to thank Warren Kilby Senior Clinical Research Scientist of Accuray Inc. for reviewing the current chapter.

REFERENCES

1. Adler JR, Murphy MJ, Chang SD et al. Image-guided robotic radiosurgery. *Neurosurgery* 1999;44:1299–1306.
2. Adler JR, Chang SD, Murphy MJ et al. The CyberKnife: A frameless robotic system for radiosurg. *Stereotact Funct Neurosurg* 1997;69:124–128.
3. Kilby W, Dooley JR, Kuduvalli G et al. The CyberKnife robotic eadiosurgery system in 2010. *Technol Cancer Res Treat* 2010;9:433–452.
4. Chang SD, Veeravagu A (Eds.) *CyberKnife Stereotactic Radiosurgery: Brain,* Vol. 1. New York: Nova Biomedical, 2014.

5. Chang SD, Veeravagu A (Eds.). *CyberKnife Stereotactic Radiosurgery: Spine*. Vol. 2. New York: Nova Biomedical, 2015.

6. Urschel HC, Kresl JJ, Luketich JD et al. (Eds.). *Robotic Radiosurgery. Treating Tumors that Move with Respiration*. Springer Science & Business Media, Berlin, Germany, 2007.

7. Dieterich S, Cavedon C, Chuang CF et al. Report of AAPM TG 135: Quality assurance for robotic radiosurgery. *Med Phys* 2011;38:2914–2936.

8. Antypas C, Pantelis E. Performance evaluation of a CyberKnife G4 image-guided robotic stereotactic radiosurgery system. *Phys Med Biol* 2008;53:4697–4718.

9. Das IJ, McGee KP, Desobrey GE et al. *A Practical Guide to CT Simulation*, Coia LR, Schultheiss TE, Hanks GE (Eds.) Madison, WI: Advanced Medical Publishing, 1995.

10. Araki F. Monte Carlo study of a Cyberknife stereotactic radiosurgery system. *Med Phys* 2006;33:2955.

11. Echner GG, Kilby W, Lee M et al. The design, physical properties and clinical utility of an iris collimator for robotic radiosurgery. *Phys Med Biol* 2009;54:5359–5380.

12. Asmerom G, Bourne D, Chappelow J et al. The design and physical characterization of a multileaf collimator for robotic radiosurgery. *Biomed Phys Eng Express* 2016;2:17003.

13. McGuinness CM, Gottschalk AR, Lessard E et al. Investigating the clinical advantages of a robotic linac equipped with a multileaf collimator in the treatment of brain and prostate cancer patients. *J Appl Clin Med Phys* 2015;16(5):284–295.

14. Goggin LM, Descovich M, McGuinness C et al. Dosimetric comparison between 3-dimensional conformal and robotic SBRT treatment plans for accelerated partial breast radiotherapy. *Technol Cancer Res Treat* 2015. doi:10.1177/1533034615601280.

15. Sharma SC, Ott JT, Williams JB et al. Commissioning and acceptance testing of a CyberKnife linear accelerator. *J Appl Clin Med Phys* 2007;8:2473.

16. Ho AK, Fu D, Cotrutz C et al. A study of the accuracy of cyberknife spinal radiosurgery using skeletal structure tracking. *Neurosurgery* 2007;60:ONS147–ONS156.

17. Chang SD, Main W, Martin DP et al. An analysis of the accuracy of the CyberKnife: A robotic frameless stereotactic radiosurgical system. *Neurosurgery* 2003;52:140-6-7.

18. Yu C, Main W, Taylor D et al. An anthropomorphic phantom study of the accuracy of CyberKnife spinal radiosurgery. *Neurosurgery* 2004;55:1138–1149.

19. Ho AK, Antony J, Adler JR. The targeting accuracy of the CyberKnife as measured with EBT2 film inside a Ball-Cube-II film cassette and head phantom. *Cureus* 2009;1:1–5.

20. Wilcox EE, Daskalov GM. Evaluation of GAFCHROMIC® EBT film for CyberKnife® dosimetry. *Med Phys* 2007;34:1967.

21. Pantelis E, Antypas C, Petrokokkinos L et al. Dosimetric characterization of CyberKnife radiosurgical photon beams using polymer gels. *Med Phys* 2008;35:2312.

22. Pantelis E, Moutsatsos A, Zourari K et al. On the implementation of a recently proposed dosimetric formalism to a robotic radiosurgery system. *Med Phys* 2010;37:2369.

23. Huet C, Dagois S, Derreumaux S et al. Characterization and optimization of EBT2 radiochromic films dosimetry system for precise measurements of output factors in small fields used in radiotherapy. *Radiat Meas* 2012;47:40–49.

24. Pantelis E, Moutsatsos A, Zourari K et al. On the output factor measurements of the CyberKnife iris collimator small fields: Experimental determination of the \kclinMSR correction factors for microchamber and diode detectors. *Med Phys* 2012;39:4875–4885.

25. Bassinet C, Huet C, Derreumaux S et al. Small fields output factors measurements and correction factors determination for several detectors for a CyberKnife® and linear accelerators equipped with microMLC and circular cones. *Med Phys* 2013;40:71725.

26. Moignier C, Huet C, Makovicka L. Determination of the \kclinMSR correction factors for detectors used with an 800 MU/min CyberKnife(\textregistered) system equipped with fixed collimators and a study of detector response to small photon beams using a Monte Carlo method. *Med Phys* 2014;41:71702.

27. Wilcox EE, Daskalov GM. Accuracy of dose measurements and calculations within and beyond heterogeneous tissues for 6 MV photon fields smaller than 4 cm produced by Cyberknife. *Med Phys* 2008;35:2259.

28. Wilcox EE, Daskalov GM, Lincoln H et al. Comparison of planned dose distributions calculated by Monte Carlo and ray-trace algorithms for the treatment of lung tumors with CyberKnife: A preliminary study in 33 Patients. *Int J Radiat Oncol Biol Phys* 2010;77:277–284.

29. Das IJ, Cheng C-W, Watts RJ et al. Accelerator beam data commissioning equipment and procedures: Report of the TG-106 of the therapy physics committee of the AAPM. *Med Phys* 2008;35:4186–4215.

30. Francescon P, Kilby W, Satariano N et al. Monte Carlo simulated correction factors for machine specific reference field dose calibration and output factor measurement using fixed and iris collimators on the CyberKnife system. *Phys Med Biol* 2012;57:3741–3758.

31. Dieterich S, Sherouse GW. Experimental comparison of seven commercial dosimetry diodes for measurement of stereotactic radiosurgery cone factors. *Med Phys* 2011;38:4166.

32. Francescon P, Cora S, Cavedon C. Total scatter factors of small beams: A multidetector and Monte Carlo study. *Med Phys* 2008;35:504.

33. Alfonso R, Andreo P, Capote R et al. A new formalism for reference dosimetry of small and nonstandard fields. *Med Phys* 2008;35:5179.

34. IAEA TRS 483. Dosimetry of small static fields used in external beam radiotherapy: An IAEA-AAPM International Code of Practice for reference and relative dose determination. Technical Report Series No. 483. Vienna, Austria, 2016.

35. Das IJ, Ding GX, Ahnesjö A. Small fields: Nonequilibrium radiation dosimetry. *Med Phys* 2008;35:206.

36. Scott AJD, Nahum AE, Fenwick JD. Using a Monte Carlo model to predict dosimetric properties of small radiotherapy photon fields. *Med Phys* 2008;35:4671.

37. Bouchard H, Kamio Y, Palmans H et al. Detector dose response in megavoltage small photon beams. I. Theoretical concepts. *Med Phys Med Phys* 2015;42:6033–6047.

38. Bouchard H, Kamio Y, Palmans H ct al. Detector dose response in megavoltage small photon beams. II. Pencil beam perturbation effects. *Med Phys* 2015;42:6048–6061.

39. Underwood TSA, Winter HC, Hill MA et al. Detector density and small field dosimetry: Integral versus point dose measurement schemes. *Med Phys* 2013;40:82102.

40. Underwood TSA, Winter HC, Hill MA et al. Mass-density compensation can improve the performance of a range of different detectors under non-equilibrium conditions. *Phys Med Biol* 2013;58:8295–310.

41. Das IJ, Morales J, Francescon P. Small field dosimetry: What have we learnt? *AIP Conference Proceedings* 2016;1747:60001.

42. Vandervoort E, La Russa D, Ploquin N et al. Sci—Thur AM: Planning—10: Improved dosimetric accuracy for patient specific quality assurance using a dual-detector measurement method for cyberknife output factors. *Med Phys* 2012;39:4621.

43. Bouchard H, Seuntjens J, Kawrakow I. A Monte Carlo method to evaluate the impact of positioning errors on detector response and quality correction factors in nonstandard beams. *Med Phys* 2011;56:2617–2634.

44. Das I. TH-A-213-01: Small field dosimetry: Overview of AAPM TG-155. *Med Phys* 2015;42:3700–3700.

45. Gago-Arias A, Antolin E, Fayos-Ferrer F et al. Correction factors for ionization chamber dosimetry in CyberKnife: Machine-specific, plan-class, and clinical fields. *Med Phys* 2013;40:11721.

46. Morales JE, Butson M, Crowe SB et al. An experimental extrapolation technique using the Gafchromic EBT3 film for relative output factor measurements in small x-ray fields. *Med Phys* 2016;43:4687–4692.

47. Francescon P, Beddar S, Satariano N et al. Variation of k Q clin, Q msr f clin, f msr for the small-field dosimetric parameters percentage depth dose, tissue-maximum ratio, and off-axis ratio. *Med Phys* 2014;41:101708.

48. Morin J, Beliveau-Nadeau D, Chung E et al. A comparative study of small field total scatter factors and dose profiles using plastic scintillation detectors and other stereotactic dosimeters: The case of the CyberKnife. *Med Phys* 2013;40:11719.

49. Sharma S, Ott J, Williams J et al. Dose calculation accuracy of the Monte Carlo algorithm for cyberknife compared with other commercially available dose calculation algorithms. *Med Dosim* 2011;36:347–350.

50. Accuray Physics Essentials Guide. *CyberKnife System Manual v.10*, Sunnyvale, CA: Accuray.

51. Kawachi T, Saitoh H, Inoue M et al. Reference dosimetry condition and beam quality correction factor for CyberKnife beam. *Med Phys* 2008;35(10):4591–4598.

52. Verhaegen F, Das IJ, Palmans H. Monte Carlo dosimetry study of a 6 MV stereotactic radiosurgery unit. *Phys Med Biol* 1998;43:2755–2768.

53. Andrés C, del Castillo A, Tortosa R et al. A comprehensive study of the Gafchromic EBT2 radiochromic film. A comparison with EBT. *Med Phys* 2010;37:6271.

54. Hardcastle N, Basavatia A, Bayliss A et al. High dose per fraction dosimetry of small fields with Gafchromic EBT2 film. *Med Phys* 2011;38:4081–4085.

55. Baart V, Devillers M, Lenaerts E. Validation and use of the "Iris quality assurance tool." *The SRS/SBRT Scientific Meeting 2013*, The Radiosurgery Society, San Mateo, CA, 2013.

56. Boyer A, Biggs P, Galvin J et al. Basic applications of multileaf collimators. AAPM Radiation Therapy Committee Task Group No. 50 Report No. 72, Madison, WI: Medical Physics Publishing, 2001.

57. Bayouth JE, Wendt D, Morrill SM. MLC quality assurance techniques for IMRT applications. *Med Phys* 2003;30:743–750.

58. Lutz W, Winston KR, Maleki N. A system for stereotactic radiosurgery with a linear accelerator. *Int J Radiat Oncol Biol Phys* 1988;14:373–381.

59. Devic S, Seuntjens J, Sham E et al. Precise radiochromic film dosimetry using a flat-bed document scanner. *Med Phys* 2005;32:2245.

60. Devic S, Aldelaijan S, Mohammed H et al. Absorption spectra time evolution of EBT-2 model GAFCHROMIC™ film. *Med Phys* 2010;37:3687.

61. Micke A, Lewis DF, Yu X. Multichannel film dosimetry with nonuniformity correction. *Med Phys* 2011;38:2523–2534.

62. Méndez I, Peterlin P, Hudej R et al. On multichannel film dosimetry with channel-independent perturbations. *Med Phys* 2014;41:11705.

63. Ma C-M, Li J. Dose specification for radiation therapy: Dose to water or dose to medium? *Phys Med Biol* 2011;56:3073–3089.

64. Siebers JV, Keall PJ, Nahum AE et al. Converting absorbed dose to medium to absorbed dose to water for Monte Carlo based photon beam dose calculations. *Phys Med Biol* 2000;45:983–995.

65. Jones AO, Das IJ. Comparison of inhomogeneity correction algorithms in small photon fields. *Med Phys* 2005;32:766–776.

66. Low DA, Harms WB, Mutic S et al. A technique for the quantitative evaluation of dose distributions. *Med Phys* 1998;25:656–661.

67. Low DA, Dempsey JF. Evaluation of the gamma dose distribution comparison method. *Med Phys* 2003;30:2455.

68. Low DA, Moran JM, Dempsey JF et al. Dosimetry tools and techniques for IMRT. *Med Phys* 2011;38:1313.

Application of GAFchromic™ films for TomoTherapy®

EDMOND STERPIN AND DAMIEN DUMONT

12.1 GENERALITIES ON FILM DOSIMETRY WITH THE TOMOTHERAPY® SYSTEM

Helical TomoTherapy® delivers intensity-modulated treatments through a unique architecture that resembles to a CT scanner. Highly conformal treatments are achieved by the simultaneous motion of the treatment couch, the gantry, and a binary multileaf collimator. The CT-like design of the TomoTherapy system enables the acquisition of CT images with a mega-voltage (MV) fan beam and the patient in treatment position. This system was first described by Mackie et al. [1] and is manufactured by Accuray Inc. In the first commercial versions of the machine, the field size in the longitudinal direction was constant during the entire treatment (5, 2.5, or 1 cm widths). The field size can now change dynamically during the treatment for faster delivery and improved dose conformation at the superior and inferior edges of the target(s) [2,3].

Radiographic films (and later GAFchromic™) have been used right from the clinical introduction of the system for various quality assurances procedures. Their high resolution is a very favorable property for TomoTherapy because of the high-dose gradients delivered and the tight geometrical tolerances related to the mechanical alignment of beam shaping devices (jaws, multileaf collimator (MLC), etc.). Several quality assurance (QA) procedures and research studies using films have been described in literature [4–11]. Many of them have been discussed and regrouped in

American Association of Physicists in Medicine (AAPM) Task Group 148 report [12]. Radiographic films were used in general. But the procedures described can be readily adapted to GAFchromic films, especially mechanical tests that typically do not rely on the quantification of the optical density into dosimetric quantities.

For patient-specific quality assurance, Thomas et al. [6] described in 2005 a thorough QA procedure making use of the vendor film analysis software that is still the standard procedure for most installed units (Figure 12.1). Dose distributions resulting from the planned fluence are computed

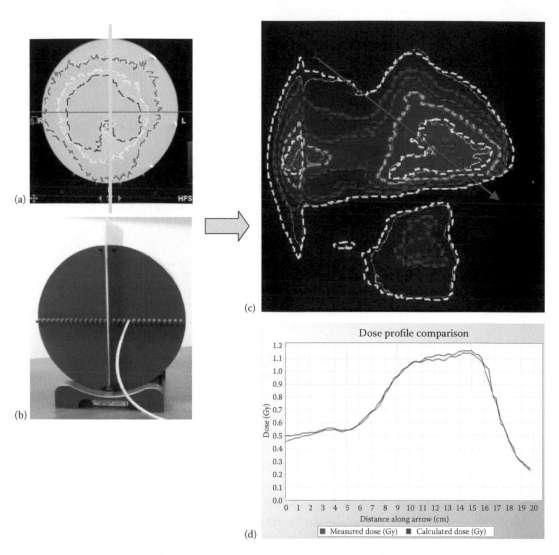

(a)

(b)

(c)

(d)

Figure 12.1 Illustrations of a typical patient quality assurance workflow using the quality assurance software provided with the Tomotherapy system (Accuray Inc.). (a), axial dose distribution of a patient on a circular phantom; (b) solid water phantom with film; (c) recomputed dose distribution in the phantom; (d) correspondence between computed and measured dose distributions in the direction of arrow indicated in (c).

in a cylindrical phantom supplied by the vendor (Virtual Water™ phantom). The phantom is separated along its main axis into two identical parts, within which a film can be inserted. After phantom setup and verification with the mega-voltage computed tomography (MVCT), the planned fluence can be delivered. The resulting dose distributions measured by the film can then be compared with the dose distribution computed in the same plane using gamma analysis software provided by the vendor (or a third-party software). Typically, tolerances for dose difference and distance to agreement criteria are 3% and 3 mm, respectively. It is worth mentioning that ion chambers can also be placed at several positions in the phantom for absolute point-dose measurements.

12.2 SPECIFICITIES OF GAFCHROMIC™ FILM DOSIMETRY FOR THE TOMOTHERAPY SYSTEM

12.2.1 FILM CALIBRATION

Films can be directly calibrated in the TomoTherapy beam quality. However, the specificities of the TomoTherapy technology make this procedure less practical than with a C-arm Linac. Some studies assume negligible dependence of the calibration curve with minor changes of beam quality [13,14], which has been shown quantitatively valid by Marrazzo et al. [14]. However, another study found deviations in gamma analysis results depending on the beam modality used for calibration [15].

Finally, GAFchromic films are ideally suited for high, hypofractionated dose delivery schemes because of their wider dynamic range (superior to 40 Gy), whereas EDR2 films saturate at doses higher than 8 Gy as it can be observed in Figure 12.2.

12.2.2 FILM SETUP

In the Virtual Water phantom, film must be set up with care if the film is in sagittal orientation. Artifacts may appear if the film is not properly clamped (Figure 12.3).

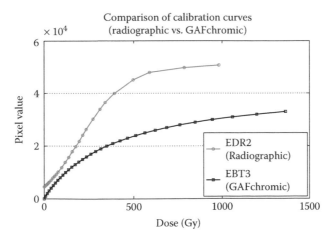

Figure 12.2 Comparison between the dynamic dose ranges of radiographic (EDR-2) and GAFchromic™ (EBT-3) films. Pixel values (PV) are plotted with respect to measured dose.

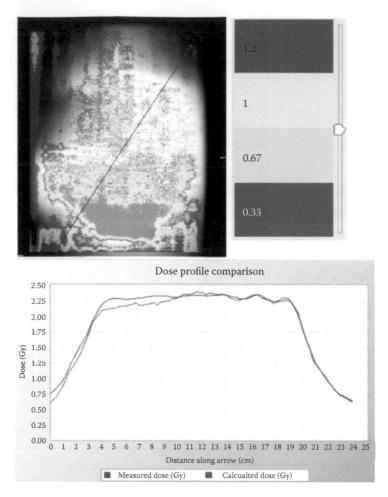

Figure 12.3 Effect of the clamping of the film on the dose distributions. The effect of an insufficient clamping can be seen on the left part of the dose profile.

12.3 GAFCHROMIC FILM APPLICATIONS FOR TOMOTHERAPY

12.3.1 PUBLISHED REPORTS ON THE USE OF GAFCHROMIC FILMS FOR PATIENT SPECIFIC QA

Marrazzo et al. [14] have established a comprehensive procedure for implementing external beam therapy (EBT) film-based patient specific QA. Using a multichannel approach, passing rates above 98% were achieved for TomoTherapy plans for a 3%/3 mm gamma criteria (EBT3 films). Using only the red channel, passing rates diminished down to 84%. Kairn et al. [13] have also reported an excellent passing rate of 99% (EBT2) for a prostate treatment and the same criteria. Tessonier et al. [15] have reported an average gamma-index of 95% for a 3%/3 mm criterion and a calibration in a TomoTherapy beam (EBT3). However, the passing rate was only of 75% on average using the film calibrated in an Elekta Synergy S (6 MV beam).

These studies generally use third-party software to perform the gamma analysis. However, the delivery quality assurance (DQA) module of the TomoTherapy planning system has an embedded gamma-analysis feature. This solution has some limitations, among others: (1) the dose criterion is global (in percentage of a dose given by the user); (2) there is no possibility to setup a threshold dose (the entire film is analyzed); and (3) the registration process is manual.

12.3.2 CORRECTION FACTORS FOR REFERENCE DOSIMETRY

Zeverino et al. [16] have considered EBT2 GAFchromic films to derive machine-specific correction factors for reference dosimetry of TomoTherapy beams with ion chambers. According to the authors, EBT2 films were suitable because they have no beam quality dependence between a TomoTherapy beam and a 6-MV beam from a conventional linear accelerator (here Varian 2100C/D). Uncertainties on the determined dose were computed and equaled 0.6%. The obtained quality correction factors with respect to a cobalt beam for Exradin A1SL, PTW Semiflex, and PTW PinPoint are in agreement with a more recent study based on comprehensive Monte Carlo simulations [17].

12.3.3 OTHER APPLICATIONS OF GAFCHROMIC FILM DOSIMETRY

Nobah et al. [18] have also reported an original study with GAFchromic EBT3 film in which they measure the dose delivered by the imaging system of the TomoTherapy unit, which is a MVCT with the imaging beam energy reduced compared with the treatment beam (3.5 MV nominal energy compared with 6 MV for the therapeutic beam). Films were calibrated in a C-arm Linac (4 MV photon beam). The red channel of the scanner was used. Dose ranges between 0.5 and 1.8 cGy were measured for various tumor locations and found consistent with published data.

Avanzo et al. [19] measured the skin dose delivered by the TomoTherapy system using EBT films. EBT films were calibrated in the TomoTherapy beam. In general, they observed an overestimation of the dose at the skin by the TomoTherapy treatment planning system (TPS) (9.2% ± 2.6% [one standard deviation]).

12.4 CONCLUSIONS

GAFchromic films are well suited for TomoTherapy devices. They can be used for any commissioning or mechanical check previously developed using radiographic films. However, the tissue-equivalence and wider dynamic range of GAFchromic films such as EBT-2 give them a strong advantage for dosimetry, including applications in reference dosimetry.

REFERENCES

1. Mackie TR, Holmes T, Swerdloff S et al. Tomotherapy: A new concept for the delivery of dynamic conformal radiotherapy. *Med Phys* 1993;20:1709–1719.
2. Sterzing F, Uhl M, Hauswald H et al. Dynamic jaws and dynamic couch in helical tomotherapy. *Int J Radiat Oncol Biol Phys* 2010;76:1266–1273.
3. Chen Y, Chen Q, Chen M et al. Dynamic tomotherapy delivery. *Med Phys* 2011;38:3013.
4. Kapatoes JM, Olivera GH, Ruchala KJ et al. On the verification of the incident energy fluence in tomotherapy IMRT. *Phys Med Biol* 2001;46:2953–2965.
5. Fenwick JD, Tomé WA, Jaradat HA et al. Quality assurance of a helical tomotherapy machine. *Phys Med Biol* 2004;49:2933–2953.
6. Thomas SD, Mackenzie M, Field GC et al. Patient specific treatment verifications for helical tomotherapy treatment plans. *Med Phys* 2005;32:3793.

7. Sterpin E, Salvat F, Olivera G et al. Monte Carlo evaluation of the convolution/superposition algorithm of Hi-Art™ tomotherapy in heterogeneous phantoms and clinical cases. *Med Phys* 2009;36:1566.

8. Balog J, Mackie TR, Pearson D et al. Benchmarking beam alignment for a clinical helical tomotherapy device. *Med Phys* 2003;30:1118.

9. Balog JP, Mackie TR, Wenman DL et al. Multileaf collimator interleaf transmission. *Med Phys* 1999;26:176.

10. Ardu V, Broggi S, Cattaneo GM et al. Dosimetric accuracy of tomotherapy dose calculation in thorax lesions. *Radiat Oncol* 2011;6:14.

11. Ramsey CR, Seibert RM, Robison B et al. Helical tomotherapy superficial dose measurements. *Med Phys* 2007;34:3286.

12. Langen KM, Papanikolaou N, Balog J et al. QA for helical tomotherapy: Report of the AAPM Task Group 148. *Med Phys* 2010;37:4817.

13. Kairn T, Hardcastle N, Kenny J et al. EBT2 radiochromic film for quality assurance of complex IMRT treatments of the prostate: Micro-collimated IMRT, RapidArc, and TomoTherapy. *Australas Phys Eng Sci Med* 2011;34:333–343.

14. Marrazzo L, Zani M, Pallotta S et al. Gafchromic® EBT3 films for patient specific IMRT QA using a multichannel approach. *Phys Med* 2015; 31(8):1035–1042.

15. Tessonnier T, Dorenlot A, Nomikossoff N. Patient quality controls with Gafchromic EBT3 for tomotherapy HI-ART 2: Analysis of the calibration's condition to achieve an absolute dosimetry. *Phys Med* 2013;29:e45.

16. Zeverino M, Agostinelli S, Pupillo F et al. Determination of the correction factors for different ionization chambers used for the calibration of the helical tomotherapy static beam. *Radiother Oncol* 2011;100:424–428.

17. Sterpin E, Mackie TR, Vynckier S. Monte Carlo computed machine-specific correction factors for reference dosimetry of TomoTherapy static beam for several ion chambers. *Med Phys* 2012;39:4066–4072.

18. Nobah A, Aldelaijan S, Devic S et al. Radiochromic film based dosimetry of image-guidance procedures on different radiotherapy modalities. *J Appl Clin Med Phys* 2014;15:229–239.

19. Avanzo M, Drigo A, Ren Kaiser S et al. Dose to the skin in helical tomotherapy: Results of in vivo measurements with radiochromic films. *Phys Med* 2013;29:304–311.

Small-field dosimetry in megavoltage beams

INDRA J. DAS AND JOHNNY MORALES

13.1 INTRODUCTION

Small fields are hallmarks of advanced radiation treatment such as stereotactic radiosurgery (SRS) involving the delivery of a high radiation dose to the lesions within the brain as well as extracranial lesions [1–3]. Stereotactic body radiotherapy has become one of the most used techniques for small lesions of lung, liver, pancreas, spine, nodes, prostate, and many other nonoperable tumors [4–10]. There are a number of radiotherapy devices that have been used to deliver small radiation fields. The most common are high-energy linear accelerators, CyberKnife®, Gamma Knife®, and TomoTherapy® units. Detailed information on these devices is not provided here; rather discussion on small fields produced from these devices is elaborated.

The dosimetry of very small X-ray fields is challenging for many reasons including lack of lateral electronic equilibrium, source occlusion, large-dose gradients, the size of detector in respect to the field size that leads to volume averaging effects and the nonwater equivalence of the particular detectors [11–13]. There have been many investigations into the choice of appropriate radiation dosimeters for relative dosimetry measurements such as depth doses, profiles, and output factors (OFs) in very small X-ray fields [14–16]. The detectors studied have included very small ionization chambers (pinpoint chambers), diodes, diamond detectors, plastic scintillator dosimeters, and radiochromic film (RCF) [14,17,18]. The incorrect choice of detector can result in up to 30%–60%

difference in relative OF leading to radiation accidents [19,20]. These detectors require significant amount of correction factors that have been reported particularly for very small field sizes [21–23].

RCF is one of the choices for dosimetry that has great potential for small-field dosimetry. This is due to very high spatial resolution, near tissue equivalence, readily available in radiation oncology clinics, cost effectiveness, and requires minimal equipment for readout. For this reason, RCF has been the subject of significant investigation for small-field dosimetry. This chapter provides an overview of RCF in small fields used in advanced radiation treatments such as SRS and stereotactic body radiotherapy.

13.2 INTERNATIONAL APPROACH—INTERNATIONAL ATOMIC ENERGY AGENCY

The difficulty of dosimetry in small fields has been realized by most organizations [11,24,25]. International Atomic Energy Agency (IAEA) provided a frame work for dosimetry in small fields [11] with speculation of providing code of practice and detector-specific data [26]. This provided method for the determination of reference dose in situations when the standard reference field 10×10 cm^2 condition cannot be realized as the case for modern technological machine that has machine-specific reference field (f_{msr}).

$$D_{w,Q_{msr}}^{f_{msr}} = M_{Q_{msr}}^{f_{msr}} \bullet N_{D,w,Q_0} \bullet k_{Q,Q_0} \bullet k_{Q_{msr},Q}^{f_{msr},f_{ref}} \qquad (13.1)$$

where:

$D_{w,Q_{msr}}^{f_{msr}}$ is the absorbed dose at a reference depth in water in the absence of the detector at its point of measurement in a field size specified by f_{msr} and beam quality Q_{msr} msr stands for machine-specific reference

M is the measurement reading by the detector (corrected for variations in environmental conditions, polarity, leakage, stem correction, and ion recombination corrections)

f_{ref} denotes the conventional reference field in dosimetry protocols for which the calibration coefficient of an ionization chamber in terms of absorbed dose to water is provided by a standard laboratory

Q is the beam quality of f_{ref}

f_{msr} is the machine-specific-reference field

$N_{D,w}$ is the chamber-specific calibration coefficient in terms of absorbed dose to water for ^{60}Co

k_{Q,Q_0} is the chamber-specific beam quality calibration factor

On the contrary, the last terms in Equation 13.1 being the k values are not readily available and more so there are variability in the literature data.

For small fields, the ratio of the detector readings is not equivalent to the ratio of doses and so a correction factor, k, is needed that was introduced by Alfonso et al. [11], which was the first small dosimetry field formalism by IAEA. The ratio of the readings for the particular small field and the reference field can be described as shown in Equation 13.2:

$$\frac{D_{w,Q_{clin}}^{f_{clin}}}{D_{w,Q_{msr}}^{f_{msr}}} = \left[\frac{M_{Q_{clin}}^{f_{clin}}}{M_{Q_{msr}}^{f_{msr}}} \right] k_{Q_{clin},Q_{msr}}^{f_{clin},f_{msr}} \qquad (13.2)$$

where:

D is the dose

M is the detector reading

f_{clin} is the clinical field different from f_{msr}

k is the correction factor that depends on Q, f, detector, and machine (focal spot)

For the sake of simplicity, Francescon et al. [23] defined correction factor, k_Ω, as follows:

$$k_\Omega \equiv k_{Q_{\text{clin}},Q_{\text{msr}}}^{f_{\text{clin}},f_{\text{msr}}} = \left(\frac{D_{w,Q_{\text{clin}}}^{f_{\text{clin}}}}{D_{w,Q_{\text{msr}}}^{f_{\text{msr}}}} \right) \left(\frac{M_{Q_{\text{msr}}}^{f_{\text{msr}}}}{M_{Q_{\text{clin}}}^{f_{\text{clin}}}} \right) \tag{13.3}$$

Equations 13.2 and 13.3 are the same except simplified to k_Ω as defined by Francescon et al. [23] due to many functional forms with dependence on many parameters as shown in the following that had been a subject of intense research over a decade. Hence, acquiring the values for k_Ω(*machine, focal spot, detector, detector orientation, depth, and f_{clin}*) is the subject of many research papers and international task groups, and work is constantly being reported and modified.

Ionization chambers had been the backbone for radiation dosimetry. However, when field size decreases to smaller values compared with the range of secondary electrons, electronic equilibrium cannot be established and Bragg–Cavity theory cannot be used. There had been surge of literature [27–34] searching the solution in such situation. For ionization chambers, the correction factor k_Ω can be derived as follows:

$$k_\Omega \equiv k_{Q_{\text{clin}},Q_{\text{msr}}}^{f_{\text{clin}},f_{\text{msr}}} = \frac{\left[\left(\dfrac{\bar{L}}{\rho} \right)_{\text{air}}^{w} \cdot P_{\text{fl}} \cdot P_{\text{grad}} \cdot P_{\text{stem}} \cdot P_{\text{cell}} \cdot P_{\text{wall}} \right]_{f_{\text{clin}}}}{\left[\left(\dfrac{\bar{L}}{\rho} \right)_{\text{air}}^{w} \cdot P_{\text{fl}} \cdot P_{\text{grad}} \cdot P_{\text{stem}} \cdot P_{\text{cell}} \cdot P_{\text{wall}} \right]_{f_{\text{msr}}}} \tag{13.4}$$

where L/ρ is the restricted stopping power ratio of water-to-air, and the other parameters are perturbation corrections due to the different components of the ionization chamber. It is noted that the parameters $P_{\text{grad}} \equiv P_\rho \cdot P_{\text{vol}}$ accounts for perturbations due to density and volume effects [35].

There are a large number of publications that have emerged over a decade providing k_Ω values for most detectors especially microdetectors that can provide meaningful dosimetry. These references and data are compiled in IAEA technical report series (TRS)-483 [26]. On the contrary, this report does not address RCF dosimetry that might be best suited for small fields. The k_Ω data for RCF are not available in published papers [18,36–40]. These papers discussed only the relative dose; the profiles, and depth dose data compared with other detectors.

13.3 RADIOCHROMIC FILM FOR SMALL-FIELD DOSIMETRY

Commissioning and periodic quality assurance of SRS, Gamma Knife, CyberKnife, TomoTherapy, intensity modulated radiation therapy (IMRT), and volumetric-modulated arc therapy (VMAT) that use small fields require radiological parameters that include percent depth dose (PDD), tissue maximum ratio (TMR), profiles at various depths for off-axis ratio, and OF need to be verified along with isodose. Detectors as described in TG-106 [41] may be too large for small field and hence RCF is a good selection. As compared with detectors and microdetectors, RCF data have paucity. In fact, IAEA TRS-483 [26] even does not provide any reference to RCF for small field. The performance of the GAFchromic™ external beam therapy (EBT3) film, thermoluminescent dosimeter (TLD) chips, the microdiamond, and other solid-state detectors for small-field dosimetry was performed by an intercomparison across 30 Italian radiotherapy centers [42] and concluded that EBT3 film, TLD chips, and microdiamond detector were all suitable for small-field dosimetry. This is consistent with a number of other experimental and Monte Carlo studies when the various detector dose measurements are compared with dose to water [43–47].

Rather than describing individually how to acquire PDD, TMR, off-center ratio (OCR), and OF, a generic approach is taken and references are provided. Francescon et al. [23] provided a detailed approach for various detectors for these parameters that can be used for comparison with RCF. Caution is taken that RCF should be placed in water to eliminate air gap. Films should be dried and scanned with every precaution as discussed in Chapter 3. Here, we provide some seminal work on RCF regarding these parameters that can be adopted for accurate dosimetry.

13.3.1 MEXICAN WORK—NATIONAL LEAD IN MEXICO

One of the first publications using GAFchromic film to take measurements for small field was that of Garcia-Garduno et al. [37]. This work reported on the use of GAFchromic EBT film to commission a Monte Carlo model for a dedicated Novalis® linear accelerator (Varian Medical Systems, Palo Alto, and United States of America) for relative dose distribution. Monte Carlo simulations of radiation transport were performed using the BEAMnrc code. GAFchromic EBT, first generation of this type of RCF film was used. The Novalis linear accelerator was equipped with Brainlab conical collimators with diameters in the range of 4–20 mm with 6-MV X-ray beam. This work deserves particular mention because it showed the ability to commission a Monte Carlo model using solely RCF as its main radiation detector for very small X-ray beams. By using RCF film, PDD, TMR, profiles, and OF were measured and compared with various other detectors for small fields with a very high degree of accuracy. They also compared the calculation grid size from a planning system and corresponding values of dose-volume histogram (DVH) from other detectors. Comparison is made throughout this chapter with the work published by this leading team.

13.3.2 FRENCH WORK—NATIONAL STANDARD IN FRANCE

The first national study that compared a wide range of commonly used detectors for the dosimetry of SRS was from the French study published by Bassinet et al. [14]. The relevance of this work was underpinned by support from the French national standards lab and was set up as there was no metrological reference dosimetry for these small beams. In this work, relative OFs were determined using a range of active and passive detectors with radiation beams generated by a CyberKnife unit, a Novalis linear accelerator equipped with the Brainlab M3 microMLC and circular cones and a Varian Clinac 2100 linear accelerator equipped with the M3 microMLC attachment. Two types of passive dosimeters were used being GAFchromic EBT2 film and lithium fluoride (LiF) microcubes (1 mm cube). Small field sizes were defined by either microMLC down to 6×6 mm^2 or circular cones down to 4 mm in diameter.

There was excellent agreement found between the relative OFs measured with the EBT2 and LiF microcubes for all field sizes. They determined that both the EBT2 RCF and lithium fluoride (LiF) microcubes did not require any correction factors and as such were selected as the reference dosimeters. Correction factors for the active detectors (solid-state diodes, miniature ionization chambers, and natural diamond detectors) were then calculated using the doses measured with the EBT2 film and LiF microcubes. Correction factors varied between detectors ranging from up to 4% for some of the diode detectors and up to 20% for the PTW PinPoint ionization chamber as has been noted in references from various detectors [26].

13.3.3 AUSTRALIAN WORK—EXTRAPOLATION TECHNIQUE

As discussed earlier, volume averaging in detector is one of the biggest problems in small-field dosimetry providing significant large k_Ω factor. This has been attempted to be resolved using RCF as shown by several investigators [14,48]. On the contrary, the region of interest (ROI) in scanning and analysis has significant impact on the optical density and thus dose that gets even harder in small fields. Due to high resolution, RCF-generated beam profiles could be taken as standard to account for the shape of the small field or by performing full geometry calculations using Monte Carlo methods to obtain correction factors [21,49]. However, Monte Carlo simulations can be used as has been performed by various investigators [21,42,49–52]. However, it is an arduous task requiring a full model of the treatment machine, a process not readily available to many radiation oncology departments and thus experimental approaches are more palatable in which RCF can play important role. In this context, various studies have been performed using film for small-field dosimetry [45,53,54]. An experimental extrapolation technique using GAFchromic EBT3 film that can potentially eliminate the volume averaging effect in small-field dosimetry has been proposed by Morales et al. [55]. This technique consists of varying the size of the ROI within the scanned area of the film. A method is presented which uses a zero-area extrapolation technique utilizing high-resolution scanning to determine the final relative OFs for very small fields that will be elaborated here.

A single lot RCF EBT3 film was used for all relative OFs measurements in this work. All pieces of film were used and handled in the process outlined in the AAPM TG-55 report [56] and as described in Chapter 3 of this book. It has been shown that EBT3 film possesses a minimal X-ray energy dependence [57,58] and therefore should have a minimal impact on OF assessment at small field sizes using 6-MV X-rays. All films were analyzed using a PC desktop scanner and ImageJ (National Institutes of Health, USA) software on a PC workstation at least 24 h after irradiation to minimize effects from postirradiation coloration [59]. An Epson 10000XL dual lens system desktop scanner (Epson, NSW, and Australia) using a scanning resolution of 1200 pixels per inch was used. The images produced were 48 bit RGB color images and analyzed with the red component of the signal making the final pixel density values 16 bit information [60]. A control film was scanned with every experimental film in the same position for each measurement. The resulting scans were then corrected for any interscan variations based on the control films' result compared with the average result in a technique similar to that used by Lewis et al. [61].

The dose delivered to each experimental film was calculated by creation of a calibration dose response curve for the GAFchromic EBT3 film using standard fields of 10×10 cm^2 at given applied dose levels. This was performed because of the known, nonlinear relationship of net optical density to dose response of EBT3 GAFchromic film when scanned using an Epson10000XL desktop scanner. The net-optical densities were calculated for each film piece using circular ROI in ImageJ software. The circular ROI was centered on the cone-produced radiation field. The relative OFs versus diameter size were plotted in a curve. A best fit extrapolation technique was applied to the results to determine the zero-area OF utilizing a second-order polynomial or linear function based on cone size and data requirements. Figure 13.1 shows the measured values for relative OF when different circular ROI ranging from 0.1 to 1.8 mm are used for the analysis in ImageJ.

The figure also shows the extrapolation estimate of the relative OF for a zero volume or area calculation that was found to be 0.651 ± 0.018 and 0.971 ± 0.017 for the 4 and 25 mm Brainlab cones, respectively. The extrapolation was performed using a second-order polynomial line of best fit to provide the best estimate at zero-volume OF. The variation in measured OF with different ROI areas is expected to change because of a number of reasons including profile shape of the beam and light-scattering properties within the scanner and film. The uncertainty values quoted in the

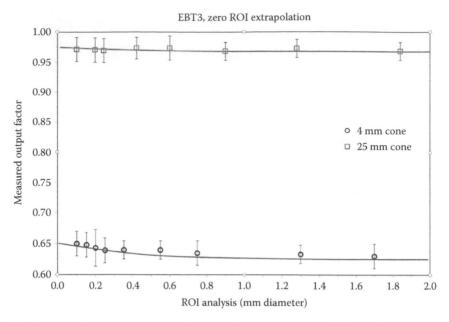

Figure 13.1 Measured and extrapolated output factor of a 4 and 25 mm Brainlab cone using GAFchromic™ EBT3 film with various sizes of ROI analysis areas. As the ROI decreases, slowly output increases, which is more visible in small cone compared with the large cone. (Redrawn from Morales, J.E. et al., *Med. Phys.*, 43, 4687–4692, 2016.)

figure and text are the standard deviation in measured results comparing the nine experimental films assessed for each cone size measured. These values combine both type A and type B errors associated with setup and dose delivery uncertainty along with experimental film-analysis errors combined. These results are in agreement with data provided by Garcia-Gurduno et al. [62] using four other detectors along with EBT and EBT2 films.

Effect of ROI can also be visualized in beam profile as shown in Figure 13.2. It shows an example of net-optical density profile for one 4-mm cone as measured by EBT3 film at a resolution of 1200 DPI that is the average of nine EBT3 film measurements. The central axis region of the profile is also shown in more detail to highlight both the variation in net-optical density and the uncertainty or noise level with the film scan for this typical measurement. In terms of very small ROI analysis, the user should make multiple measurements using different films to minimize the impact of selecting a small ROI for analysis around either a noise peak or trough, thus skewing the measured dose level either high or low by the magnitude of the noise which in our case was found to be between 0.5% and 1.5%.

The EBT film ROI extrapolation technique is compared with the data from microdiamond that has been advocated for small-field dosimetry in many publications [21,42,43,64–67]. Figure 13.3 shows the comparison, and inset data show percent difference indicating that EBT data with ROI extrapolation provide nearly identical data to within ± 1.5% over a range of the cones used in Brainlab.

For small fields that are almost entirely penumbral and thus nonflat area at the center of the field, the selection of ROI size in planar measurements can significantly impact the measurement result. Therefore, obtaining a series of ROIs of different diameters will enable extrapolation to the zero volume to obtain the true OF as shown in Figures 13.1 and 13.3.

The importance of dosimetric accuracy for very small-field relative OF that lies both in planning data as well as experimental dose verification. Although planning computer grid sizes are

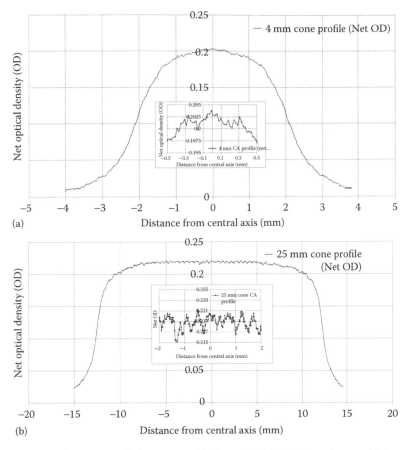

Figure 13.2 Measured net-optical density profile for a 6-MV SRS X-ray beam: (a) 4-mm cone and (b) 25-mm cone. These profiles were taken as the average of nine EBT3 film measurements. The inset in the figure includes details of the center of the beam for both cone. (Adapted from Morales, J.E. et al., *Med. Phys.*, 43, 4687–4692, 2016.)

often larger than the submillimeter measurement size in this work, they still require an accurate relative dose factor for each cone as their defined value. For example, the Brainlab iPlan treatment planning system utilizes an adaptive grid resolution down to 0.5 mm. By utilizing the extrapolation technique, as shown by Morales et al. [55], the relative output factor for the very small fields such as a 4 mm cone can be accurately measured and used for dose calculation in the treatment planning system.

13.3.4 SCANNING RESOLUTION

RCF has been used more extensively for the determination of small field size relative OFs as shown in Table 13.1 from various investigators over a long span of time. It shows the different DPI resolutions used in various small-field dosimetry studies [37,39,68–72]. The maximum DPI so far has been 150 (Table 13.1), and none of them used extrapolation techniques except [55]. The 150 DPI resolutions may be appropriate for cone sizes up to 10 mm and above; however, X-ray beams with dimensions less than 10 mm and microbeams require higher resolution. With caution, multiple film, higher DPI, and extrapolation technique can provide satisfactory results for output measurements in small field; thus, one can measure k_Ω for any detector when compared with RCF.

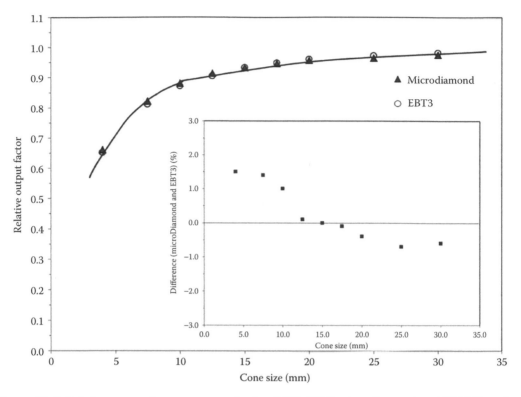

Figure 13.3 Relative output factor measured using PTW-60019 microdiamond and EBT3 films data extrapolated to zero ROI for Brainlab cones. Differences between these two detectors are very small (inset curve). (Data from Morales, J.E. et al., *Med. Phys.*, 43, 4687–4692, 2016.)

Table 13.1 The scanning resolution (DPI) used across studies for small-field dosimetry

Study and year	Film type	Resolution DPI	Year
Wilcox and Daskalov [68]	EBT	75	2007
Garcia-Garduno et al. [37]	EBT	100	2010
Kairn et al. [69]	EBT2	72	2011
Aland et al. [70]	EBT2	75	2011
Huet et al. [39]	EBT2	150	2012
Fiandra et al. [71]	EBT2,EBT3	72	2013
Moignier et al. [72]	EBT3	150	2014
Morales et al. [45]	EBT3	150	2014

13.4 RADIOCHROMIC FILM APPLICATION IN SMALL-FIELD PARAMETERS RELATIVE DOSIMETRY

A number of groups have used RCF for relative dosimetry measurements of percentage depth doses, TMRs, and off-axis profiles. Garcia-Garduno et al. [37] used GAFchromic EBT film within a water phantom for the dosimetry of small X-ray beams with diameters ranging from 4 to 20 mm on a Novalis linear accelerator. The film dose measurements consisted off-axis ratios, TMRs, and total scatter factors measurements and were compared with full Monte Carlo calculations using the BEAMnrc/EGSnrc Monte Carlo code. They found that the agreement between the EBT dose data

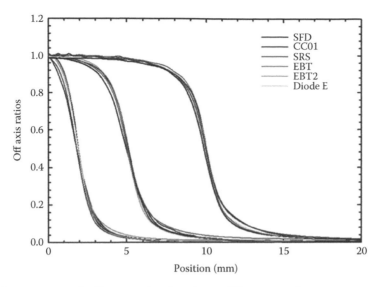

Figure 13.4 Comparison of off-axis ratios for three different small cones measured with four different detectors. GAFchromic EBT and EBT2 data are also shown that are indistinguishable from other data. (Adapted from Garcia-Garduno, O.A. et al., *Med. Phys.*, 41, 092101, 2014.) Similar data compared to Monte Carlo simulation and EBT can be found in Garcia-Garduno, O.A. et al., *Radiother Oncol.*, 96, 250–253, 2010.)

and Monte Carlo calculations was good and with differences of less than 2.7%. Figure 13.4 shows the EBT measured and Monte Carlo calculated profiles for all field diameters.

Morales et al. [73] determined surface doses for small-size circular collimators as used on a Novalis linear accelerator. Surface doses were measured using GAFchromic EBT3 film positioned at the surface of a Virtual Water phantom for circular collimators with diameters ranging from 4 to 30 mm. The structure of the EBT3 film is such that the active layer is very close to the International Commission on Radiological Protection (ICRP) skin dose depth of 70 μm. The film measurements were compared with Monte Carlo calculations performed using BEAMnrc/EGSnrc with a detailed geometric model of the linear accelerator. The agreement between EBT3 measured and calculated was better than 2.0%. From these results, they concluded that GAFchromic EBT3 film was an accurate dosimeter for surface dosimetry.

Chan et al. [74] used GAFchromic EBT2 film for measurement of depth doses and profiles for commissioning of iPlan in a Novalis system for field sizes ranging from 1×1 to 10×10 cm^2. All measurements were performed in polystyrene phantoms. These doses were compared with those measured with an EDGE diode detector and calculations in the treatment planning system. They concluded that EBT2 was an excellent dosimeter; however, such validity using edge detector without proper k_Ω is questionable. The measured planar dose distributions had pass rates of better than 98% passing rates and 95% with a set of 2%/2-mm dose and distance to agreement (DTA) criteria for all square fields and all patient treatment fields less than 5 cm.

13.5 DOSIMETRIC VERIFICATION—END-TO-END TESTING

The study by Esparza-Moreno et al. [75] looked at the dosimetry of very small SRS fields as used for the treatment of trigeminal neuralgia. This work studies the commissioning dosimetry data set obtained from Kodak X-OMAT V2 film and from GAFchromic EBT2 film. A total of 23 patients had dose calculations measured using both types of film and the relative isodoses compared. They found that there were differences in the full width at half maximum (FWHM) of the 35 and 40-Gy isodose lines that was of importance for doses to the brainstem. They concluded that there were statistically significant differences between the two calculated dose distributions that different types of film will small but significant differences in dose distributions.

Kim et al. [76] used GAFchromic EBT2 film for end-to-end testing during the commissioning of a Novalis Trilogy linear accelerator. In this report on commissioning, they found that measured

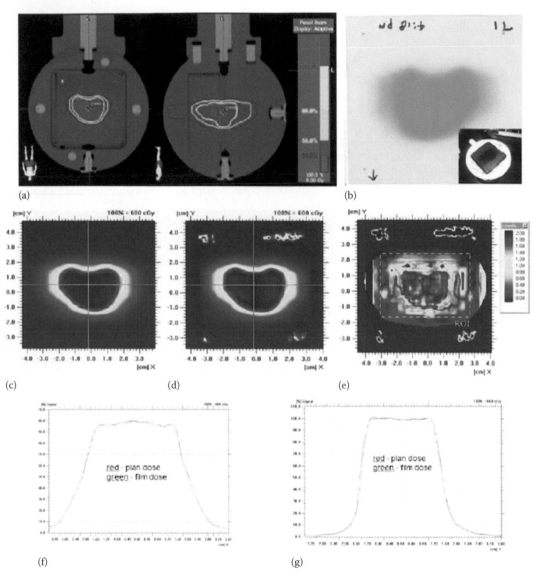

Figure 13.5 End-to-end test using Lucy 3D QA phantom: (a) treatment plan developed on phantom CT scan, (b) an exposed irradiated GAFchromic EBT film, (c) plan dose distribution, (d) film dose distribution, (e) gamma-index image with 3% dose difference and 1-mm distance to agreement (95% of pixels passed within ROI), (f) horizontal dose profiles, and (g) vertical profiles. (Redrawn from Kim, J. et al., *J. Appl. Clin. Med. Phys.*, 13, 124–151, 2012.)

doses at the isocenter had differences from the planned values of 0.8% and 0.4%, respectively, when measured with a small volume ionization chamber and GAFchromic EBT2 film. The gamma pass rate, measured by EBT2 film, was 95% (3% and 1-mm DTA) for their analysis. Figure 13.5 shows the results of the film measurements when performed inside of the Lucy phantom for a set of test SRS plans.

13.6 SUMMARY

RCF provides unsurpassed high resolution and superior 2D dosimetric information compared with any other types of detectors. If proper precautions for handling as discussed in Chapters 2–3 are adapted, RCF could provide suitable PDD, TMR, profiles for use in any SRS system or devices for small fields. The PDD and TMR can be satisfactorily derived on the basis of the precautions as discussed in extrapolation section. However, there are still inconsistencies in OF measurements as we lack the better understanding of the correction factor k_Ω. Nonetheless, being a tissue-equivalent detector, RCF should provide accurate dose in small fields which many of the publications [46,47,78–80] have advocated for several type of devices. It is our conviction that k_Ω is nearly unity exactly similar to other tissue equivalent detectors such as plastic scintillator [46,81–89] and microdiamond [21,43,44,64,65,67,90]. Additional research is needed to eliminate conflicting results and remarks as shown by several researchers [37,38,91] about the use of film.

REFERENCES

1. De Salles AA, Pedroso AG, Medin P et al. Spinal lesions treated with Novalis shaped beam intensity-modulated radiosurgery and stereotactic radiotherapy. *J Neurosurg* 2004; 101(3):435–440.
2. Lunsford LD, Kondziolka D, Flickinger JC et al. Stereotactic radiosurgery for arteriovenous malformations of the brain. *J Neurosurg* 1991;75:512–524.
3. Kondziolka D, Lunsford LD, Coffey RJ, Flickinger JC. Stereotactic radiosurgery of meningiomas. *J Neurosurg* 1991;74:552–559.
4. Eaton DJ, Naismith OF, Henry AM. Need for consensus when prescribing stereotactic body radiation therapy for prostate cancer. *Int J Radiat Oncol Biol Phys* 2015;91:239–241.
5. Garg AK, Wang XS, Shiu AS et al. Prospective evaluation of spinal reirradiation by using stereotactic body radiation therapy: The University of Texas MD Anderson Cancer Center experience. *Cancer* 2011;117:3509–3516.
6. Mahadevan A, Floyd S, Wong E et al. Stereotactic body radiotherapy reirradiation for recurrent epidural spinal metastases. *Int J Radiat Oncol Biol Phys* 2011;81:1500–1505.
7. Martin A, Gaya A. Streotactic body radiotherapy: A review. *Clin Oncol (R Coll Radiol)* 2010;22:157–172.
8. Potters L, Kavanagh B, Galvin JM et al. American society for therapeutic radiology and oncology (ASTRO) and American college of radiology (ACR) practice guideline for the performance of stereotactic body radiation therapy. *Int J Radiat Oncol Biol Phys* 2010;76:326–332.
9. Zheng X, Schipper M, Kidwell K et al. Survival outcome after stereotactic body radiation therapy and surgery for stage I non-small cell lung cancer: A meta-analysis. *Int J Radiat Oncol Biol Phys* 2014;90:603–611.
10. Benedict SH, Yenice KM, Followill D et al. Stereotactic body radiation therapy: The report of AAPM Task Group 101. *Med Phys* 2010;37:4078–4101.

11. Alfonso P, Andreo P, Capote R et al. A new formalism for reference dosimetry of small and nonstandard fields. *Med Phys* 2008;35:5179–5186.
12. Das IJ, Ding GX, Ahnesjö A. Small fields: Non-equilibrium radiation dosimetry. *Med Phys* 2008;35:206–215.
13. McKerracher C, Thwaites DI. Assessment of new small-field detectors against standard-field detectors for practical stereotactic beam data acquisition. *Phys Med Biol* 1999;44:2143–2160.
14. Bassinet C, Huet C, Derreumaux S et al. Small fields output factors measurements and correction factors determination for several detectors for a CyberKnife(R) and linear accelerators equipped with microMLC and circular cones. *Med Phys* 2013;40:071725.
15. McKerracher C, Thwaites DI. Head scatter factors for small MV photon fields. Part I: A comparison of phantom types and methodologies. *Radiot Oncol* 2007;85:277–285.
16. Charles PH, Cranmer-Sargison G, Thwaites DI et al. A practical and theoretical definition of very small field size for radiotherapy output factor measurements. *Med Phys* 2014;41:041707.
17. Martens C, De Wagter C, De Neve W. The value of the PinPoint ion chamber for characterization of small field segments used in intensity-modulated radiotherapy. *Phys Med Biol* 2000;45:2519–2530.
18. Ralston A, Liu P, Warrener K et al. Small field diode correction factors derived using an air core fibre optic scintillation dosimeter and EBT2 film. *Phys Med Biol* 2012;57:2587–2602.
19. Bogdanich W, Rebelo K. A pinpoint beam strays invisibly, harming instead of healing. *The New York Times*. http://www.nytimes.com/2010/12/29/health/29radiation.html?_r=0. New York, 2010.
20. Bogdanich W, Ruiz RR. Radiation errors reported in Missouri. *The New York Times*. http://www.nytimes.com/2010/02/25/us/25radiation.html. New York, 2010.
21. Benmakhlouf H, Sempau J, Andreo P. Output correction factors for nine small field detectors in 6 MV radiation therapy photon beams: A PENELOPE Monte Carlo study. *Med Phys* 2014;41:041711.
22. Francescon P, Cora S, Cavedon C. Total scatter factors of small beams: A multidetector and Monte Carlo study. *Med Phys* 2008;35:504–513.
23. Francescon P, Beddar S, Satariano N, Das IJ. Variation of k(fclin, fmsr, Qclin, Qmsr) for the small-field dosimetric parameters percentage depth dose, tissue-maximum ratio, and off-axis ratio. *Med Phys* 2014;41:101708.
24. Aspradakis MM, Bryne JP, Palmans H et al. Report No 103: Small Field MV Dosimetry. York, UK: Institute of Physics and Engineering in Medicine, IPEM; 2010.
25. Das IJ, Francescon P, Ahnesjö A et al. Small fields and non-equilibrium condition photon beam dosimetry: AAPM Task Group Report 155. *Med Phys* 2017; (in review).
26. IAEA TRS 483. Dosimetry of small static fields used in external beam radiotherapy: An IAEA-AAPM international code of practice for reference and relative dose determination, Technical Report Series No. 483. Vienna, Austria: International Atomic Energy Agency; 2017.
27. Bouchard H, Seuntjens J. Ionization chamber-based reference dosimetry of intensity modulated radiation beams. *Med Phys* 2004;31:2454–2465.
28. Bouchard H, Seuntjens J, Kawrakow I. A Monte Carlo method to evaluate the impact of positioning errors on detector response and quality correction factors in nonstandard beams. *Phys Med Biol* 2011;56:2617–2634.
29. Bouchard H, Seuntjens J, Palmans H. On charged particle equilibrium violation in external photon fields. *Med Phys* 2012;39:1473–1480.
30. Bouchard H. A theoretical re-examination of Spencer-Attix cavity theory. *Phys Med Biol* 2012;57:3333–3358.
31. Bouchard H, Seuntjens J, Duane S et al. Detector dose response in megavoltage small photon beams. I. Theoretical concepts. *Med Phys* 2015;42:6033–6047.

32. Bouchard H, Kamio Y, Palmans H et al. Detector dose response in megavoltage small photon beams. II. Pencil beam perturbation effects. *Med Phys* 2015;42:6048–6061.

33. Fenwick JD, Kumar S, Scott AJ, Nahum AE. Using cavity theory to describe the dependence on detector density of dosimeter response in non-equilibrium small fields. *Phys Med Biol* 2013;58:2901–2923.

34. Kumar S, Fenwick JD, Underwood TS et al. Breakdown of Bragg-Gray behaviour for low-density detectors under electronic disequilibrium conditions in small megavoltage photon fields. *Phys Med Biol* 2015;60:8187–8212.

35. Bouchard H, Seuntjens J, Carrier JF, Kawrakow I. Ionization chamber gradient effects in non-standard beam configurations. *Med Phys* 2009;36:4654–4663.

36. Zeidan OA, Li JG, Low DA, Dempsey JF. Comparison of small photon beams measured using radiochromic and silver-halide films in solid water phantoms. *Med Phys* 2004;31:2730–2737.

37. Garcia-Garduno OA, Larraga-Gutierrez JM, Rodriguez-Villafuerte M et al. Small photon beam measurements using radiochromic film and Monte Carlo simulations in a water phantom. *Radiother Oncol* 2010;96:250–253.

38. Larraga-Gutierrez JM, Garcia-Hernandez D, Garcia-Garduno OA et al. Evaluation of the Gafchromic((R)) EBT2 film for the dosimetry of radiosurgical beams. *Med Phys* 2012;39:6111–6117.

39. Huet C, Dagois S, Derreumaux S et al. Characterization and optimization of EBT2 radiochromic film dosimetry system for precise measurements of output factors in small fields used in radiotherapy. *Radiat Meas* 2012;47:40–49.

40. Massillon JLG, Cueva-Procel D, Diaz-Aguirre P et al. Dosimetry for small fields in stereotactic radiosurgery using Gafchromic MD-V2-55 film, TLD-100 and alanine dosimeters. *PloS One* 2013;8:e63418.

41. Das IJ, Cheng CW, Watts RJ, Ahnesjo A, Gibbons J, Li XA, Lowenstein J, Mitra RK, Simon WE, Zhu TC. Accelerator beam data commissioning equipment and procedures: Report of the TG-106 of the therapy physics committee of the AAPM. *Med Phys* 2008;35:4186–4215.

42. Russo S, Masi L, Francescon P et al. Multicenter evaluation of a synthetic single-crystal diamond detector for CyberKnife small field size output factors. *Phys Med* 2016;32:575–581.

43. Chalkley A, Heyes G. Evaluation of a synthetic single-crystal diamond detector for relative dosimetry measurements on a CyberKnife. *Br J Radiol* 2014;87:20130768.

44. Larraga-Gutierrez JM, Ballesteros-Zebadua P, Rodriguez-Ponce M et al. Properties of a commercial PTW-60019 synthetic diamond detector for the dosimetry of small radiotherapy beams. *Phys Med Biol.* 2015;60:905–924.

45. Morales JE, Crowe SB, Hill R et al. Dosimetry of cone-defined stereotactic radiosurgery fields with a commercial synthetic diamond detector. *Med Phys* 2014;41:111702.

46. Papaconstadopoulos P, Archambault L, Seuntjens J. Experimental investigation on the accuracy of plastic scintillators and of the spectrum discrimination method in small photon fields. *Med Phys* 2017;44:654–664.

47. Shukaili KA, Petasecca M, Newall M et al. A 2D silicon detector array for quality assurance in small field dosimetry: DUO. *Med Phys* 2017;44:628–636.

48. Azangwe G, Grochowska P, Georg D et al. Detector to detector corrections: A comprehensive study of detector specific correction factors for beam output measurements for small radiotherapy beams. *Med Phys* 2014;41:072103.

49. Francescon P, Kilby W, Satariano N. Monte Carlo simulated correction factors or output factor measurement with the CyberKnife system—results for new detectors and correction factor dependence on measurement distance and detector orientation. *Phys Med Biol* 2014;59:N11–N17.

50. Francescon P, Kilby W, Satariano N, Cora S. Monte Carlo simulated correction factors for machines specific reference field dose calibration and output factor measurement using fixed and iris collimators on the CyberKnife system. *Phys Med Biol* 2012;57:3741–3258.

51. Cranmer-Sargison G, Weston S, Evans JA et al. Implementing a newly proposed Monte Carlo based small field dosimetry formalism for a comprehensive set of diode detectors. *Med Phys* 2011;38:6592–6602.

52. Cranmer-Sargison G, Weston S, Evans JA et al. Monte Carlo modelling of diode detectors for small field MV photon dosimetry: Detector model simplification and the sensitivity of correction factors to source parameterization. *Phys Med Biol* 2012;57:5141–5153.

53. Gonzalez-Lopez A, Vera-Sanchez JA, Lago-Martin JD. Small fields measurements with radiochromic films. *J Med Phys* 2015;40:61–67.

54. Reinhardt S, Hillbrand M, Wilkens JJ, Assmann W. Comparison of Gafchromic EBT2 and EBT3 films for clinical photon and proton beams. *Med Phys* 2012;39:5257–5562.

55. Morales JE, Butson M, Crowe SB et al. An experimental extrapolation technique using the Gafchromic EBT3 film for relative output factor measurements in small X-ray fields. *Med Phys* 2016;43:4687.

56. Niroomand-Rad A, Blackwell CR, Coursey BM et al. Radiochromic film dosimetry: Recommendations of AAPM radiation therapy committee task group 55. American Association of Physicists in Medicine. *Med Phys* 1998;25:2093–2115.

57. Hill R, Healy B, Holloway L et al. Advances in kilovoltage X-ray beam dosimetry. *Phys Med Biol* 2014;59:R183–231.

58. Brown TA, Hogstrom KR, Alvarez D et al. Dose-response curve of EBT, EBT2, and EBT3 radiochromic films to synchrotron-produced monochromatic x-ray beams. *Med Phys* 2012;39:7412–7417.

59. Cheung T, Butson MJ, Yu PK. Post-irradiation colouration of Gafchromic EBT radiochromic film. *Phys Med Biol* 2005;50:N281–N285.

60. Butson MJ, Cheung T, Yu PK. Absorption spectra variations of EBT radiochromic film from radiation exposure. *Phys Med Biol* 2005;50:N135–N140.

61. Lewis D, Chan MF. Correcting lateral response artifacts from flatbed scanners for radiochromic film dosimetry. *Med Phys* 2015;42:416–429.

62. Garcia-Garduno OA, Rodriguez-Ponce M, Gamboa-deBuen I et al. Effect of dosimeter type for commissioning small photon beams on calculated dose distribution in stereotactic radiosurgery. *Med Phys* 2014;41:092101.

63. Morales JE, Butson M, Crowe SB et al. An experimental extrapolation technique using the Gafchromic EBT3 film for relative output factor measurements in small x-ray fields. *Med Phys* 2016;43:4687–4692.

64. Andreo P, Palmans H, Marteinsdottir M et al. On the Monte Carlo simulation of small-field micro-diamond detectors for megavoltage photon dosimetry. *Phys Med Biol* 2016;61:L1–L10.

65. Lechner W, Palmans H, Solkner L et al. Detector comparison for small field output factor measurements in flattening filter free photon beams. *Radiot Oncol* 2013;109:356–360.

66. Mancosu P, Reggiori G, Stravato A et al. Evaluation of a synthetic single-crystal diamond detector for relative dosimetry on the Leksell Gamma Knife Perfexion radiosurgery system. *Med Phys* 2015;42:5035–5041.

67. Marsolat F, Tromson D, Tranchant N et al. A new single crystal diamond dosimeter for small beam: Comparison with different commercial active detectors. *Phys Med Biol* 2013;58:7647–7660.

68. Wilcox EE, Daskalov GM. Evaluation of GAFCHROMIC EBT film for CyberKnife dosimetry. *Med Phys* 2007;34:1967–1974.

69. Kairn T, Hardcastle N, Kenny J et al. EBT2 radiochromic film for quality assurance of complex IMRT treatments of the prostate: Micro-collimated IMRT, RapidArc, and TomoTherapy. *Aust Phys & Eng Sci Med* 2011;34:333–343.

70. Aland T, Kairn T, Kenny J. Evaluation of a Gafchromic EBT2 film dosimetry system for radiotherapy quality assurance. *Aust Phys & Eng Sci Med* 2011;34:251–260.

71. Fiandra C, Fusella M, Giglioli FR et al. Comparison of Gafchromic EBT2 and EBT3 for patient-specific quality assurance: Cranial stereotactic radiosurgery using volumetric modulated arc therapy with multiple noncoplanar arcs. *Med Phys* 2013;40:082105.

72. Moignier C, Huet C, Makovicka L. Determination of the KQclinfclin, Qmsr fmsr correction factors for detectors used with an 800 MU/min CyberKnife((R)) system equipped with fixed collimators and a study of detector response to small photon beams using a Monte Carlo method. *Med Phys* 2014;41:071702.

73. Morales JE, Hill R, Crowe SB et al. A comparison of surface doses for very small field size X-ray beams: Monte Carlo calculations and radiochromic film measurements. *Australas Phys Eng Sci Med* 2014;37:303–309.

74. Chan MF, Zhang Q, Li J et al. The verification of iPlan commissioning by radiochromic EBT2 films. *Int J Med Phys Clin Eng Radiat Oncol* 2012;1:1–7.

75. Esparza-Moreno KP, Garcia-Garduno OA, Ballesteros-Zebadua P et al. Comparison of trigeminal neuralgia radiosurgery plans using two film detectors for the commissioning of small photon beams. *J Appl Clin Med Phys* 2013;14:3824.

76. Kim J, Yoon M, Kim S et al. Three-dimensional radiochromic film dosimetry of proton clinical beams using a Gafchromic EBT2 film array. *Radiat Prot Dosimetry* 2012;151:272–277.

77. Kim J, Wen N, Jin JY et al. Clinical commissioning and use of the Novalis Tx linear accelerator for SRS and SBRT. *J Appl Clin Med Phys* 2012;13:124–151.

78. Ploquin N, Kertzscher G, Vandervoort E et al. Use of novel fibre-coupled radioluminescence and RADPOS dosimetry systems for total scatter factor measurements in small fields. *Phys Med Biol* 2015;60:1–14.

79. Hassani H, Nedaie HA, Zahmatkesh MH, Shirani K. A dosimetric study of small photon fields using polymer gel and Gafchromic EBT films. *Med Dosim* 2014;39:102–107.

80. Pantelis E, Moutsatsos A, Zourari K et al. On the output factor measurements of the CyberKnife iris collimator small fields: Experimental determination of the k(Q(clin),Q(msr))(f(clin),f(msr)) correction factors for microchamber and diode detectors. *Med Phys* 2012;39:4875–4885.

81. Masi L, Russo S, Francescon P et al. CyberKnife beam output factor measurements: A multi-site and multi-detector study. *Phys Med* 2016;32:1637–1643.

82. Tanny S, Sperling N, Parsai EI. Correction factor measurements for multiple detectors used in small field dosimetry on the Varian Edge radiosurgery system. *Med Phys* 2015;42:5370–5376.

83. Underwood TS, Rowland BC, Ferrand R, Vieillevigne L. Application of the exradin W1 scintillator to determine Ediode 60017 and microDiamond 60019 correction factors for relative dosimetry within small MV and FFF fields. *Phys Med Biol* 2015;60:6669–6683.

84. Carrasco P, Jornet N, Jordi O et al. Characterization of the Exradin W1 scintillator for use in radiotherapy. *Med Phys* 2015;42:297–304.

85. Morin J, Beliveau-Nadeau D, Chung E et al. A comparative study of small field total scatter factors and dose profiles using plastic scintillation detectors and other stereotactic dosimeters: The case of the CyberKnife. *Med Phys* 2013;40:011719.

86. Klein DM, Tailor RC, Archambault L et al. Measuring output factors of small fields formed by collimator jaws and multileaf collimator using plastic scintillation detectors. *Med Phys* 2010;37:5541–5549.

87. Wang LL, Beddar S. Study of the response of plastic scintillation detectors in small-field 6 MV photon beams by Monte Carlo simulations. *Med Phys* 2011;38:1596–1599.

88. Gagnon JC, Theriault D, Guillot M et al. Dosimetric performance and array assessment of plastic scintillation detectors for stereotactic radiosurgery quality assurance. *Med Phys* 2012;39:429–436.

89. Cho SH, Vassiliev ON, Lee S et al. Reference photon dosimetry data and reference phase space data for the 6 MV photon beam from Varian Clinac 2100 series linear accelerators. *Med Phys* 2005;32:137–148.

90. Papaconstadopoulos P, Tessier F, Seuntjens J. On the correction, perturbation and modification of small field detectors in relative dosimetry. *Phys Med Biol* 2014;59:5937–5952.

91. Tyler M, Liu PZ, Chan KW et al. Characterization of small-field stereotactic radiosurgery beams with modern detectors. *Phys Med Biol* 2013;58:7595–7608.

Use of radiochromic film in electron beam radiation therapy

THOMAS R. MAZUR AND H. HAROLD LI

14.1 INTRODUCTION

Electron beams have found various clinical applications including treatment of superficial diseases. The limited penetration depth of high-energy electrons enables uniform coverage of shallow targets while simultaneously sparing normal tissues underlying the treatment site. Although electron beams have many promising applications, accurate dosimetry can be challenging primarily due to steep dose gradients near a patient's surface. Radiochromic film has proven to be highly valuable in electron-beam dosimetry for characterizing dose distributions largely due to its ability to provide high spatial resolution among other characteristics.

Dose distributions can be mapped in many cases using point dosimeters such as thermoluminescent dosimeters (TLDs) and diode detectors; however, tracing distributions point-by-point can be labor intensive and time consuming in certain applications in which many sets of treatment parameters are investigated. Moreover, certain applications of electron beams require dosimetric measurements in highly atypical conditions in which conventional dosimeters and protocols may no longer be applicable. Film is appealing for many of these applications as it provides visualization of dose distributions across a large dynamic range with high spatial resolution while also providing a permanent record.

Beyond these general features, radiochromic film has been largely shown to be only moderately energy dependent for high energy (>6 MeV) electron beams [1–5]. For example, Su et al. [1] irradiated GAFchromic™ EBT films to 150 cGy for electron energies spanning from 6 to 22 MeV. Mean optical density (OD) values within the irradiated films deviated by less than 4% relative to the mean-value spanning all films for each energy. In addition, they measured limited response dependence on fractionation (<3% variation for different fractionations) and dose rate (<2% variation for dose rates up to 600 MU per min), among other parameters including measurement depth and cone size. Beyond these desirable response characteristics, film can also be readily embedded into both phantoms (including water) and treatments. Moreover, handling radiochromic film is relatively straightforward, not requiring chemical processing as in the case of radiographic film.

Various investigators have successfully shown that GAFchromic films can be used satisfactorily and uniquely in electron-beam dosimetry [5,6]. In this chapter, we provide an overview of several exciting applications of electron beams in therapy in which radiochromic film has contributed substantially to dosimetry. We first discuss applications of radiochromic film in investigating dose calculation platforms for electron beams. For each of these applications (in which dosimetry must be performed under highly unique conditions), film contributes significantly to diverse aspects of the treatments ranging from commissioning to in vivo dosimetry. We then summarize the role of radiochromic film in facilitating reliable delivery of several types of treatments including the following: (1) total skin electron radiation therapy, (2) modulated electron radiation therapy (MERT), and (3) intraoperative electron radiation therapy (IOERT).

14.2 DOSE CALCULATION VALIDATION

Verification of dose-calculation platforms is a pervasive application for many types of dosimeters. Radiochromic film, however, has been particularly useful in evaluating dose calculations of electron beams in which spatial resolution is highly valuable for probing steep-dose gradients. For example, Coleman et al. [7] used film as an experimental benchmark in comparing electron beam dose calculations performed both via Monte Carlo (MC) and the Fermi–Eyges algorithm. They particularly investigated dose calculation of electron-boost treatments delivered to breast-cancer patients following whole-breast photon irradiation (with interest in assessing variation in calculated dose to the ipsilateral lung and left ventricle). These boosts delivered 16 Gy—in addition to the 45 Gy during the whole-breast treatment—in 2 Gy fractions prescribed to the 80% isodose level. Planning systems at the time largely relied upon the Fermi–Eyges algorithm as implemented by Hogstrom [8] for its simplicity despite MC being more realistic by modeling-relevant physical interactions.

Although dose calculations could be directly compared, film measurements were also performed. Older-generation film (GAFchromic MD-55) was inserted between slabs defining the chest of an anthropomorphic phantom (Alderson RANDO) with the film edge cut to conform to the chest surface. The phantom itself was positioned on a breast board at an angle of 10.5° relative to the couch. Likewise, the gantry was rotated to ensure normal incidence of the electron beam on the phantom surface. A computed tomography (CT) scan was performed prior to delivery to orient the field relative to a set of radio-opaque markers within the phantom. The left chest wall of the phantom was ultimately irradiated to 70 Gy (maximum) with a $5 \times 7 \text{ cm}^2$ electron beam at 12 MeV. Figure 14.1 shows depth–dose curves along the central axis of the beam obtained from both MC simulation and film measurement. Film measurement and MC calculation results agreed within 3%/3 mm gamma criteria across the entire film, thus providing confidence (particularly when supplemented by diode measurements also shown in Figure 14.1) in the MC calculations for subsequent evaluation of the Fermi–Eyges algorithm.

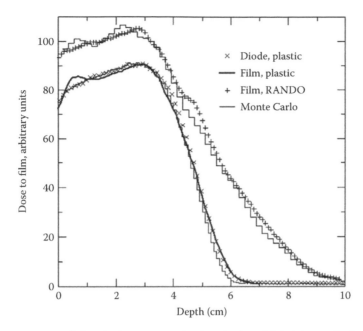

Figure 14.1 Comparison of depth-dose curves both calculated (via Monte Carlo simulation) and measured by Coleman et al. along the central axis of water-equivalent plastic and an anthropomorphic phantom for 12 MeV electrons. (Adapted from Coleman, J. et al., *Int. J. Radiat. Oncol.*, 621–628, 2005. With Permission.)

More recently, Vandervoort et al. [9] used film heavily for evaluating a commercial MC algorithm for electron-beam calculation. In particular, film (GAFchromic EBT) supplemented diode measurements performed in water for benchmarking the performance of the XiO electron MC treatment planning software developed by Elekta in a heterogeneous phantom. The phantom developed for these measurements was designed to model the geometry of the trachea and spine [10]. The phantom consists of a 4.7 cm thick slab of muscle-substitute material (MS-11) that is 15 × 15 cm² in size. A 2.5 cm hole runs along the length of the slab through its center just below its surface (to simulate the trachea). A set of four hard-bone equivalent plastic disks (1 cm thick and 2.5 cm in diameter) spaced by 0.5 cm were placed underneath the hole to simulate the spine. The composite phantom was placed on 12 cm thick solid water slabs that were 30 × 30 cm² in size.

Films were placed both directly between the composite phantom and slabs and 1 cm below the composite phantom. Films were irradiated by 9 and 17 MeV electron beams, whereas XiO electron MC was used for calculating dose to the anatomy simulated by the phantom. Figure 14.2 compares isodose plots produced both from calculation and measurement directly below the spinal disks for 17 MeV electrons. The distributions agree reasonably across most of the films with failure in gamma analysis—using 3%/2 mm criteria—observed mostly at the boundaries between different materials. The good agreement between calculation and film (with over 98% of pixels deemed passing using 3%/2 mm criteria for all measurements) provides confidence in the dose calculation.

For electron-therapy applications discussed in the following, film allows for verification of planned dose (both in vitro and in vivo). Film particularly facilitates dose validation by allowing fast and accurate data acquisition with high-spatial resolution.

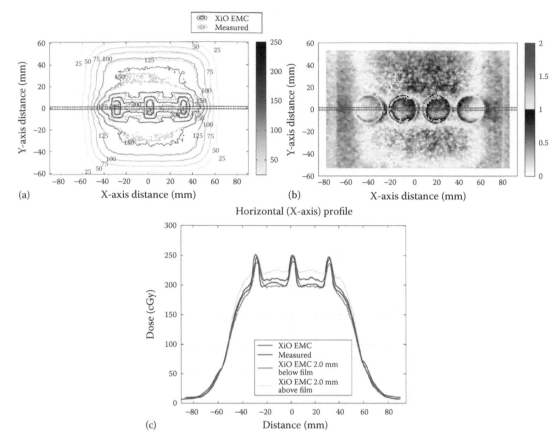

Figure 14.2 (a) Comparison of calculated (XiO EMC) and measured (radiochromic film) isodose lines following irradiation by a 9 MeV electron beam to 200 MU immediately below a set of hard-bone equivalent disks in a phantom modeling the spine and trachea. (b) Gamma analysis obtained using 3%/2 mm criteria for the data in (a). (c) Average field profiles within the dashed-blue lines are shown in (a) and (b). (Vandervoort, E.J. et al., *Med. Phys.*, 41, 21711, 2014. With Permission.)

14.3 TOTAL SKIN ELECTRON THERAPY

Total skin electron therapy (TSET) is a technique that is most frequently applied to the treatment of mycosis fungoides, a cutaneous T-cell lymphoma. Given that lesions are predominantly confined to the first few millimeters of the skin, low-energy electron beams (~2–9 MeV) are highly effective for treatment due to their limited penetration depth. A challenge in applying TSET toward treatment of mycosis fungoides, among other whole-body, superficial diseases, is identifying a suitable patient and beam setup for ensuring uniform application of an electron beam across the entire area to be treated (which in the case of mycosis fungoides can include the entire skin). In addition, quality assurance must be rigorously performed to verify the repeatability of irradiations across the large treatment areas.

Large-area beam performance must be first characterized prior to implementing a linear accelerator for application to TSET, and then monitored periodically for constancy to be confident in the dosimetric results of treatments. As outlined in Task Group 30 of the American Association of Physicists in Medicine (AAPM TG-30) [11], given the highly extended source-to-skin distance (SSD) typically used for these treatments (~300 cm) and the concomitant beam current required to maintain adequate dose rates in this setup, identifying suitable dosimeters and protocols for beam calibration and monitoring requires serious consideration. Although institutions have adapted dosimetric protocols—such as TG-51 [12]—that rely on (plane-parallel) ionization chamber measurements for TSET dosimetry, radiochromic film has enabled more comprehensive two-dimensional measurements.

For example, in 2010 Schiapparelli et al. [13] developed absolute dosimetry for an implementation of TSET on a linear accelerator (Varian 2100 C/D) by combining measurements from a plane-parallel ionization chamber and radiochromic film (GAFchromic EBT-1). Similarly to other TSET implementations, the authors produced 6 MeV electron beams with a 36 × 36 cm² field size at isocenter, ultimately yielding an approximate dose rate of 2500 cGy per min at isocenter in a high dose rate mode of operation. Irradiation was performed with the gantry rotated to 90° (i.e. beam line parallel to floor) to allow adequate distance (SSD > 350 cm) for geometrically broadening the field to uniformly cover a patient. A patient would stand on a wooden rotating platform behind a 1 cm thick 200 × 100 cm² acrylic slab that degraded the electron beam energy from its nominal value of 6 MeV to between 3 and 5 MeV.

Patients are typically treated with TSET using a variation of a technique—referred to as the Stanford technique in deference to the origin of TSET at Stanford University—that combines multiple beam angles in an effort to achieve more uniform irradiation [14]. Most commonly, two beams are used with one directed in the patient's superior direction and the other toward the inferior direction. Meanwhile, the patient rotates on the treatment platform either continuously or in discrete steps. Figure 14.3 indicates the core components of the TSET setup implemented by Schiapparelli et al. (among other institutions) and identifies a coordinate system for reference.

Schiapparelli et al. used radiochromic film for both identifying an optimal-beam geometry and performing thorough quality assurance for ensuring uniformity and repeatability. Figure 14.4 summarizes film measurements from their optimization experience. Film strips (2 × 2.5 cm²) lining a plastic sheet at the treatment plane along the patient's axis (i.e., y-axis) were first irradiated with a dual-beam setup at various tilt angles for the beams. Relative dose measured along films was then compared for the angles considered. As shown in Figure 14.4a, the optimal tilt angle was measured to be 19°. Shallower angles produced maximal dose along the beam axis with measurably less dose at ±80 cm (i.e., typical patient height) in the superior and inferior directions.

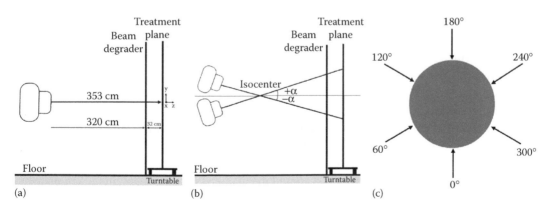

Figure 14.3 (a) Simplified side view (with coordinate definitions) of TSET geometry used by Schiapparelli et al. among others. (b) Schematic of dual-beam arrangement implemented for ensuring uniform dose within a field size comparable to a patient's dimensions. (c) Top-down view of turntable indicating six incident beam angles (achieved by rotating turntable) for uniformly irradiating a patient's surface. (Adapted from Schiapparelli, P. et al., *Med. Phys.*, 37, 3510, 2010. With Permission.)

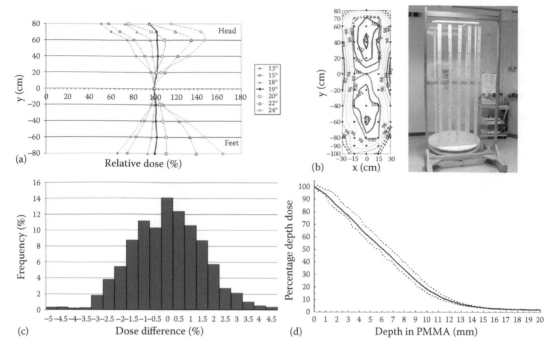

Figure 14.4 (a) Film-measured dose profiles along the y-axis (as indicated in Figure 14.3) for different angles α used in dual-beam irradiations. (b) Isodose lines (left) measured on the treatment plane using radiochromic film strips (as shown on the right). (c) Histogram summarizing percent differences between film-measured doses on the surface of a cylindrical phantom (with its proximal surface on the treatment plane) and the mean dose applied to the surface. (d) Variation (indicated by dashed lines) in percent depth-dose curves measured using films pressed between two cylindrical phantoms that were irradiated by a dual-beam, six-position TSET procedure. (Adapted from Schiapparelli, P. et al., *Med. Phys.*, 37, 3510, 2010. With Permission.)

In contrast, steeper angles yielded a pair of measured maxima offset from the central axis (with the apparent asymmetry being a consequence of scatter from the floor in the inferior direction).

After configuring an optimal dual-beam geometry, the in-plane dose uniformity was measured at the treatment distance using radiochromic film. Film strips (2×2.5 cm²) were arranged in a grid on a 180×60 cm² surface at the treatment position. As shown in Figure 14.4b, measured dose was highly uniform across the plane with a mean relative value (normalized to the measured value at the origin along the beam axis) of 97% ± 5%. Film was likewise used for measuring percent depth dose (PDD) curves in both single- and dual-beam geometries for the sake of comparison with percent depth ionization curves measured using an ionization chamber in plastic (PMMA). Depths where measured curves reached 50% of their maximal values (R_{50}) were in agreement for both ionization chamber and film.

Film was lastly used for characterizing a Stanford-type irradiation combining two beams in six phantom geometries. Films were wrapped around the diameter of a CT phantom (32 cm diameter) to measure surface dose. The phantom was setup coaxially with the treatment platform and then offset away from the beam to position its proximal surface on the treatment plane. In addition, films were sandwiched between identical cylinders at various angles relative to the central axis to measure cumulative PDD curves for a typical treatment. Figure 14.4c summarizes the distribution of dose differences between measured values on the phantom surface and the mean dose. Figure 14.4d lastly shows the range of PDD curves measured across all films.

Of critical importance, radiochromic film confirmed calibration factors that were determined using ionization chamber measurements for relating skin dose (defined as the mean dose along a circle near a cylindrical phantom surface) following irradiation by dual beams in the six patients' orientations to dose measured at a calibration point for a single, horizontal beam. First, dose at the calibration point in dual-beam geometry was related to the dose in a geometry using a single horizontal beam by a factor A. This factor was determined by the ratio between PDD values for the two configurations at the calibration point. Then, the dose at the calibration point for the dual-beam geometry was related to the skin dose by another multiplicative factor B. This factor was derived by dividing the cumulative skin dose measured (using film) in the six configurations to that measured for a static dual-beam geometry.

The TSET geometry and delivery protocol varies by institution, partially because linear accelerators must be adapted to the application (from more routine external beam treatments). For instance, rather than using a multibeam configuration (à la Schiapparelli et al. [13]) for ensuring uniform coverage of a patient, Reynard et al. [15] integrated a custom flattening filter consisting of aluminum, lead, and plastic into the gantry head for a Varian Clinac 21EX. Moreover, rather than positioning a patient in discrete orientations relative to the beam, Reynard et al. continuously rotated their treatment platform in an effort to ensure more uniform coverage of the skin. In addition, boost electron fields were applied for supplementing the dose to areas that might be shielded from the prescription dose (such as the palms of the hands and soles of the feet). These institution-specific approaches to TSET require thorough, multidimensional dosimetric characterization.

Again, film (GAFchromic EBT) was used for measuring surface dose and PDDs, and results were compared with plane-parallel ionization chamber measurements (and in turn used for identifying calibration factors). Films were also placed at various positions on a humanoid phantom (Rando phantom) to search for locations in which the applied dose deviated substantially from the prescription. These measurements showed that delivered dose agreed with prescription (200 cGy) to within ±10% for all locations except an anterior point on the neck.

Beyond phantom-based characterization of under-dosed areas like the work of Reynard et al. [15], film has served as an in vivo dosimeter for TSET treatments. For example, Gamble et al. [16] used film to measure dose in the inframammary folds of patients being treated for inflammatory and multinodular breast cancer (using the six-position Stanford-based technique) with long-term interests in applying film toward assessing dose in areas that might be under dosed during treatment of mycosis fungoides. Although treatments are designed to optimize uniformity in electron fluence, certain areas such as the inframammary folds could be under dosed because of self-shielding. By monitoring dose to these problematic areas across a fractionated regimen, supplemental fields could be subsequently prescribed accordingly. Prior work has suggested that these supplemental fields could be critically important toward minimizing the likelihood of disease relapse [17].

Gamble et al. [16] adhered double-layer GAFchromic MD-55 film (~50 × 110 mm^2) to a pair of patients for three fractions of a thirty fraction regimen, being sure to include a set of irradiations spanning all six configurations. Patients were treated to 35 Gy on the skin using a degraded 6 MeV beam from a linear accelerator (Varian 2100C). Lithium fluoride TLDs were placed on the films for the sake of comparing measurements. TLDs had been more conventionally used for in vivo dosimetry of TSET treatments; however, film should provide more comprehensive two-dimensional information [18]. As disease was limited to the contralateral skin for both patients, large dynamic ranges were measured. Absolute film dose was inferred by constructing a calibration curve for optical densities by exposing a set of films to known doses.

Figure 14.5 shows isodose curves obtained from film measurements for both patients (with percentage dose normalized to prescription). The gradient in dose from the inframammary fold

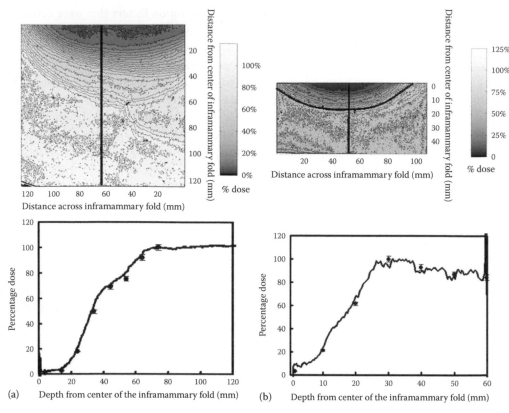

Figure 14.5 Sample radiochromic film measurements recorded by Gamble et al. with films placed in the inframammary folds of two patients being treated for breast cancer using a TSET protocol. Upper images show isodose curves, whereas lower figures show percent-dose curves along vertical lines indicated on isodose plots. (a) Patient 1, (b) Patient 2 (Adapted from Gamble, L.M. et al., *Int. J. Radiat. Oncol.*, 62, 920–924, 2005. With Permission.)

to the skin beyond the fold is immediately evident from two-dimensional data. Percentage dose profiles extending outward from the folds show that substantial areas receive less than 50% of the prescription dose. For example, the isodose line in bold for the second patient defines the area in proximity to the fold that receives less than 40% of the prescribed dose. This area indicated by film measurements could subsequently be used for defining a boost field to the fold. Data points on the percentage dose curves show TLD measurements at discrete locations. Nearly, all points agree to within indicated uncertainties with the profiles extracted from film.

Bufacchi et al. [19] similarly positioned films (GAFchromic EBT) on four patients undergoing TSET to monitor dose uniformity. Twenty positions were considered, and TLDs were again positioned on top of films for the sake of comparison. Treatments were performed at a highly extended SSD using a PMMA-degraded 6 MeV electron beam from a linear accelerator (Varian 2100 C/D). The Stanford-based dual-beam, six-position protocol was used.

Figure 14.6a indicates the points of measurement on the patients. Beyond just allowing for dose monitoring, the average measurement spanning the first six locations (across the belt) at the beginning of treatments was used to accurately determine the monitor units required for delivering the prescribed dose. Figure 14.6b summarizes percentage difference between film and TLD measurements for all positions considered. In contrast to the work of Gamble et al. [16], substantially higher deviation is observed between TLD and film measurements, although percentage

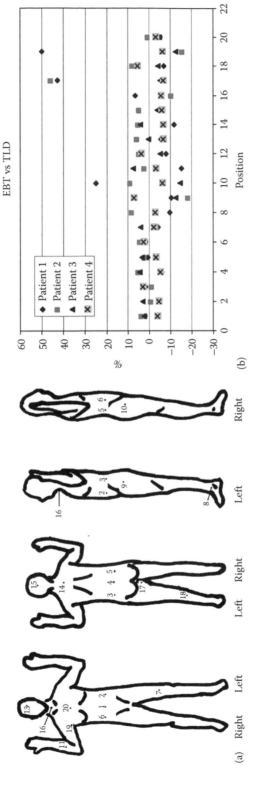

Figure 14.6 (a) Summary of film and thermoluminescent dosimeter (TLD) placements used by Bufacchi et al. for in vivo dosimetry of TSET treatments. (b) Percent differences between film and TLD measurements for all positions considered among a set of four patients. (Adapted from Bufacchi, A. et al., *Phys. Medica.*, 23, 67–72, 2007. With Permission.)

differences are within 10% for 85% of measurements. These deviations are largely attributed to the variations in patients' surfaces. Such variations should arguably be better captured by film that can be conformed accordingly to a patient's surface.

14.4 MODULATED ELECTRON RADIATION THERAPY

MERT is a promising approach for producing more conformal dose distributions than conventional electron therapy while maintaining the advantages of electron therapy over photon-based treatments for treating superficial targets. Energy modulation allows for better localization of dose deposition at the target volume, whereas intensity modulation ensures dose uniformity across the volume. Electron beam modulation (both in energy and intensity) has historically been accomplished using patient-specific bolus that is positioned in proximity to a patient's skin [20]. Largely because of the inconvenience of using bolus—both in terms of fabrication and implementation during treatment—modulation using bolus is typically limited to several segments per treatment.

More recently, automated devices have been investigated for application of MERT. Several designs have been considered for electron-specific multileaf collimators (MLCs). These devices are typically appended to the gantry head of a linear accelerator in the applicator tray and positioned closely (within a few centimeters) to the patient's surface. Alternative approaches have adapted existing MLCs for photon therapy to electron beam treatments. Utilizing these MLCs simplifies delivery by not requiring additional hardware and allows for the combined application of electrons and photons in treatment. On the other hand, machine constraints associated with photon-specific MLCs—such as a minimum distance from MLC to patient—can hinder modulation effectiveness for electron therapy. Radiochromic film has been instrumental in developing both of these approaches to MERT in which two-dimensional dosimetry with high spatial resolution is of the utmost importance for developing confidence in the dose delivered by a highly modulated beam.

For example, Salguero et al. [21] applied film toward assessing the feasibility of adapting photon MLCs toward the application of MERT to the irradiation of the chest wall in the treatment of post-mastectomy breast cancer patients. The authors particularly applied MERT using a linear accelerator (Siemens PRIMUS) that included an MLC with 29 opposed leaf pairs. The distal distance to the MLC from the radiation focus was approximately 36 cm, and the width of the shadow cast by most pairs at the isocenter was 1 cm. The minimum SSD allowed by the apparatus was 50 cm. Simulation of electron interactions in the MLC—which had not been implemented in commercial-treatment planning systems—required the authors to develop an MC platform for dose calculation. This platform (1) identifies beamlet-based electron fluences for considered energies, (2) calculates dose contributions from each beamlet to all voxels within a volume of interest and configures beamlet weights according to delivery constraints, (3) sequences fluences into segments, and then (4) reoptimizes weights upon re-calculating dose using fully defined segments.

To verify deliveries prepared by the in-house planning system, both an ionization chamber array and radiochromic film were implemented for mapping dose distributions. The array consisted of 27×27 plane-parallel ionization chambers with 5 mm spacing between each chamber. Array measurements were compared with simulated dose of the air cavities within the chambers. As conversion of dose-to-air to dose-to-water is energy dependent and the beam spectrum varies spatially chamber-to-chamber, determination of dose-to-water would be prohibitively time consuming for quality assurance. In contrast, two-dimensional distributions at higher spatial resolution could be directly measured using film (GAFchromic EBT) embedded in a solid water phantom irradiated perpendicularly by the electron beam. Films were irradiated at depths of 0.5,

1.5, and 2.0 cm with absolute dosimetry being performed by a single plane-parallel ionization chamber positioned at 1.6 cm depth. These combined measurements then allowed for quality assurance through the assessment of gamma analysis (3 mm/3%) applied to relative dose distributions and differences in absolute point dose measurements (<3%). Dose calculation in solid water was approximated by identifying stopping powers corresponding to the mean energies at the phantom surface for each segment.

Figure 14.7a shows relative isodose lines measured at 1.5 cm depth using both the ionization chamber array and radiochromic film for a plan considered. Film measurements provide finer detail due to film's higher spatial resolution and also allow for gamma comparison (right) with dose calculations generated by the planning system. Figure 14.7b compares DVH curves generated by dose calculation for MERT for the plan shown in Figure 14.7a with the more conventional approach of using shielding blocks (or a static MLC configuration) for modulation. For four patients considered, MERT generally produced more uniform coverage of planning target volumes (PTV) while also reducing dose to the heart and lung in some cases. With plan verification provided by film measurements, MERT delivered using photon MLCs presents a promising approach for improving chest-wall irradiation in breast cancer treatment.

Salguero et al. [22] later demonstrated the application of radiochromic film to patient-specific quality assurance of MERT applied to the treatment of shallow head and neck tumors. As in their prior work, photon MLCs were used for modulation, and treatment planning was performed using an in-house system. Initial validation of their MERT planning system in application to head and heck targets again used a commercial two-dimensional ionization chamber array. A plan was constructed for treating a synthetic PTV embedded within a pair of adjacent organs-at-risk as illustrated in Figure 14.8a. This artificial treatment volume required four segments including three energies (9, 12, and 18 MeV). Figure 14.8b outlines planned and delivered dose distributions at two depths with delivered measurements being performed again using an ionization chamber array.

Although chamber array measurements demonstrated the feasibility of delivering MERT using photon MLC segments defined using an in-house dose calculation engine, the spatial resolution provided by the array was insufficient for quality assurance of actual clinical plans. The isodose comparisons in Figure 14.8b were constructed by overlapping measurements and calculations to minimize local error in the region of highest dose. Measurements with higher spatial resolution might indicate non-negligible deviation in this region that could have meaningful clinical impact. Especially for clinical cases including steep dose gradients and highly irregular geometries, higher spatial resolution is critically important for quality assurance. In addition, ionization chambers might not be applicable for clinical cases including steep dose gradients in which lateral electronic equilibrium might not be satisfied.

Four plans were considered for film-based delivery verification including cases of (1) recurring squamous cell carcinoma in the oronasal region following radical surgery, (2) parotid mucoepidermoid carninoma with infiltration in the right preauricular region, (3) basal cell carcinoma in the right ear, and (4) diffuse large B-cell lymphoma of the parotid gland. Figure 14.9a outlines calculated isodose distributions for the third case considered. Isodose lines from film measurements at three depths within a solid water phantom are shown in Figure 14.9b. General agreement is evident in comparing calculated and measured distributions; however, some deviations are apparent that are hidden by the coarser resolution provided by the ionization chamber array. Film measurements enabled quantitative verification through gamma analysis (3 mm/3%). In combination with absolute point dose measurement using a single chamber, radiochromic film measurements thus allowed for quality assurance of the highly modulated fields delivered for MERT by photon MLCs.

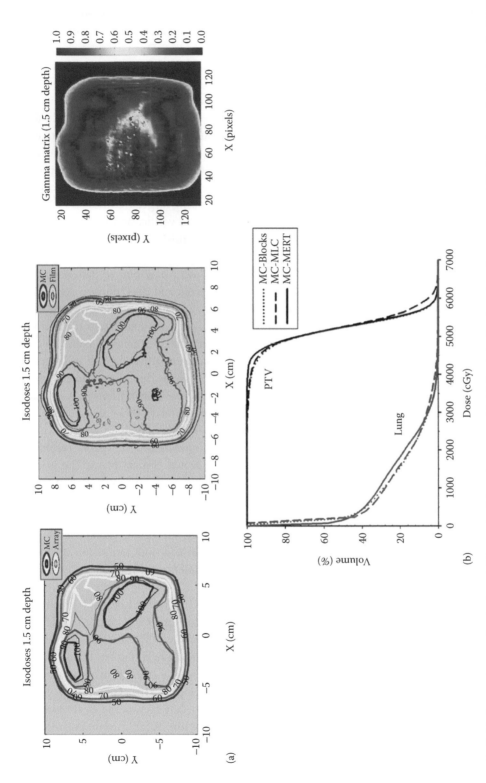

Figure 14.7 (a) Comparisons of calculated (Monte Carlo) and measured (with both ionization chamber arrays and film) isodose distributions for MERT plans to be delivered using photon MLCs for postmastectomy chest wall irradiation. Displayed measurements were performed at 1.5 cm depth in a solid water phantom. (b) Comparison of calculated lung and PTV DVHs for the patient shown in (a) for conditions in which treatment is to be delivered using conventional blocks (MC-Blocks), a static MLC configuration (MC-MLC), or MERT based on photon MLCs (MC-MERT). (Adapted from Salguero, F.J. et al., *Radiother. Oncol.,* 93, 625–632, 2009. With Permission.)

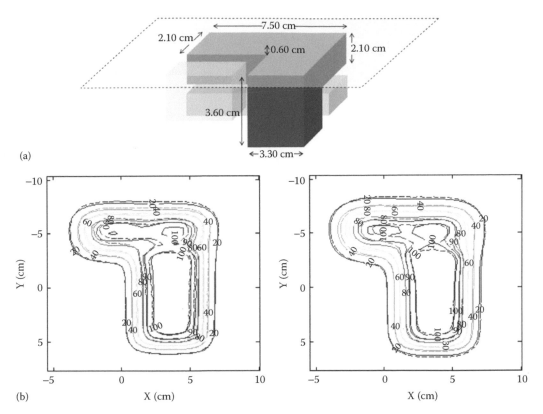

Figure 14.8 (a) Geometry considered by Salguero et al. for evaluating the feasibility of delivering MERT using photon MLCs to head and neck targets. The L-shaped red volume represents the PTV, whereas the other shapes simulate organs-at-risk where dose should be minimized. (b) Comparisons between calculated and measured (using ionization chamber arrays) dose distributions at two depths for a MERT plan derived using the geometry in (a). (Adapted from Salguero, F.J. et al., *Phys. Med. Biol.*, 55, 1413–1427, 2010. With Permission.)

Likewise, radiochromic film has been critically important in verifying plan deliveries for MERT applied using an electron-specific MLC (more specifically a few-leaf electron collimator or FLEC). The FLEC was designed as an accessory to be appended to a linear accelerator in the tray at the base of an electron applicator. A first-generation device developed by Al-Yahya et al. [23], as shown in Figure 14.10, consisted of two pairs of copper leaves oriented perpendicularly. The leaf thickness was chosen to be 1.2 cm to balance electron stopping power and bremsstrahlung yield. Although the photon jaws are configured to provide beam collimation, the leaf width was chosen to be suitably wide (3.0 cm) to suppress electron leakage beyond the field size defined by the leaves. The device was configured to fit in a 15×15 cm^2 Varian applicator, ultimately allowing for a peak field opening of 8×8 cm^2 and positioning of the opening within a few centimeters of a patient's surface. Each leaf can be individually positioned via computer control of stepper motors, with encoders being implemented for closed-loop feedback of position.

Prior to integrating the FLEC device into treatment planning and actual treatments, device parameters are needed to be optimized as a function of FLEC configuration (with each opening being deemed a fieldlet). For every fieldlet (1) output was maximized while (2) the width of the dose penumbra was minimized, and (3) the percentage dose beyond the defined field edge was minimized. MC simulation allowed for these characteristics to be predicted for a given geometry, and radiochromic film measurements enabled high-resolution confirmation of simulated results. Beam

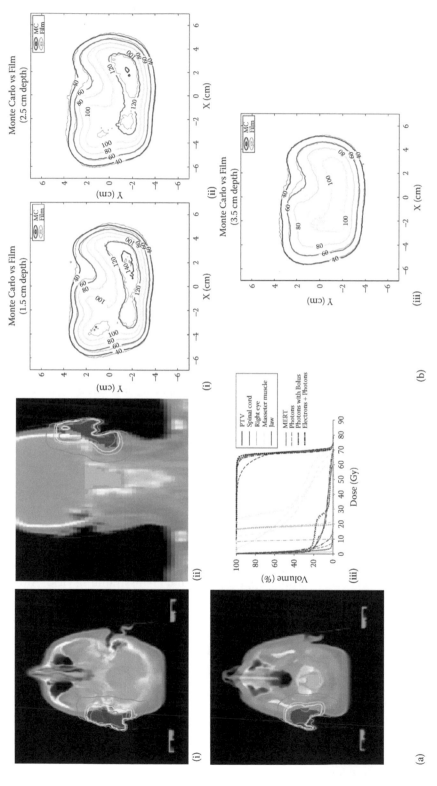

Figure 14.9 (a) Example of MERT-based plan calculated using the planning system described by Salguero et al. for treating basal-cell carcinoma in the right ear, with 110% (red), 110% (orange), 90% (yellow), 80% (green), and 40% (blue) isodose levels relative to prescription being displayed. Bottom-right compares calculated DVH curves for MERT with alternative treatment modalities. (b) Comparison between calculated and film-measured isodose lines at 2.5 cm depth in a water phantom for the MERT-plan outlined in (a). (Adapted from Salguero, F.J. et al., *Phys. Med. Biol.*, 55, 1413–1427, 2010. With Permission.)

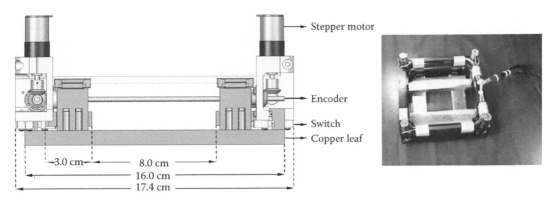

Figure 14.10 Schematic (left) with photograph (right) of FLEC device developed by Al-Yahya et al. for MERT delivery. (Adapted from Al-Yahya, K. et al., *Med. Phys.*, 34, 4782, 2007. With Permission.)

profiles for various photon jaw openings and FLEC-defined field sizes were measured using film (GAFchromic HS) at 2.8 and 1.5 cm depths in solid water for 18 and 9 MeV beams, respectively. Absolute dose was inferred from film measurements using the well-established protocol of Devic et al. [24] where a calibration curve is constructed from a polynomial fit to a set of films irradiated to known doses.

Figure 14.11a shows beam profile comparisons between film measurements and MC calculations at 9 and 18 MeV beam energies for a square FLEC opening of 6 cm with jaw openings of 6, 7, and 8 cm. Al-Yahya et al. [23] devised a notation in which beam energy, jaw opening, and FLEC size are specified in a format such as 18J8C6 for an 18 MeV beam with 8 cm jaw opening and 6 cm FLEC size. For both energies considered, beam profiles show that larger jaw openings provide higher outputs and sharper penumbra, but worse off-axis leakage. Again, radiochromic-film measurements showed good agreement with MC calculation and importantly contributed to identifying suitable choices of jaw openings for given FLEC-defined field sizes. Figure 14.11b compares planar measurements and calculations for both several individual FLEC fieldlets at 2.8-cm depth and a composite distribution for the combined fieldlets at 3- and 5-cm depths. Measured and calculated isodose lines were generally in agreement, except near the periphery of the film in which leakage dose is likely problematic.

Radiochromic film (EBT3) has been recently used to evaluate the delivery accuracy of a clinical plan using the FLEC device [25]. An in-house treatment planning system was used for devising a FLEC-based MERT treatment for a patient who had undergone whole-breast irradiation for invasive ductal carcinoma. The prior treatment first irradiated the left breast to 42.56 Gy in 16 fractions and then applied a conventional electron boost to 10 Gy in four fractions. Although the MERT plan was constructed for this patient simply for the sake of evaluating delivery accuracy, in principal MERT could be considered as a candidate for more effective administration of the boost. The beamlet-based inverse planning constructed a plan for delivering a prescription dose of 2.78 Gy to 95% of the PTV.

Planned and delivered dose distributions were compared using a $30 \times 30 \times 12$ cm^3 solid water phantom. The planned delivery was first recalculated onto the phantom, and then the plan was delivered to a film placed at 2 cm depth in the phantom. Of note, the multichannel dosimetry technique developed by Micke et al. [26] was applied to measurements to extract any nondose-dependent contributions from measured signals. The plan included 196 FLEC segments at energies of 9, 12, 16, and 20 MeV. Figure 14.12 compares planned and delivered dose both for a constituent energy in the plan (16 MeV) and the cumulative plan. Film measurements were used for calculating

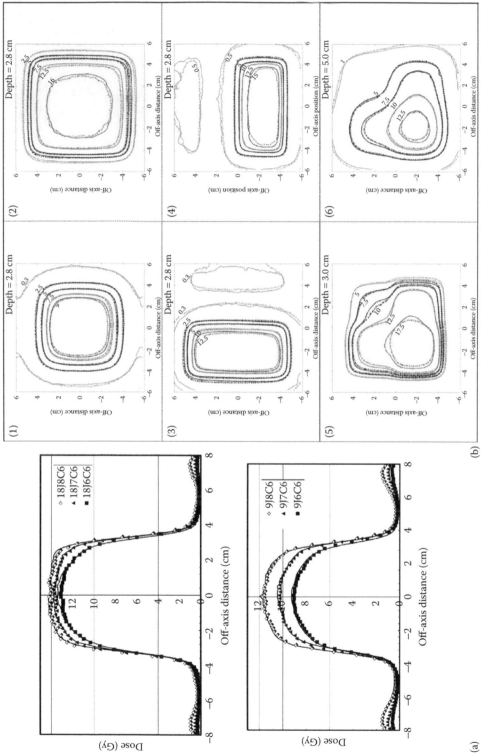

Figure 14.11 (a) Calculated (symbols) and measured (lines) transverse profiles for several FLEC configurations using 18 MeV (upper) and 9 MeV (lower) electrons. (b) Calculated (symbols) and measured (lines) dose distributions for various single (1—4) and composite (5—6) fieldlet deliveries. (Adapted from Al-Yahya, K. et al., *Med. Phys.*, 34, 4782, 2007. With Permission.)

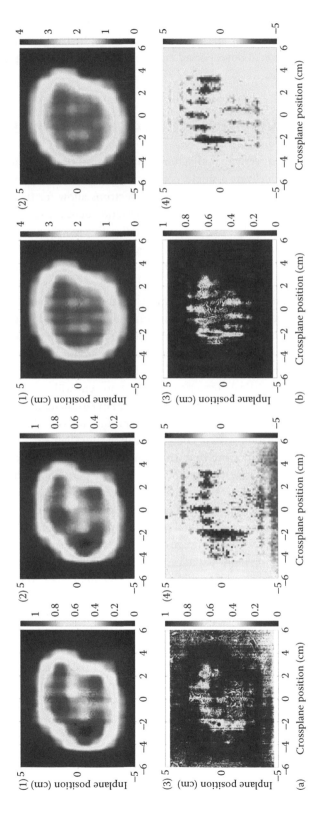

Figure 14.12 (a) Measured (1) and planned (2) dose distributions at 2 cm depth in a solid water phantom for the 16 MeV contribution of a 196 segment FLEC-based MERT treatment along with a 5%/5 mm gamma map (3) and a map of the percent difference between the planned and measured dose (4). (b) The same as (a), except now showing the cumulative, 196 segment delivery. (Adapted from Connell, T. et al., *Med. Phys.*, 41, 61715, 2014. With Permission.)

a gamma passing rate at 2 cm depth. Gamma passing rates for 3 mm/3% were above 97% except at the lowest energy considered (9 MeV) in which the applied dose was substantially lower than other energies. Vector shifts were identified for all films to produce a qualitatively minimal subtraction image between planned and delivered dose. For all energies (and the composite plan), these shifts were less than 2 mm in magnitude.

14.5 INTRAOPERATIVE ELECTRON RADIATION THERAPY

IOERT applies ionizing radiation directly to both a tumor site and resection margins during a surgical intervention. IOERT typically delivers a single, concentrated dose to a site either just before or after surgical removal of a tumor. In contrast to photons, electrons allow for the dose to be more precisely delivered to the tumor site depending upon the penetration depth. To date, IOERT has been predominantly applied to breast and pelvic treatments either directly preceding or following surgery. In fact, breast and rectal cancers are the diseases most frequently treated by IOERT in Europe [27].

Given the high doses typically applied during IOERT treatments (~25 Gy), in vivo dosimetry is critically important for verifying dose delivery. The significance of in vivo dosimetry for IOERT is enhanced by the fact that treatment planning for most treatments relies solely on manual calculations. These calculations depend on measurements—including size and shape of the region to be treated—performed upon visualizing the target during surgery. Although preoperative CT scans have been applied to IOERT planning, these images do not reflect anatomical changes that occur during surgery [28]. Intraoperative imaging has been proposed for more realistic planning, although CT imaging during surgery presents logistical obstacles that might prevent widespread implementation in the foreseeable future. In the absence of planning images, manual calculations typically specify dose prescription to the 90% isodose level as determined by profiles measured in a water phantom under various beam conditions (including beam energy and applicator diameter and bevel angle).

A challenge in performing in vivo dosimetry during IOERT is identifying a dosimeter that can accurately characterize dose across irregular geometries without perturbing the radiation field while maintaining sterility in proximity to the surgical area. The high-dose rates administered during IOERT treatments also complicate calibration dosimetry. Conventional protocols using plane-parallel ionization chambers often cannot be applied because of uncertainties in ion-recombination factors associated with high dose rates [29]. Radiochromic films are well suited to the unique challenges presented by IOERT, including high spatial resolution and dynamic range, low dependence on both electron energy and dose rate, and low dependence on incident beam angle. Consequently, film has been applied to both in vitro and in vivo measurements for IOERT.

Ciocca et al. [30] applied radiochromic film to in vivo dosimetry for IOERT in treating early-stage breast cancer. In particular, film was used to assess delivered dose to a majority of patients (35 of 54) who were treated for early-stage breast unifocal carcinoma exclusively by IOERT immediately following quadrantectomy or wide excision. Prescription for these patients was 21 Gy at the depth of 90% isodose. Dose was delivered at nominal beam energies of 5, 7, and 9 MeV by a mobile linear accelerator (Hitesys Novac7) that was capable of producing up to 0.09 Gy per pulse at 5 Hz, thus enabling prescription delivery within 2 min. Round acrylic applicators were used for hard-docking beam collimation to the tumor bed. Applicator diameters varied between 4 and 8 cm, and applicator ends were either perpendicular to the collimation axis or beveled at 22.5°.

Two radiochromic films (GAFchromic MD-55-2) cut into 1.5×1.5 cm^2 pieces were wrapped in a sterile envelope and placed on top of the surgical bed at its center prior to docking the applicator. Surface dose was extracted from these films by first determining the mean OD within a

1×1 cm^2 window centered on the films. Dose was then obtained using a calibration curve that was constructed from measurements in a solid phantom at the depth of maximum dose for 6 MeV electrons. Entrance dose—defined here as the dose at depth of maximum dose—was ultimately determined by applying an experimentally determined correction factor to the measured surface dose. Ciocca et al. [30] estimated the overall uncertainty in dose measurements to be approximately 4%. Contributions to uncertainty in the measurements included film calibration and various correction factors applied to OD measurements. Figure 14.13 summarizes the observed differences between measured and expected doses for all patients considered. The mean difference in entrance dose spanning all patients was 1.8% ± 4.7%. From the measured distribution, Ciocca et al. [30] defined thresholds beyond which certain actions should be taken for quality assurance. For example, deviations of 7 and 10% were identified as thresholds for confirming film calibration curves and electron beam calibration, respectively.

More recently, Severgnini [31] et al. performed in vivo film dosimetry with radiochromic film (GAFchromic EBT-3) for verifying the alignment of shielding disks in breast IOERT treatments. In this study, 37 breast cancer patients were prescribed an IOERT-based boost of 10 Gy in a single fraction prior to conventional whole-breast radiotherapy (50 Gy in 25 fractions). For most patients, 8 cm diameter shielding disks—consisting of a 3 mm layer of copper sandwiched between two PMMA layers—were positioned between the residual breast and the pectoralis fascia to protect tissues underneath the target volume (including the heart and lungs).

Films were fixed to both faces of the disk, allowing for dose distributions to be measured both just below the target volume and immediately above the tissues that should be protected from irradiation by the disk. More specifically, the average absolute dose was estimated within a 2 cm diameter in the center of the exposed area of the upper film. Likewise, images of the lower film were used for measuring transmission across the shield. Calibration curves were generated for the film at 6 and 9 MeV beam energies in a water-equivalent phantom with a 10 cm collimator perpendicular to the phantom surface. Measurements were performed at a reference depth both with 5×5 cm^2 film strips and an ionization chamber.

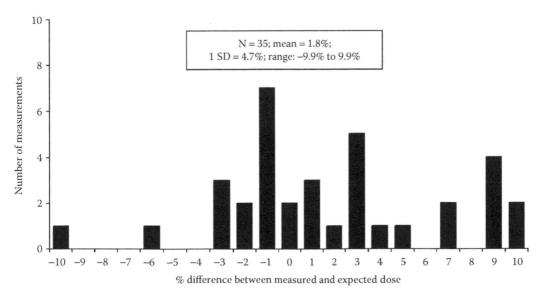

Figure 14.13 Histogram of observed deviations between measured and expected entrance doses delivered during IOERT in the treatment of 35 breast cancer patients. (Adapted from Ciocca, M. et al., *Radiother. Oncol.*, 69, 285–289, 2003. With Permission.)

Figure 14.14a shows an example image generated by measurement of the upper film for a single patient. In this case where the film dimensions exceed the field size (6.5 cm), the shield is seen to be misaligned with the treatment field as evident by overlapping a 6.5 cm diameter circle with the measured field. In fact, beam area beyond the shield can be inferred from the image to be 4.9 cm². Moreover, the average dose inferred from the upper film is measured to be 4% less than the expected dose. Excluding two patients, the measured average dose spanning the cohort agreed to within 4% of the prescribed dose. In addition, the dose measured on the lower film was less than 40 cGy for all patients (confirming the effectiveness of the shield when properly aligned). Based on observed misalignment for several patients, Severgnini et al. [31] modified the shielding setup to include an elastic band that is wrapped around the assembly and subsequently stitched to the pectoralis fascia prior to treatment to prevent slippage of the disk. For 21 patients in which this modified setup was implemented, the field area beyond the shield was reduced (relative to the other 16 patients) from a mean value of 7.3 cm² to just 3.6 cm². For eight of those patients (with an example being shown in Figure 14.14b), the shield was setup such that the incident field was contained entirely within the shield.

Film has similarly been instrumental for in vivo dosimetry of IOERT applied to pelvic treatments. For instance, Krengli et al. [32] used radiochromic film for assessing rectal dose in a study investigating the feasibility of IOERT for prostate cancer treatment. Between 2005 and 2008, 38 patients with locally advanced prostate cancer were treated with IOERT prior to radical prostatectomy and lymph node dissection. Radiation was delivered by a linear accelerator (IntraOp Mobetron) using electron energies between 9 and 12 MeV. Prescription for all patients except three was 12 Gy to the 90% isodose level. Any patient whose pathology indicated extracapsular extension or positive surgical margins were to receive postoperative radiation therapy.

The target volume was irradiated with beveled collimators (at angles between 15°and 30°) with diameters ranging between 4.5 and 6.5 cm. To identify a suitable choice for beam parameters, an urologist first exposed the prostate and placed a stitch on the bladder neck for guiding the positioning of the collimator to the target volume. Ultrasound imaging of the prostate—particularly

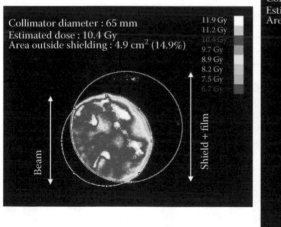
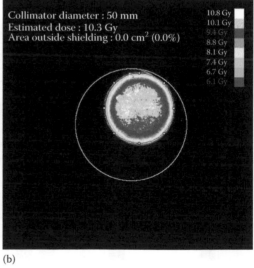

(a) (b)

Figure 14.14 (a) Example of film-measured misalignment between copper shield and electron beam during breast treatment via IOERT. (b) Example of good alignment between shield and beam as evident from the beam being completely captured by film measurement. (Adapted from Severgnini, M. et al., *J. Appl. Clin. Med. Phys.*, 16, 1–9, 2015. With Permission.)

the diameter in the anterior–posterior direction and distance between the prostate surface and anterior rectal wall—was used for choosing a specific collimator diameter and bevel angle.

An underlying rationale for performing IOERT prior to prostatectomy is to allow for prostate tissue to shield the anterior rectal wall from being directly irradiated. To evaluate dose to the rectum, four radiochromic films were affixed to the surface of a rectal probe that was 12 cm in length and 2.5 cm in diameter. Mean doses to these films were interpreted as doses to the anterior rectal wall (3.9 Gy), left and right rectal walls (1.1 and 1.4 Gy), and posterior rectal wall (0.1 Gy). The range in measured dose to the anterior rectal wall spanned from 0.4 to 8.9 Gy. The large difference between measured doses to the anterior and posterior walls indicated that the rectum was within the steep dose fall-off of the incident field.

As a final example of the application of film to IOERT, Costa et al. [33] used radiochromic film to thoroughly assess variation in dose distributions in pelvic IOERT treatments for an assortment of treatment parameters. Pelvic IOERT treatments typically require beveled applicators, with 30°and 45° bevels being the most commonly used. As the irradiated surface is often somewhat irregular, the applicator sometimes cannot be docked flush with the surface. Costa et al. [33] explored film measurements in a solid phantom to assess changes in dose distributions that result from irregular treatment surfaces (leading to poor applicator coupling). Radiochromic film (GAFchromic EBT3) allowed for dose distributions to be feasibly measured with high spatial resolution for various combinations of treatment parameters while simulating poor docking between applicator and surface.

A challenge in measuring beam profiles in a solid phantom along the field direction (or more precisely, perpendicular to the phantom surface first encountered by the incident radiation) is that air gaps between the film and phantom can induce artifacts in dose distributions. To minimize air gaps, films were pressed at the center of a set of 16 solid phantom slabs. Clamps applied pressure to acrylic plates that sandwiched the slabs. In one configuration, an old film was placed between the center-most slabs. This film included a slot that defined a gap for loading a film to be irradiated for a test measurement. Both films were fixed to the slabs by adhesive tape. In a second configuration, the slotted film for inserting films for measurement between slabs was replaced by a 1 cm thick bolus (30×30 cm^2) whose elasticity was hypothesized to enable better conformity between an inserted film and adjacent surfaces.

Measurements were performed on a linear accelerator (Varian Clinac 2100 C/D) that included a hard-docking system for cylindrical applicators. Various applicator diameters (6, 7, and 8 cm), bevel angles (0°, 30°, and 45°), and beam energies (6, 9, and 12 MeV) were considered. Percent depth dose curves were measured along a line perpendicular to the phantom surface that intersects the central axis of a given applicator. Figure 14.15a compares PDD curves for measurements performed with a nonbeveled 10×10 cm^2 applicator using the different setups outlined earlier. Poor agreement was observed (particularly at shallow depth) between diode measurements in water and film simply pressed between phantom slabs. In contrast, good agreement was obtained at all depths for the configuration using bolus, whereas dose measured at some shallow depths is slightly low for the setup using slotted film presumably because not all air gaps could be suppressed. As exemplified in Figure 14.15b, deviation between film and diode measurements could be quantified via gamma analysis. For all applicator diameters, bevel angles, and beam energies considered, gamma values were less than one using 2%/2 mm criteria at all depths between 2 mm and the depth in which dose reaches 10% of its maximum value. In addition, the percent difference between film and ionization chamber (in water) doses obtained along the central axis at the depth of maximum dose was at most 3%.

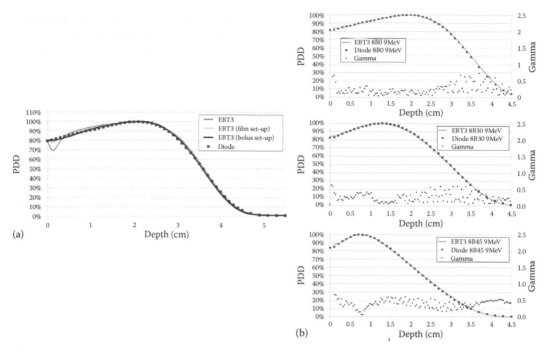

Figure 14.15 (a) Comparison of PDD curves measured (for a conventional applicator) by a diode in a water tank and film in a solid water phantom with various arrangements, including being directly pressed between phantom slabs (EBT3), sliding between slabs via a cutout in a film that is sandwiched between the slabs (film set-up), and being clamped between slabs with flat bolus on one face of the film (bolus set-up). (b) PDD curves (with associated gamma values for comparison to diode measurements) recorded with the bolus set up in "a" for 0°, 30°, and 45° bevel angles on an 8 cm applicator. (Adapted from Costa, F. et al., *Phys. Medica.*, 31, 692–701, 2015. With Permission.)

Two-dimensional distributions could ultimately be imaged from films irradiated for PDD measurements. Figure 14.16a shows reference distributions, whereas Figure 14.16b shows distributions measured for cases designed to simulate clinically realistic scenarios. For the more realistic measurements, two pieces of bolus were prepared on top of the slabs. A concave hole was prepared between the bolus pieces to simulate a male sacral bone. Film was positioned between these additional pieces of bolus with a cut-out that conformed to the hole. The particular configurations shown in Figure 14.16b include (1) two beveled applicators positioned over the hole and (2) a beveled applicator oriented at an incline by resting on an additional, adjacent piece of bolus. All shown distributions were normalized to the dose at the depth of maximum dose measured on corresponding reference distributions. Comparison of distributions in Figure 14.16 provides clinical insights into the consequences of irregular treatment surfaces. For example, the uppermost distribution in Figure 14.16b appear to be shifted forward and lengthened relative to its counterpart in Figure 14.16a. Similar effects appear to be present for the other distributions as shown in Figure 14.16b.

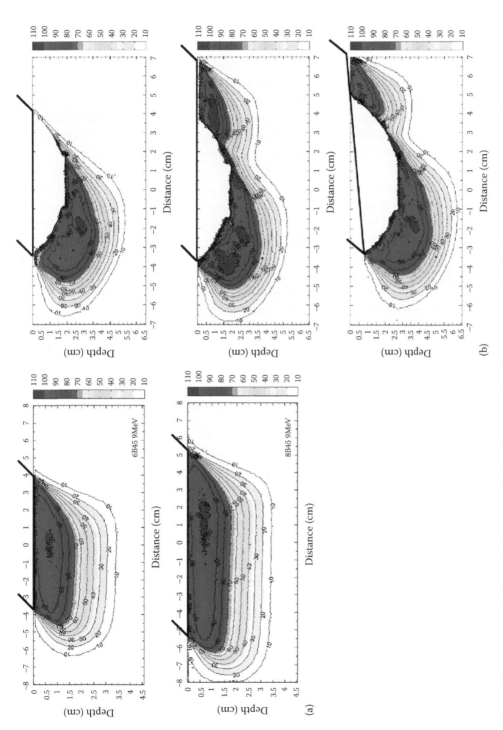

Figure 14.16 (a) Film-measured dose distributions for irradiation by 9-MeV electrons with 45° bevels and 6 (upper) and 8 (lower) cm applicators. (b) Dose distributions with similar, beveled applicator, but with a convex cutout introduced into bolus positions on top of the solid phantom. (Adapted from Costa, F. et al., *Phys. Medica.*, 31, 692–701, 2015. With Permission.)

14.6 SUMMARY

Radiochromic film has contributed extensively to the development and implementation of diverse electron-beam therapies. In this chapter, we discussed applications of radiochromic film to dose calculation validation for electron beams, TSET, MERT, and IOERT. Radiochromic film has served important roles in all of these applications.

Film measurements spanning large areas support commissioning of TSET. In addition, radiochromic film is well suited for in vivo dosimetry during TSET as it can be adhered to surfaces that might be underdosed following treatment. Radiochromic film supports quality assurance of MERT treatments in which modulation can produce sharp dose gradients. Likewise, radiochromic film supports quality assurance of IOERT through in vivo measurement of extremely high, single-fraction irradiations.

In these applications among others, radiochromic film can provide PDDs and other profiles in irregular fields with minimal corrections for beam energy, dose rate, and angle of incidence. Moreover, radiochromic film has unique applications in characterizing complex electron fields in which other dosimeters might be impractical.

REFERENCES

1. Su F-C, Liu Y, Stathakis S et al. Dosimetry characteristics of GAFCHROMIC® EBT film responding to therapeutic electron beams. *Appl Radiat Isot* 2007;65:1187–1192.
2. Richter C, Pawelke J, Karsch L et al. Energy dependence of EBT-1 radiochromic film response for photon (10 kVp–15 MVp) and electron beams (6–18 MeV) readout by a flatbed scanner. *Med Phys* 2009;36:5506.
3. Arjomandy B, Tailor R, Anand A et al. Energy dependence and dose response of Gafchromic EBT2 film over a wide range of photon, electron, and proton beam energies. *Med Phys* 2010;37:1942.
4. Sorriaux J, Kacperek A, Rossomme S et al. Evaluation of Gafchromic® EBT3 films characteristics in therapy photon, electron and proton beams. *Med Phys.* 2013;29:599–606.
5. Petri S, Ojala J, Kaijaluoto S et al. Gafchromic EBT3 film dosimetry in electron beams—Energy dependence and improved film read-out. *J Appl Clin Med Phys* 2016;17:360–373.
6. Chan E, Lydon J, Kron T. On the use of Gafchromic EBT3 films for validating a commercial electron Monte Carlo dose calculation algorithm. *Phys Med Biol* 2015;60:2091–2102.
7. Coleman J, Park C, Villarreal-Barajas JE et al. A comparison of Monte Carlo and Fermi-Eyges-Hogstrom estimates of heart and lung dose from breast electron boost treatment. *Int J Radiat Oncol Biol Phys* 2005;61:621–628.
8. Hogstrom KR, Mills MD, Almond PR. Electron beam dose calculations. *Phys Med Biol* 1981;26:445–459.
9. Vandervoort EJ, Tchistiakova E, La Russa DJ et al. Evaluation of a new commercial Monte Carlo dose calculation algorithm for electron beams. *Med Phys* 2014;41:21711.
10. Cygler JE, Daskalov GM, Chan GH et al. Evaluation of the first commercial Monte Carlo dose calculation engine for electron beam treatment planning. *Med Phys* 2004;31:142.
11. Karzmark CJ, Anderson J, Buffa A et al. *Total Skin Electron Therapy: Technique and Dosimetry.* American Institute of Physics: College Park, MD; 1987.
12. Gossman MS, Sharma SC. Total skin high-dose-rate electron therapy dosimetry using TG-51. *Med Dosim* 2004;29:285–287.
13. Schiapparelli P, Zefiro D, Massone F et al. Total skin electron therapy (TSET): A reimplementation using radiochromic films and IAEA TRS-398 code of practice. *Med Phys* 2010;37:3510.

14. Karzmark CJ, Loevinger R, Steele RE et al. A technique for large-field, superficial electron therapy 1. *Radiology* 1960;74:633–644.
15. Reynard EP, Evans MDC, Devic S et al. Rotational total skin electron irradiation (RTSEI) with a linear accelerator. *J Appl Clin Med Phys* 2008;9.
16. Gamble LM, Farrell TJ, Jones GW et al. Two-dimensional mapping of underdosed areas using radiochromic film for patients undergoing total skin electron beam radiotherapy. *Int J Radiat Oncol Biol Phys* 2005;62:920–924.
17. Jones G., Wong R, Kastikainen J et al. 15-year results of total skin electron beam (TSEB) radiation as first-line monotherapy for newly diagnosed stage IA–IB mycosis fungoides (MF). *Int J Radiat Oncol Biol Phys* 2003;57:S291.
18. Anacak Y, Arican Z, Bar-Deroma R et al. Total skin electron irradiation: Evaluation of dose uniformity throughout the skin surface. *Med Dosim* 2003;28:31–34.
19. Bufacchi A, Carosi A, Adorante N et al. In vivo EBT radiochromic film dosimetry of electron beam for total skin electron therapy (TSET). *Med Phys* 2007;23:67–72.
20. Kudchadker RJ, Hogstrom KR, Garden AS et al. Electron conformal radiotherapy using bolus and intensity modulation. *Int J Radiat Oncol Biol Phys* 2002;53:1023–1037.
21. Salguero FJ, Palma B, Arrans R et al. Modulated electron radiotherapy treatment planning using a photon multileaf collimator for post-mastectomized chest walls. *Radiother Oncol* 2009;93:625–632.
22. Salguero FJ, Arráns R, Palma BA et al. Intensity- and energy-modulated electron radiotherapy by means of an xMLC for head and neck shallow tumors. *Phys Med Biol* 2010;55:1413–1427.
23. Al-Yahya K, Verhaegen F, Seuntjens J. Design and dosimetry of a few leaf electron collimator for energy modulated electron therapy. *Med Phys* 2007;34:4782.
24. Devic S, Seuntjens J, Sham E et al. Precise radiochromic film dosimetry using a flat-bed document scanner. *Med Phys* 2005;32:2245.
25. Connell T, Alexander A, Papaconstadopoulos P et al. Delivery validation of an automated modulated electron radiotherapy plan. *Med Phys* 2014;41:61715.
26. Micke A, Lewis DF, Yu X. Multichannel film dosimetry with nonuniformity correction. *Med Phys* 2011;38:2523.
27. Krengli M, Calvo FA, Sedlmayer F et al. Clinical and technical characteristics of intraoperative radiotherapy. *Strahlenther Onkol* 2013;189:729–737.
28. Pascau J, Santos Miranda JA, Calvo FA et al. An innovative tool for intraoperative electron beam radiotherapy simulation and planning: Description and initial evaluation by radiation oncologists. *Int J Radiat Oncol Biol Phys* 2012;83:e287–e295.
29. Piermattei A, delle Canne S, Azario L et al. The saturation loss for plane parallel ionization chambers at high dose per pulse values. *Phys Med Biol* 2000;45:1869.
30. Ciocca M, Orecchia R, Garibaldi C et al. In vivo dosimetry using radiochromic films during intraoperative electron beam radiation therapy in early-stage breast cancer. *Radiother Oncol* 2003;69:285–289.
31. Severgnini M, de Denaro M, Bortul M et al. In vivo dosimetry and shielding disk alignment verification by EBT3 GAFCHROMIC film in breast IOERT treatment. *J Appl Clin Med Phys* 2015;16:1–9.
32. Krengli M, Terrone C, Ballarè A et al. Intraoperative radiotherapy during radical prostatectomy for locally advanced prostate cancer: Technical and dosimetric aspects. *Int J Radiat Oncol Biol Phys* 2010;76:1073–1077.
33. Costa F, Sarmento S, Sousa O. Assessment of clinically relevant dose distributions in pelvic IOERT using Gafchromic EBT3 films. *Med Phys* 2015;31:692–701.

Proton beam dosimetry using radiochromic film

INDRA J. DAS

15.1 INTRODUCTION

Particle beam started at Berkley California in late thirties, but its clinical implication was not realized till the concept was proposed by Wilson [1] in 1946. Its full scope did not realize till the beginning of this century when multiple centers started operating clinical proton beam in the United States of America. Due to particle beam unique characteristics with Bragg peak and finite range, it provides attractive choice for patient treatment. In general, the skin dose (entrance dose) is relatively low and no dose beyond the range of the particles (Figure 15.1). The first center in the United States of America was built soon after World War II at the Harvard Cyclotron Laboratory in Cambridge, Massachusetts. A detailed account of the Harvard Cyclotron is presented in a book by Richard Wilson [2]. After 40 years, first synchrotron-based proton beam therapy center was built in Loma Linda Medical Center California, followed by a commercial proton device at Massachusetts General Hospital, Boston in 2001. Indiana University in Bloomington converted its cyclotron for clinical use during 2003 with addition of ion beam application (IBA) gantry thus becoming third center

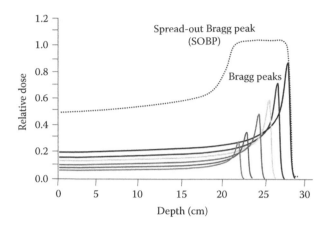

Figure 15.1 Proton beam depth doses of various energies are shown with different ranges. Note the entrance dose (surface dose) and Bragg peak (maximum dose) and sharp dropoff representing the range of the beam. When these curves are added with proper beam weights, they produce a resultant depth dose known as spread-out Bragg peak (SOBP). Most patients are treated with SOBP with variable width depending upon the tumor size.

in the country. At present time, there are closed to 35 proton centers operational treating various types of diseases. The growth of proton beam in United States of America is exponentially rising as shown from the reference [3].

Proton beams in general are divided into three classes: (1) passive scattered beam, (2) uniform scanning beam, and (3) pencil beam for spot scanning and intensity-modulated proton beam therapy. Dosimetry of scattered beam is relatively simple as beam characteristics (depth dose, profiles and output) can be measured with ion chambers. Even though ion chamber measurements are time consuming but treated as gold standard. For uniform scanning ion chamber measurements takes long time and could be prohibitive as each data point should be collected for entire sweep. This is similar to dynamic wedge in a photon beam in which a multidetector array should be used. With development of multilayer ion chamber [4], data collection becomes relatively easier; however, this detector is extremely expensive and requires periodic calibrations.

Small proton fields and pencil beams are considered a separate class in which dosimetry is even much more complex. The typical Bragg peak loses its unique characteristics as the field size is reduced [5]. Also the detector size could be comparable with field size or spot size, thus unable to measure dose from standard detectors.

Film dosimetry provides unique opportunity to map 2D dose, thus saving significant amount of time. Radiochromic film (RCF) is attractive for dosimetry as discussed in many chapters for photon beams, and thus its utility in proton beam is worth exploring which is discussed in this chapter.

15.2 BRAGG PEAK CHARACTERISTICS

Every particle as it travels has energy related to kinetic energy. As they travel in medium, they lose energy and speed. This slowing process produces high linear energy transfer (LET) and thus high dose. For heavy charged particle such as proton, the rate of LET is significant and at the end of its journey produces a peak called Bragg peak [6]. Bragg peak is unique and characteristic of heavy charged particle. It is only a few millimeters in widths, and there is no dose distal to Bragg peak

(Figure 15.1). The position of Bragg peak also defines the beam energy and range that will be discussed briefly in the following section.

15.2.1 STOPPING POWER AND RANGE

Bragg peak is unique to the particle beams and directly associated with loss of energy in medium as known as stopping power. The concept of stopping power was first introduced by Bragg and Kleeman [7,8]. Bethe and Bloch [9,10] provided a general equation for energy loss by charged particle in medium as given by Equation 15.1. This equation is also adopted by International Commission on Radiation Units and Measurements (ICRU)-49 [11].

$$\frac{dE}{\rho dx} = \frac{4\pi N_A r_e^2 m_e c^2}{\beta^2} \frac{Z}{A} z^2 \left[\ln\left(\frac{2m_e c^2 \beta^2}{I(1-\beta^2)} \right) - \beta^2 \right] \tag{15.1}$$

where dE/dx is the energy loss per unit path length known as stopping power or LET, ρ is the density of the medium, E is the particle energy (MeV), N_A is Avogadro's number, r_e is the classical electron radius, m_e is the mass of electron, c is the velocity of light, z is the charge of the particle, Z is the atomic number of medium, A is the atomic mass of the medium, β is the relative velocity (v/c) of the particle, and I is the ionization potential of the medium, that is, energy needed to ionize molecule or atom of the medium.

The rate of energy loss is directly related to energy of the particle at a point in space. There are empirical equations providing link to LET and range [12]. Thus very closed to its range, it provides significant energy loss; thus, the dose is very high. This behavior of energy deposition is known as Bragg peak as shown in Figure 15.1.

Particle range, R, is inversely proportion of stopping power (dE/dx) and more importantly given as

$$R = \int_0^E \frac{dE}{(dE/dx)} \tag{15.2}$$

Range has many definitions depending on the observation and measurements [13,14]. For clinical use, continuous slowing down approximation (CSDA) is often used as shown in Equation 15.3. For a monoenergetic proton energy, range in a specific medium is precisely known and tabulated in ICRU-49 [11] based on the integration as shown in Equation 15.3.

$$R_{CSDA} = \int_0^{E_{max}} \left(\left(\frac{S}{\rho} \right)^{-1} \right) dE \tag{15.3}$$

For billions of protons used in clinical use, the probability of interaction is statistically distributed; thus, there is energy straggling that also creates range straggling which is described elsewhere [3]. The straggling process creates fuzziness in the range in which measurements could be uncertain depending upon the detectors. It is important to emphasize that the CSDA approximation is not a good model for proton beam. For simplicity, it can be calculated analytically as shown in Equation 15.4 using a quadratic equation. Various other models also have been suggested that are also empirical in nature [12–14].

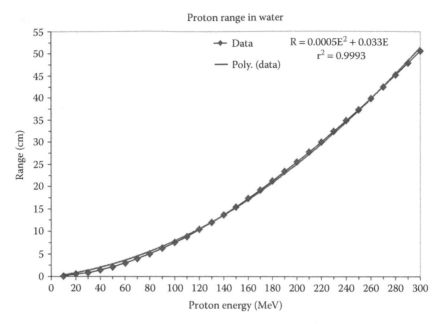

Figure 15.2 A polynomial fit as shown in Equation 15.4 for range energy relationship in water.

$$R = aE + bE^2 \qquad (15.4)$$

where a and b are fitting parameters equal to 0.033 cm/MeV and 0.0005 cm/MeV^2 respectively as shown in Figure 15.2.

15.2.2 SPREAD OUT BRAGG PEAK

Due to narrow curve (Figure 15.1) of the Bragg peak, it is not suitable for clinical use unless pencil or spot scanning is used. Currently, most centers use broadened beam by scattering the pencil beam. Such broadening is performed by passive scattering using single or double scatterer. The range broadening is performed by modulating wheel by adding many layers of different energy proton beam or mixing many Bragg peaks with selective weighting as shown in Figure 15.1. The spread out Bragg peak (SOBP) provides tumor coverage in depth. One should appreciate that depending upon the modulation for creating broader SOBP; the surface dose is significantly increased.

15.3 DOSIMETRIC DIFFICULTIES

Characterizing beams for clinical use rely on the measured depth dose and profiles that form the isodose curves. Due to permutations of beam energy and SOBP, these data sets could become extremely large in which point-by-point measurements become labor intensive and time consuming in a costly proton center. In such situation, film provides unique opportunity for quick 2D dose map. However, selection of the film and irradiation techniques along with readout must be carefully addressed.

15.4 OUTPUT

Output is a general term defined as dose/monitor unit (MU) for a beam and field. Output is calibrated for a reference condition in which typically it is maintained to be 1.0 cGy/MU. The calibration is provided with reference ion chamber that is periodically calibrated using a standard protocol such as ICRU-78 [15]. The doses in every other energies, fields, and depths are related to the reference condition and are called relative output factor. The calculation of MU is then performed empirically and verified by measurements [16,17]. For broad beams, this method works well; however, for small fields, the shape of depth dose changes rapidly, and there is no well-defined Bragg peak [18,19]. This creates unique problem in which dosimetry is uncertain. RCF can be used to quantify dose in small fields with caution [20].

15.5 RADIOCHROMIC FILM DOSIMETRY OF PROTON BEAM

Use of RCF in proton beams has been cited since 1997. The earlier version of RCF,MD has been attempted by several investigators [21–24] and MD-2 by Piermattei et al. [22] for low energy proton beam used for ocular melanoma treatment. When HD-2 film became available, it was attempted for use in proton beam dosimetry [25]. Recently external beam therapy (EBT) series of films have been used with limited success [26–40]. Many difficulties need to be overcome, which is discussed for accurate dosimetry in proton beam.

15.5.1 FILM ORIENTATION

Film orientation plays an important role in radiation dosimetry due to radiation interaction causing polymerization in RCF. Zhao and Das [31] studied this effect in the context of depth dose measurements. Recent study [40] on EBT and RTQA films performed with proton beam provides technical challenges and its solution for 2D dosimetry. In general, RCF can be exposed in two orientations, perpendicular and parallel. Due to cost associated with RCF film, perpendicular exposure requires a lot of film and not preferred. It is shown that the parallel orientation is superior as one can get entire 2D data set from one single film; however, the air column creates artifact when a solid phantom is used as shown in Figure 15.3. To avoid such problem beam can be angled 2°–3° or the film can be kept in water; however, Borowicz et al. [40] recommended 5°. Ajomandy et al. [27] provided method for exposing these films with one set of film holder that can be used in water. RCF dosimetry in proton beam is more sensitive to experimental setup and hence care has to be taken as discussed in literature [27,31,40].

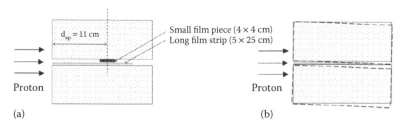

Figure 15.3 EBT film exposure in parallel orientation in two different beam angles. In parallel, orientation exposure minimum air gap (a) as well as at least 2°–3° gantry angle should be maintained (b). (Adopted from Zhao, L. and Das, I.J., *Phys. Med. Biol.*, 55, N291–N301, 2010. With Permission IOP.)

Figure 15.4 Streaking effect due to air column in circle indicating distal tail area behind Bragg peak. (Courtesy of Aswin Hoffman Dresden, Germany, 2016.)

Effect of air gap in solid phantom has significant problem in proton beams. The air provides channel for proton beam to travel freely even with multiple scattering, and it creates fuzziness and broader Bragg peak as shown in Figure 15.4. Such interpretation in complex geometry could be difficult; hence, for accurate dosimetry, mainly in water phantom should be preferred.

15.5.2 Air gap

Like radiographic films that create large perturbation [41] in electron beam due to variation in fluence, RCF gives similar response when used in solid phantom. Zhao and Das [31] showed that air column significantly distorts proton beam depth dose as shown in Figure 15.5. The entrance as well as Bragg peak gets distorted and so SOBP is skewed up. To remove such distortion, air from the phantom should be eliminated, which can be achieved by putting the film in water with aid of

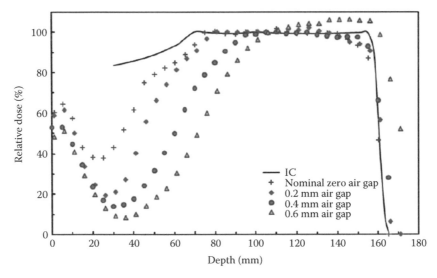

Figure 15.5 Depth dose measured with various amounts of air in the path compared to ion chamber measurement. Note that a large air gap provides significant underdose in the entrance side of the depth dose curve. A minimum air gap or exposure in water is preferred. (Adopted from Zhao, L. and Das, I.J., *Phys. Med. Biol.*, 55, N291–N301, 2010. With Permission IOP.)

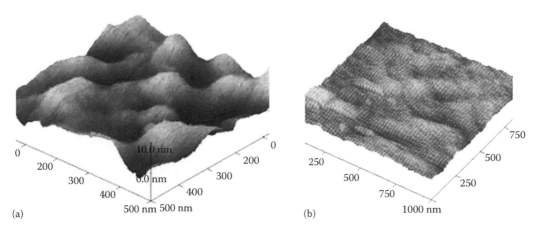

0

200

300

400

500 nm 500 nm

10.0 nm

0.0 nm

0

200

300

400

(a)

750

500

250

250

500

750

1000 nm

(b)

Figure 15.6 Scanning electron micrograph of EBT2 film (a) before and (b) after 5-Gy irradiation. The polymerization is clearly visible with disappearance of trough and valley. (Adapted from Zhao, L. et al., *IFMBE Proceedings*, 39, 1164–1167, 2012.)

film folder. Care also should be taken to give enough width so that water leakage in film should not cause additional perturbation.

15.5.3 TEMPORAL RESPONSE

Radiochromic film response is dependent on the polymerization of the chemical bonds producing color, the process is time dependent. This has been studied in photon beam by Andres et al. [42] and in proton beam by Zhao et al. [30] The film polymerization is a slow process that takes long time. Zhao et al. [30] provided two-exponential function for a full saturation reading. Alternatively, at least a minimum of 12 h should be given before film is scanned. It is also a good practice to allow the same time between calibration and experimental film postirradiation readout. Figure 15.6 shows the impact of polymerization on the molecular bond.

15.5.4 LINEAR ENERGY TRANSFER DEPENDENCE

LET is positional dependence in the particle beam. As the energy reduces, LET increases several order of magnitude as shown by Anferov and Das [43]. Figure 15.7 shows the LET versus depth information in a pristine Bragg peak, and Figure 15.8 shows the LET variation in broad beam (SOBP) and with different ranges. It shows clearly that LET effect is important. These observations have been also observed by many on cell survival [44–46]. On the contrary, these cannot be measured with ion chambers. Additionally, using RCF dosimetry these observations get lost due to LET dependence.

Polymerization of molecular bonds in the RCF is energy dependence. In proton beam, energy of the particle is variable and assumed to be losing energy continuously based on continuous slowing down approximation. In such situation, energy is highest at the surface and zero at range of the particle. Thus at Bragg peak the energy is lowest and may have highest effect. This was clearly shown by Zhao and Das [31] that energy response is dependent on the position of the Bragg peak. Figure 15.9a shows the variation of film and ion chamber measured depth dose in various energy beams. One could correct the LET dependence by empirical equation. By using correction, depth dose is in good agreement with ion chamber data as shown in Figure 15.9b.

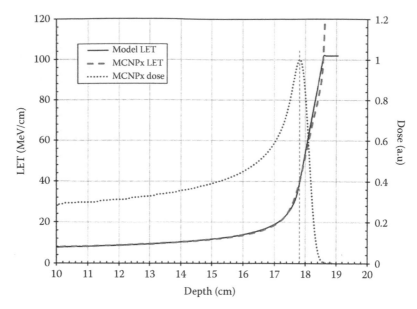

Figure 15.7 Monte Carlo simulated Bragg-peak and corresponding LET. (Adopted from Anferov, V. and Das, I.J., *Int. J. Med. Phys. Clin. Eng. Radiat. Oncol.*, 4, 149–161, 2015.)

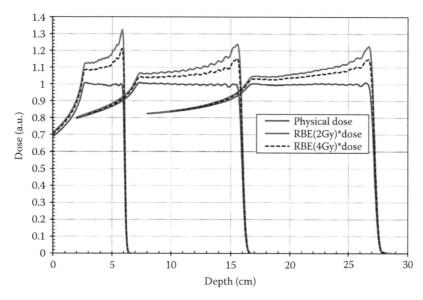

Figure 15.8 Physical and RBE evaluated depth doses are shown for various SOBP and range of the proton beam. Note that dashed line is not constant rather much higher at shallower range and smaller SOBP. (Adopted from Anferov, V. and Das, I.J., *Int. J. Med. Phys. Clin. Eng. Radiat. Oncol.*, 4, 149–161, 2015.)

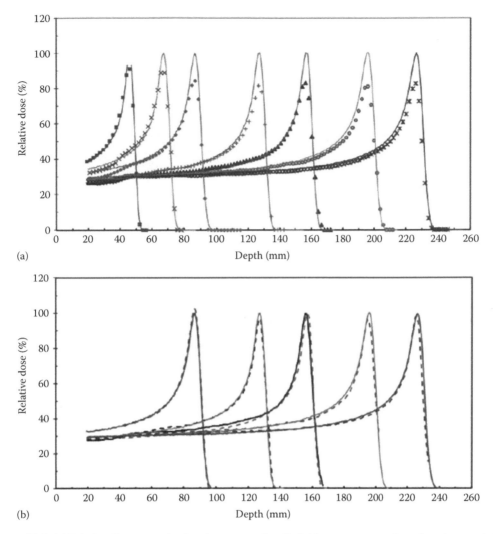

Figure 15.9 (a) Pristine Bragg peak of various energies. Solid line represents ion chamber and data points represent EBT2 film. (b) When LET correction is applied to the EBT2 films, ion chamber and film data are in good agreement. (Adopted from Zhao, L. and Das, I.J., *Phys. Med. Biol.*, 55, N291–N301, 2010. With Permission IOP.)

15.6 CONCLUSIONS

Radiochromic films can be used for complex dosimetry in proton beam, either to save time for commissioning beam data for depth and profiles or for small fields; however, the following precautions must be taken.

- Standard precautions as discussed in previous section for variability in batch, age, lot size should be properly taken.
- For readout, frequency, position, and orientation on scanner should be maintained.

- Air gap between film and phantom during irradiation should be avoided. The best choice is to expose film in water.
- Parallel versus perpendicular film orientation provides different response and hence calibration and measurement should be performed in the same orientation.
- Polymerization is time dependent. At least 12 postirradiation should be given or mathematical expression described by Zhao et al. [30] should be applied for proton beam.
- LET dependence should be properly accounted which is energy dependence as described by several investigators [6,12,31,36].

REFERENCES

1. Wilson RR. Radiological use of fast protons. *Radiology* 1946;47:487–491.
2. Wilson RR. *A Brief History of Harvard University Cyclotrons*. Cambridge, MA: Harvard University Press; 2004.
3. Das IJ, Paganetti H. (Eds.). *Principle and Practice of Proton Beam Therapy*. Madison, WI: Medical Physics Publishing; 2015.
4. Nichiporov D, Solberg K, Hsi W et al. Multichannel detectors for profile measurements in clinical proton fields. *Med Phys* 2007;34:2683–2690.
5. Moskvin VP, Estabrook NC, Cheng CW et al. Effect of scanning beam for superficial dose in proton therapy. *Technol Cancer Res Treat* 2014;14(5):643–652.
6. Kempe J, Gudowska I, Brahme A. Depth absorbed dose and LET distributions of therapeutic 1H, 4He, 7Li, and 12C beams. *Med Phys* 2007;34:183–192.
7. Bragg WH, Kleeman R. On the α particles of radium and their loss of range in passing through various atoms and molecules. *Phil Mag J Sci* 1905;10:318–340.
8. Bragg WH, Kleeman R. On the ionization curves of radium. *Phil Mag J Sci* 1904;8:726–738.
9. Bethe HA. Zur Theorie des Durchgangs schneller Korpuskularstrahlen durch Materie. *Ann Phys (Leipzig)* 1930;5:324–400.
10. Bloch F. Zur Bremsung rasch bewegter Teilchen beim Durchgang durch Materie. *Ann Phys (Leipzig)* 1933;16:285–320.
11. ICRU Report 49. Stopping powers and ranges for protons and alpha Particles. Bethesda, MD: International Commission on Radiation Units and Measurements; 1993.
12. Kempe J, Brahme A. Energy-range relation and mean energy variation in therapeutic particle beams. *Med Phys* 2008;35:159–170.
13. Attix FH. *Introduction to Radiological Physics and Radiation Dosimetry*. New York: John Wiley & Sons; 1986.
14. Bichsel H. Passage of charged particles through matter. In *American Institute of Physics Handbook*. New York: McGraw-Hill; 1972.
15. ICRU Report 78. Prescribing, recording, and reporting proton beam therapy. Bethesda, MD: International Commission on Radiation Units and Measurements; 2007.
16. Kooy HM, Rosenthal SJ, Engelsman M, Mazal A, Slopsema RL, Paganetti H, Flanz JB. The prediction of output factors for spread-out proton Bragg peak fields in clinical practice. *Phys Med Biol* 2005;50:5847–5856.
17. Zhao Q, Wu H, Wolanski M et al. A sector-integration method for dose/MU calculation in a uniform scanning proton beam. *Phys Med Biol* 2010;55:N87–N95.
18. Bednarz B, Daartz J, Paganetti H. Dosimetric accuracy of planning and delivering small proton therapy fields. *Phys Med Biol* 2010;55:7425–7438.
19. Fontenot JD, Newhauser WD, Bloch C et al. Determination of output factors for small proton therapy fields. *Med Phys* 2007;34:489–498.

20. Zhao L, Newton J, Oldham M et al. Feasibility of using PRESAGE(R) for relative 3D dosimetry of small proton fields. *Phys Med Biol* 2012;57:N431–N443.
21. Vatnitsky SM, Schulte RW, Galindo R et al. Radiochromic film dosimetry for verification of dose distributions delivered with proton-beam radiosurgery. *Phys Med Biol* 1997;42:1887–1898.
22. Piermattei A, Miceli R, Azario L. et al. Radiochromic film dosimetry of a low energy proton beam. *Med Phys* 2000;27:1655–1660.
23. Vatnitsky SM. Radiochromic film dosimetry for clinical proton beams. *Appl Radiat Isot* 1997;48:643–651.
24. Daftari I, Castenadas C, Petti PL et al. An application of GafChromic MD-55 film for 67.5 MeV clinical proton beam dosimetry. *Phys Med Biol* 1999;44:2735–2745.
25. Mercado Uribe H, Gamboa-deBuen I, Buenfil AE et al. Experimental study of the response of radiochromic films to proton radiation of low energy. *Nucl Instrum Meth Phys Res* 2009;267:1849–1851.
26. Arjomandy B, Tailor R, Anand A et al. Energy dependence and dose response of Gafchromic EBT2 film over a wide range of photon, electron, and proton beam energies. *Med Phys* 2010;37:1942–1947.
27. Arjomandy B, Tailor R, Zhao L, Devic S. EBT2 film as a depth-dose measurement tool for radiotherapy beams over a wide range of energies and modalities. *Med Phys* 2012;39:912–921.
28. Karsch L, Beyreuther E, Burris-Mog T et al. Dose rate dependence for different dosimeters and detectors: TLD, OSL, EBT films, and diamond detectors. *Med Phys* 2012;39:2447–2455.
29. Reinhardt S, Hillbrand M, Wilkens JJ, Assmann W. Comparison of Gafchromic EBT2 and EBT3 films for clinical photon and proton beams. *Med Phys* 2012;39:5257–5262.
30. Zhao L, Coutinho L, Cao N et al. Temporal response of Gafchromic EBT2 radiochromic film in proton beam irradiation. *IFMBE Proceedings* 2012;39:1164–1167.
31. Zhao L, Das IJ. Gafchromic EBT film dosimetry in proton beams. *Phys Med Biol* 2010;55:N291–N301.
32. Gomà C, Andreo P, Sempau J. Spencer–Attix water/medium stopping-power ratios for the dosimetry of proton pencil beams. *Phys Med Biol* 2013;58:2509–2522.
33. Angellier G, Gautier M, Herault J. Radiochromic EBT2 film dosimetry for low-energy proton-therapy. *Med Phys* 2011;38:6171–6177.
34. Gueli AM, De Vincolis R, Kacperek A, Troja SO. An approach to 3D dose mapping using Gafchromic film. *Radiat Prot Dosimetry* 2005;115:616–622.
35. Kim J, Yoon M, Kim S et al. Three-dimensional radiochromic film dosimetry of proton clinical beams using a GAFchromic EBT2 film array. *Radiat Prot Dosimetry* 2012;151:272–277.
36. Kirby D, Green S, Palmans H et al. LET dependence of GafChromic films and an ion chamber in low-energy proton dosimetry. *Phys Med Biol* 2010;55:417–433.
37. Martisikova M, Jakel O. Dosimetric properties of Gafchromic EBT films in monoenergetic medical ion beams. *Phys Med Biol* 2010;55:3741–3751.
38. Sorriaux J, Kacperek A, Rossomme S et al. Evaluation of Gafchromic(R) EBT3 films characteristics in therapy photon, electron and proton beams. *Phys Med* 2013;29:599–606.
39. Troja SO, Egger E, Francescon P et al. 2D and 3D dose distribution determination in proton beam radiotherapy with GafChromic film detectors. *Technol Health Care* 2000;8:155–164.
40. Borowicz DM, Malicki J, Mytsin G, Shipulin K. Dose distribution at the Bragg peak: Dose measurements using EBT and RTQA GAFchromic film set at two positions to the central beam axis. *Med Phys* 2017;44:1538–1544.
41. Dutreix J, Dutreix A. Film dosimetry of high-energy electrons. *Ann N Y Acad Sci* 1969;161:33–43.
42. Andres C, del Castillo A, Tortosa R et al. A comprehensive study of the Gafchromic EBT2 radiochromic film. A comparison with EBT. *Med Phys* 2010;37:6271–6278.

43. Anferov V, Das IJ. Biological dose estimation model for proton beam therapy. *Int J Med Phys Clin Eng Radiat Oncol* 2015;4:149–161.

44. Paganetti H, Goitein M. Radiobiological significance of beamline dependent proton energy distributions in a spread-out Bragg peak. *Med Phys* 2000;27:1119–1126.

45. Paganetti H, Niemierko A, Ancukiewicz M et al. Relative biological effectiveness (RBE) values for proton beam therapy. *Int J Radiat Oncol Biol Phys* 2002;53:407–421.

46. Kagawa K, Murakami M, Hishikawa Y et al. Preclinical biological assessment of proton and carbon ion beams at Hyogo Ion Beam Medical Center. *Int J Radiat Oncol Biol Phys* 2002;54:928–938.

Skin-dose and build-up dose assessment with EBT film dosimetry

CARLOS DE WAGTER AND ANNEMIEKE DE PUYSSELEYR

16.1 INTRODUCTION

16.1.1 PHYSICS OF DOSE BUILD-UP IN MEGAVOLTAGE PHOTON BEAMS

The build-up region and its concomitant skin-sparing effect are one of the most distinct characteristics of megavoltage photon beams. This gradual build-up of dose absorption from the patient surface with depth originates from the two-step process of dose deposition in megavoltage photon beams. In all photon beams, the absorbed dose is not deposited by the photons itself, rather by

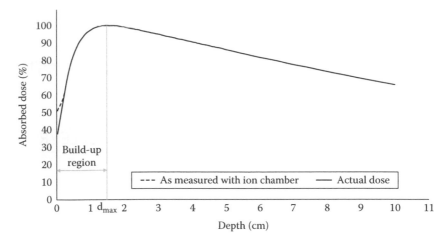

Figure 16.1 The build-up of absorbed dose with depth for a 6-MV photon beam. The build-up region and the depth of the dose maximum are depicted. The broken curve is affected by the finite active volume of the ionization chamber.

the secondary charged particles generated during photon-tissue interactions. As these secondary electrons predominantly travel forward in the beam direction while depositing their energy, the electron fluence and absorbed dose rise steeply from the tissue–air interface toward the depth of the dose maximum (d_{max}). The region between the surface and the depth of the dose maximum is generally referred to as the build-up region (Figure 16.1) and exhibits, by nature, a lack of charged particle equilibrium (CPE). The extent of this build-up region increases with the range and energy of the secondary charged particles and thus varies with the megavoltage energy spectrum.

In clinical practice, the skin-sparing effect of the build-up region is compromised by the presence of backscattered radiation from the patient and contaminant photons or electrons from outside the patient [1]. This contaminant radiation consists of head scattered photons, mainly originating from the flattening filter, and electrons generated in the treatment head or in the traversed air column [2,3].

16.1.2 THE CLINICAL RELEVANCE OF THE BUILD-UP DOSE

16.1.2.1 SKIN-DOSE DETERMINATION

First, accurate knowledge and determination of build-up doses in megavoltage photon beams are essential in the clinical management of the patients, especially the skin toxicity.

As illustrated in Figure 16.2, human skin is a complex organ composed of two layers: the epidermis and the dermis. The epidermis is the most superficial layer with a thickness between 0.020 and 0.100 mm in most body sites, though this may greatly vary with the anatomical region considered, between patients and with age [4,5]. The outermost epidermal layer primarily consists of dead cornified cells that are shed through normal desquamation. These detached cells are continuously replaced by mature skin cells originating from an underlying single layer of stem cells, generally referred to as the basal cell layer. The dermis underlies the epidermis and is considerably thicker (1–10 mm [4]). Not only it consists of intersecting

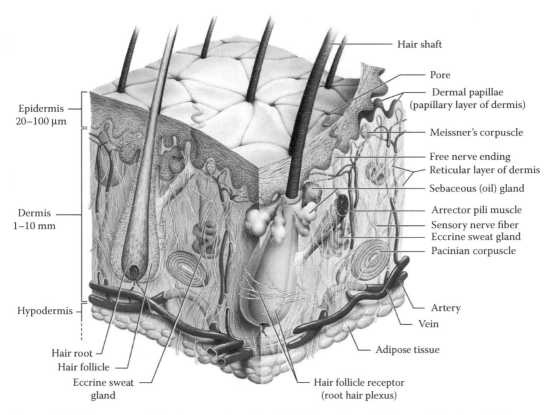

Figure 16.2 Structure of the human skin. (Modified from Marieb, E.N. and Hoehn, K., *Human Anatomy and Physiology*, Pearson Education/Benjamin Cummings, London, 2008. With Permission.)

collagen bundles but also contains hair follicles, glands, nerves, lymphatic conduits, and blood vessels supplying nutrients and support [6].

In literature, both acute and late radiation-induced skin reactions have been reported. Acute skin toxicity primarily originates from damage to the basal cell layer in the epidermis [4]. As the stem cells in this layer are sterilized, they are prevented from proliferating and repopulating the epidermal layers. This process might result in dry or moist desquamation and has been reported for absorbed doses from 20 to 25 Gy when delivered in 2 Gy daily fractions [7]. In response to the epidermal damage, erythema might develop as a secondary inflammatory response in the dermal structures [4]. At higher doses, more severe radiation skin reactions include ulceration, hemorrhage, and necrosis. Late skin reactions, on the other hand, are defined as radiation-induced changes present beyond 90 days of treatment. They might include telangiectasia, atrophy, fibrosis, edema, alopecia, and ulceration [4].

In skin dosimetry, the recommended depth for dose determination varies with the effect considered. The International Commission on Radiological Protection and the International Commission on Radiation Units and Measurements generally recommends skin-dose assessments at 0.07 and 1.0 mm depth for epidermal and dermal effects, respectively [4,8]. As previously mentioned, however, skin-layer thickness might vary considerably with the anatomical region, age and even between patients. In this respect, Devic et al. [9] suggested to measure build-up doses at a wide range of depths and to extract the relevant skin dose at a case-specific depth defined by a radiation oncologist.

16.1.2.2 SUPERFICIAL TARGET VOLUMES

In megavoltage photon radiotherapy, accurate dosimetry in the build-up region also plays a major role in the treatment of superficial target volumes. The skin/build-up dose is a complex function of many parameters: beam energy, field size, beam angle, source-to-surface distance, beam modifying devices, immobilization accessories, and couch top in the beam line. Target volumes extending close to the skin, mainly occur in the treatment of breast, chest wall, and head-and-neck cancer. In these cases, adequate target coverage requires a boost of the inherently limited build-up doses. Such an increase in superficial doses is generally achieved through the application of bolus material [10] or through tangentially incident photon beams [11].

16.2 THE VALUE OF RADIOCHROMIC FILM DOSIMETRY IN THE BUILD-UP REGION

The available knowledge on skin and superficial target doses in clinical practice strongly relies on accurate build-up dose measurements for two reasons. First, the absorbed dose in the build-up region depends in a complex manner on a wide range of treatment parameters as described earlier. Second, modern treatment planning systems often fail to accurately predict superficial doses, as dose calculations are complicated by the steep dose gradient, lack of CPE, and presence of contaminant radiation in the build-up region, and more so, the calculation grid used for dose calculations as shown by Akino et al. [12]. In literature, discrepancies of up to 25% between build-up dose measurements and treatment planning system (TPS) calculations have been reported [13–16]. Even Monte Carlo simulations rely on measured data to validate the simulations and unveil neglected particles contributing to surface doses [17].

Build-up measurements, however, are evenly complicated by the physical characteristics of the build-up region. The steep dose gradient, lack of CPE, and presence of contaminant radiation require the detector to have a high spatial resolution combined with a low-energy dependence but more so a window-less detector. For these reasons, radiochromic film has been extensively used as a build-up region dosimeter, as it has a high spatial resolution and a low spectral sensitivity in both photon and electron beams [18]. Although these investigations mainly focused on the determination of the build-up dose in single beams, radiochromic film has also been shown to be a valuable tool for entire-treatment skin-dose verifications as discussed in the following.

16.2.1 SINGLE BEAM BUILD-UP DEPTH–DOSE GRADIENT ANALYSIS

16.2.1.1 FILM POSITIONED ON A FLAT PHANTOM SURFACE

Given the tissue-equivalence of radiochromic film, it is natural to mount a film piece on top of a flat phantom surface in order to assess superficially absorbed doses as shown by Butson et al. [19] and Devic et al. [9]. In this orientation, radiochromic films allow one to measure absorbed doses at a very small effective depth of measurement. As illustrated in Figure 16.3 for various film types, the minimal effective depth of measurement depends on the film structure and composition. For megavoltage beams, the center of the active layer can be taken as the effective point of measurement. The resulting effective depth of measurement is then obtained as the radiological distance from this center point to the film entrance surface. For all types of radiochromic film, the thicknesses and mass densities of the substrate, coating, protection, adhesive, and active layers are specified by the manufacturer and as shown in Chapters 2 and 3. Note that, for the asymmetrically structured external beam therapy (EBT)2 film, the effective depth of measurement will also depend on the side of the film facing the beam source [20].

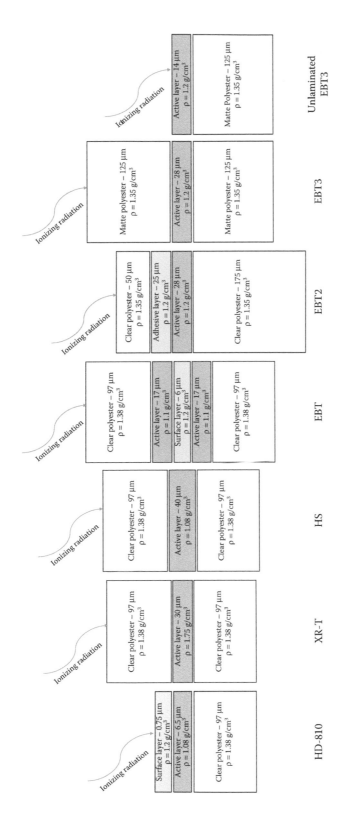

Figure 16.3 Layer geometry (not to scale) of successive types of GAFchromic™ film. (Modified from Devic, S. et al., *Med. Phys.*, 33, 1116–1124, 2006. With Permission.)

The resulting effective depths of measurement are indicated in Table 16.1 for various film types and are generally slightly higher than the nominal skin depth of 0.07 mm recommended by the International Commission on Radiological Protection and International Commission on Radiation Units and Measurements [4,8]. For this reason, Devic et al. [9] proposed field-size dependent correction factors for various film types, listed in Table 16.1, to convert the

Table 16.1 Studies assessing surface and build-up doses using radiochromic film on top of and parallel to a phantom surface. The effective depth values are as reported in the references.

Type of radiochromic film	Phantom characteristics	Orientation of film relative to beam axis	Effective depths of measurement (mm)	Reference
MD-55-2	Flat phantom	Perpendicular	0.17	[23]
MD-55-2	Flat phantom	Perpendicular	0.175, 0.525, 0.875, 1.225, 1.575 stacking 5 films	[24]
MD-55-2	Hemicylindrical chest-wall phantom	Perpendicular and oblique	0.17	[25]
MD-55-2	Flat phantom	Perpendicular	0.17 1.17 (adding solid-water sheet)	[21]
MD-55-2	Flat phantom	Perpendicular	0.181 2.54 (adding 2-mm thick sheet)	[22]
HD-810	Flat phantom	Perpendicular	0.004	[9]
XR-T	Flat phantom	Perpendicular	0.157	[9]
HS	Flat phantom	Perpendicular	0.153	[9]
EBT	Flat phantom	Perpendicular	0.153	[9,26]
EBT	Flat phantom	Perpendicular	0.153, 0.459, 0.765, 1.071, 1.377, 1.683, 1.989, 2.295 stacking eight films	[27]
EBT	Curved anthropomorphic phantom	Perpendicular and oblique	0.153	[28]
EBT2	Flat and curved phantoms	Perpendicular and oblique	0.115 1	[20]
EBT2	Curved anthropomorphic phantom	Perpendicular and oblique	0.115 3, 6, 11 (adding bolus sheets)	[12]
EBT2	Flat phantom	Perpendicular	0.115 (substrate down) 0.254 (substrate up)	[29]
EBT3	RANDO phantom	All angular incidences (kV cone-beam CT)	0.186 (not reported in Reference)	[30]

measured dose values into the dose at the nominal skin depth of 0.07 mm. These correction factors amounted to approximately 15% for 6-MV photon beams for the investigated film types (Table 16.1). As illustrated by other authors [21,22], a further increase in the effective depth of measurement can be obtained by positioning build-up material, such as solid water slabs or other films, on top of the measurement film.

As illustrated in Table 16.1, radiochromic films positioned on a flat phantom surface have been extensively used in build-up region dosimetry for various applications. Butson et al. [23], for example, employed MD55-2 radiochromic films to assess central axis, off-axis, and peripheral skin doses at 0.17 mm water equivalent depth in 6-MV, 10-MV, and 18-MV photon beams. At the central axis, skin doses amounted to 15.5 to 22% of the dose at d_{max}, depending on the beam energy. Peripheral doses were found to increase with field size. The same radiochromic film type, supplemented by an Attix Model 449 parallel plate ionization chamber (RMI, Middleton, WI), was also employed to assess the clinically relevant effect of an unintended air gap under bolus material for perpendicular and obliquely incident beams [31]. Air gaps were found to produce minor decreases in surface dose at perpendicular incidence. The largest reduction in surface dose, being 10%, was measured for a 6 MV (8×8)-cm^2 field at 60° angle of incidence with a 10-mm gap. Butson et al. [21] compared superficially absorbed doses in 6-MV photon beams with or without two types of treatment couches in the beam line. Absorbed doses were estimated at the basal cell layer (0.17 mm water equivalent depth), and at the deeper edge of the dermis (1.17 mm water equivalent depth, obtained by positioning the film at a 1-mm depth in solid water [32]). Both the absorbed doses at the basal cell layer and dermal layer were found to increase substantially for beams traversing the treatment couches. This increase was the most pronounced for the carbon fiber/Mylar treatment couch compared with the tennis string/Mylar couch. Using EBT film pieces positioned on a flat solid water phantom, Bilge et al. [26] evaluated the absorbed dose at the central axis at 0.0153-mm depth in 6- and 18-MV photon beams. These superficially absorbed doses, normalized to the dose at d_{max}, were shown to decrease with energy and increase with field size. Nakano et al. [20] not only measured the superficially absorbed dose in 6-MV photon beams using EBT2 film pieces positioned on a rectangular phantom but also extended this method to a cylindrical phantom. For films positioned on the cylinder surface at large angles with respect to the beam axis, these measurements demonstrated the important contribution to the superficially absorbed dose of laterally scattered radiation from material inside the phantom. In their phantom study using EBT2 film, Akino et al. [12] came to the same conclusion for oblique beams.

A particularly remarkable application of radiochromic films positioned on a phantom surface is the use of a radiochromic film stack as a three-dimensional build-up dose detector. This method was first introduced by Butson et al. [24], who positioned a stack of five MD-55-2 film pieces on top of a solid water phantom to measure the absorbed dose at effective depths of measurement between 0.175 and 1.575 mm. Considering the nonlinearity of the depth–dose gradient in the build-up region, a second-order polynomial extrapolation was then applied to determine the absorbed dose at more shallow depths. Using this method, the extrapolated surface dose agreed within 4% with the Attix parallel plate chamber surface ionization (normalized to the dose and ionization at d_{max}, respectively). More recently, Chiu-Tsao et al. [27] employed a stack of eight EBT radiochromic films to measure absorbed doses between 0.153 and 2.295 mm depth in 6 and 15 MV photon beams. In contrast to the study by Butson et al. [24], solid water slabs were now placed around the film stack on top of the phantom. Extrapolation of these measurements to zero depth was performed using a square root of a linear function and a linear function for 6 and 15 MV beams, respectively. The obtained results generally agreed well with previously published percentage depth–dose data from parallel plate chamber and ultrathin thermoluminescent dosimeter (TLD) measurements.

16.2.1.2 FILM PERPENDICULAR TO THE PHANTOM SURFACE

Although build-up region dosimetry using radiochromic films mounted parallel to a phantom surface has extensively been documented in literature, the opposite is true for the use of radiochromic films in a perpendicular orientation with respect to the phantom surface. In these studies, radiochromic films are generally mounted between the slabs of a stacked phantom and are positioned in a parallel orientation with respect to the beam axis. For build-up region dosimetry, the most important advantage of this orientation is the potential to measure the dose in the complete build-up region, rather than at one specified depth.

Paelinck et al. [22], for example, combined the parallel and the perpendicular orientation of the film and concluded that radiochromic film is able to measure the dose in the build-up region accurately, irrespective of its orientation with respect to the beam axis. More recently, De Puysseleyr et al. [33] positioned radiochromic films perpendicularly to the phantom surface to evaluate the build-up doses in standard and flattening filter free (FFF) megavoltage photon beams.

An important disadvantage of radiochromic films irradiated in this orientation, however, is the restriction to measurement depths greater than 1 mm. Optical densities within the first millimeter from the film edge cannot be analyzed as they might be affected by the mechanical pressure during film cutting by the user or during manufacturing [34], as discussed in Chapter 3. In addition, potential air gaps between the film and phantom slabs might result in measurement inaccuracies on the central axis. Such inaccuracies should be avoided by a meticulous phantom design or by positioning the radiochromic films slightly off the central axis [35].

16.2.2 ENTIRE-TREATMENT SKIN-DOSE VERIFICATION (PHANTOM AND IN VIVO)

Besides the evaluation of skin and build-up dose in a single beam, radiochromic film has evenly shown to be a valuable detector for entire-treatment skin-dose verification, in both phantom-based and in vivo evaluations. In these applications, radiochromic films integrally measure the superficial doses deposited by beams entering the tissue, as well as by beams leaving the patient in this region (exit dose).

Almberg et al. [28], for example, compared the superficially absorbed doses at 0.153 mm depth for 4 different breast cancer treatment techniques. To that purpose, EBT radiochromic film strips were taped on an anthropomorphic phantom. This approach demonstrated similar superficial doses for conventional and tangential intensity modulated radiation therapy (IMRT) treatments but detected much lower skin doses for a 7-field intensity-modulated radiation technique including nontangential fields. Similarly, Akino et al. [12] evaluated the superficially absorbed dose for four different breast cancer treatment techniques and compared the obtained measurements with treatment planning data. The EBT2 radiochromic film strips were positioned at various depths between superflab layers positioned on a humanoid acrylic phantom. As a consequence, superficial doses were analyzed at the surface and at 3, 6, and 11 mm depth. This approach allowed one to demonstrate that the superficially absorbed dose (at 3 mm depth) differed considerably from the treatment planning system dose calculations for all techniques.

In literature, radiochromic film-based entire-treatment skin-dose verification has also been extended to in vivo applications. The most widespread application in this respect includes the measurement of skin doses by radiochromic film strips during total skin electron therapy [36]. Additional information related to electron beam can be found in Chapter 14. For megavoltage photon beams, Rudat et al. [37] used EBT3 radiochromic film pieces to perform in vivo skin-dose measurements for patients treated with three different breast cancer treatment techniques. To that purpose, small radiochromic film pieces (3×3 cm^2) were positioned on the skin at three predefined positions (medially and laterally) on the irradiated breast or chest wall. This approach allowed one to demonstrate that a 7-field intensity-modulated radiation technique resulted in considerably smaller skin doses compared with tangential beam IMRT or 3D conformal radiotherapy.

16.2.3 ANGULAR RESPONSE

There is no evidence that radiochromic film produces an angular response [38]. Therefore, it is used as a reference in various studies that assess the angular response of other detectors, for example [39], and that measure superficial dose from oblique beams [12,20,25].

16.2.4 COMPARISON WITH OTHER DETECTORS

In build-up region dosimetry, possible alternative approaches to radiochromic film dosimetry include the use of extrapolation chambers (EPC), fixed-separation parallel-plate ionization chambers, micro-diamond detectors, metal–oxide–semiconductor field-effect transistor (MOSFET) or TLD or optically stimulated luminescence dosimeter (OSLD). However, a direct comparison between radiochromic film and other detectors is often complicated by small differences in the effective depth of measurement, which becomes of major importance in the steep dose gradient of the build-up region.

For build-up region dosimetry in megavoltage photon beams, EPC dosimetry is generally considered the golden standard [40–42]. In these chambers, the cavity ionization can be extrapolated to an infinitesimally small cavity volume by varying the distance between two parallel collecting electrodes, in order to eliminate the effects of electron fluence perturbations arising from the presence of the air cavity in the medium [43]. These chambers usually have a very thin entrance window allowing for an effective depth of measurement as small as 0.0075 mm (Böhm EPC type 23392, PTW Freiburg). However, as EPC measurements are often cumbersome and time-consuming to use, few authors have compared EPC and radiochromic film measurements. For 6-MV photon beams, Devic et al. [9] compared EPC measurements at 0.069-mm depth with radiochromic film measurements, using pieces of various types of radiochromic film positioned on top of a solid water phantom (Table 16.1). Dose differences of approximately 15% between EPC and film dosimetry were noted and were attributed to the differences in effective depth of measurements. De Puysseleyr et al. [33] compared EPC with EBT2 radiochromic film measurements in flattened and unflattened photon beams at depths beyond 1 mm. To that purpose, radiochromic films were positioned in a solid water slab phantom in a parallel orientation with respect to the beam axis. Excellent agreement between film and EPC data was found. A similar conclusion was drawn in a dosimetric study of a patient immobilization device for radiotherapy [44], as illustrated in Figure 16.4.

Fixed-separation plane-parallel chambers are more practical to use but suffer considerably from perturbations of the electron fluence in the cavity volume. These perturbations mainly originate from electrons scattered from the cavity side and back walls and result in an overestimation of the ionization of up to 15% [41–43]. These effects need to be corrected or minimized by optimizing the chamber design, featuring a small cavity height with respect to the electrode diameter and a sufficiently large guard ring [45]. In this respect, the most widespread type of fixed-separation plane-parallel chamber is the Attix ionization chamber. For this chamber, both Devic et al. [9] and Butson et al. [24] demonstrated good agreement with radiochromic film measurements when taking the differences in effective depth of measurement into account. Compared with radiochromic film, however, both extrapolation and fixed-separation plane-parallel chambers are less suitable for

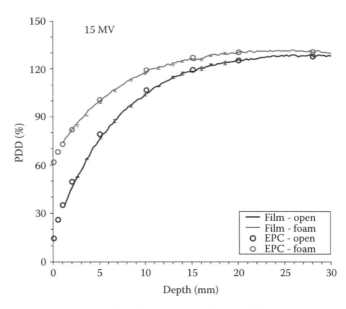

Figure 16.4 Comparison between absorbed dose in the build-up region as measured by EBT2 radiochromic film and extrapolation chamber (EPC) dosimetry for a 10 × 10 cm² 15-MV beam without (open) and with a 10-cm thick support cushion (foam) in the beam line. (From De Puysseleyr, A. et al., *Phys Med*, 32, 758–766, 2016.)

use with curved phantom surfaces or obliquely incident beams, which are both important determinants of build-up dose.

On the other hand, TLD, optically stimulated, and MOSFETs can easily be positioned on curved phantoms or patients. Therefore, they are mainly used on anthropomorphic phantoms or for in vivo skin dosimetry. Jong et al. [46], for example, compared EBT2 radiochromic film measurements at the surface of a cubic solid water phantom to measurements with a MOSFET-based detector. This latter detector consists of a radiation-sensitive MOSFET dye with submicron thickness covered in a Kapton pigtail strip. In this study, EBT2 film measurements were approximately 3%–4% (normalized to the dose at d_{max}) higher compared with the MOSFET measurements. These differences were attributed to the differences in effective depth of measurement between both detectors, equaling 0.07 vs. 0.115 mm for the MOSFET-based and EBT2 detector, respectively.

16.3 SPECIFIC PRECAUTIONS IN USING RADIOCHROMIC FILM DOSIMETRY IN THE BUILD-UP REGION

In literature, some specific precautions in using radiochromic film as a skin or build-up detector have been described along with the general precautions provided in Chapter 3.

First, it is important to note that most applications described previously use radiochromic films cut into small pieces or strips to measure absorbed doses in the build-up region. The use of films cut to small sizes is particularly relevant for in vivo applications, in which very small film pieces might be required for measurements on, for example, the eye of a patient [47]. For this reason, multiple authors have investigated whether the optical densities near the cutting edge of the film pieces are affected by the mechanical pressure during film cutting [34,47,48]. For EBT radiochromic film, Yu et al. [34] found that optical density disturbances generally remain limited to a 1-mm wide area around the film edge, even though artifacts up to 8 mm from the cut were observed in the worst-case scenario. Similarly, Avanzo et al. [48] reported artifacts up to 1.7 mm away from the edge of EBT2 film pieces. For EBT2 and EBT3 radiochromic film, Moylan et al. [47] evaluated the dosimetric accuracy of film pieces with a size between 5×5 mm^2 and 10×10 mm^2. They concluded that, for both EBT2 and EBT3 radiochromic film, a film size as small as 5×5 mm^2 provided accurate dosimetric results when averaging dose values over a region of interest (ROI) of 2.1×2.1 mm^2. The use of larger ROIs resulted in a considerable uncertainty increase, which was attributed to the inclusion of pixels near the film edge in the ROI. The use of a smaller ROI evenly resulted in less accurate dose values and increased the uncertainty. In general, most authors recommend not to analyze the optical densities within 1 to 2 mm of the film edge, depending on the film type and cutting method considered.

Second, radiochromic film dosimetry in the steep-dose gradient of the build-up region requires careful interpretation of the effective depth of measurement. As mentioned earlier, this depth depends on the film type and structure, and, for the asymmetrically layered EBT2 film, on the orientation of the film.

Finally, it is important to avoid potential disturbances caused by air gaps between the radiochromic film and surrounding tissue or phantom material. When mounting film on curved surfaces, such air gaps are generally avoided by using small film sizes and by carefully taping the film edges to the phantom surface or patient skin [28]. For films positioned in a slab phantom in a parallel orientation with respect to the beam axis, the impact of air gaps can be minimized by a meticulous phantom design or by positioning radiochromic films slightly off the central axis [35].

16.4 EXPLOITING THE ASYMMETRIC LAYER GEOMETRY OF EBT2 RADIOCHROMIC FILM FOR ASSESSMENT OF SKIN-DOSE AND STEEP-DOSE GRADIENTS

EBT2 film had to be carefully positioned on the surface of a phantom [20] due to its asymmetric layer configuration. But the asymmetry could also be exploited to measure at two different depths using the same film batch [29,49]. In a further step, De Wagter et al. [29] conceived a method to determine the superficial longitudinal dose gradient. They measured a gradient of 32%/mm in a 6-MV (10×10cm^2) field, whereas the highest measured value previously reported is 20%/mm [23].

Although EBT2 is no longer commercially available, its successor EBT3, which intrinsically has the same dosimetric properties [50], is available in an unlaminated configuration. The layer configuration concerned is illustrated in Figure 16.3. The user can add extra layer thicknesses to create multiple measuring depths with film pieces from a single-film sheet. The unlaminated EBT3 film was already used to determine the dose enhancement of gold films for kV X-rays [51].

16.5 SUMMARY

Exploiting its high spatial resolution, tissue equivalence and low-energy dependence, radiochromic film was shown to be an accurate and valuable skin and build-up dose detector in both megavoltage photon and electron beams. Although most applications focus on the determination of the build-up doses in single beams, radiochromic film has also been shown to be a valuable tool for entire-treatment for skin-dose verifications.

For build-up region dosimetry, some special precautions regarding the use of radiochromic film are required. In this respect, it is recommended to carefully consider the effective depth of measurement, not to analyze the optical densities near the film edge and to avoid the presence of air gaps around the film during irradiation.

One remarkable application of unlaminated radiochromic film is to exploit its asymmetric layer structure to measure the absorbed dose at two different depths in order to determine the superficial longitudinal dose gradient.

REFERENCES

1. Hounsell AR, Wilkinson JM. Electron contamination and build-up doses in conformal radiotherapy fields. *Phys Med Biol* 1999;44:43–55.
2. Chaney EL, Cullip TJ, Gabriel TA. Monte-Carlo study of accelerator head scatter. *Med Phys* 1994;21:1383–1390.
3. Medina AL, Teijeiro A, Garcia J, Esperon J. Characterization of electron contamination in megavoltage photon beams. *Med Phys* 2005;32:1281–1292.
4. ICRP. Publication No. 59: The biological basis for dose limitation in the skin. Annals of the ICRP (International Commission on Radiological Protection) Smith H. Oxford, UK: Pergamon, 1992.
5. Farage MA, Miller KW, Elsner P, Maibach HI. Structural characteristics of the aging skin: A review. *Cutan Ocul Toxicol* 2007;26:343–357.
6. FitzGerald TJ, Jodoin MB, Tillman G et al. Radiation therapy toxicity to the skin. *Dermatol Clin* 2008;26:161–172.
7. Archambeau JO, Pezner R, Wasserman T. Pathophysiology of irradiated skin and breast. *Int J Radiat Oncol Biol Phys* 1995;31:1171–1185.
8. ICRU. Publication 39: Determination of dose equivalents resulting from external radiation sources. Washington, DC: International Commission on Radiation Units and Measurements, 1985.
9. Devic S, Seuntjens J, Abdel-Rahman W et al. Accurate skin dose measurements using radiochromic film in clinical applications. *Med Phys* 2006;33:1116–1124.
10. Hsu SH, Roberson PL, Chen Y et al. Assessment of skin dose for breast chest wall radiotherapy as a function of bolus material. *Phys Med Biol* 2008;53:2593–2606.
11. Tournel K, Verellen D, Duchateau M et al. An assessment of the use of skin flashes in helical tomotherapy using phantom and in-vivo dosimetry. *Radiother Oncol* 2007;84:34–39.
12. Akino Y, Das IJ, Bartlett GK et al. Evaluation of superficial dosimetry between treatment planning system and measurement for several breast cancer treatment techniques. *Med Phys* 2013;40:011714.
13. Hsu SH, Moran JM, Chen Y et al. Dose discrepancies in the buildup region and their impact on dose calculations for IMRT fields. *Med Phys* 2010;37:2043–2053.
14. Court LE, Tishler RB, Allen AM et al. Experimental evaluation of the accuracy of skin dose calculation for a commercial treatment planning system. *J Appl Clin Med Phys* 2008;9:29–35.
15. Panettieri V, Barsoum P, Westermark M et al. AAA and PBC calculation accuracy in the surface build-up region in tangential beam treatments. Phantom and breast case study with the Monte Carlo code PENELOPE. *Radiother Oncol* 2009;93:94–101.
16. Chung HT, Jin HS, Dempsey JF et al. Evaluation of surface and build-up region dose for intensity-modulated radiation therapy in head and neck cancer. *Med Phys* 2005;32:2682–2689.
17. Verhaegen F, Seuntjens J. Monte Carlo modelling of external radiotherapy photon beams. *Phys Med Biol* 2003;48:R107–R164.

18. Arjomandy B, Tailor R, Anand A et al. Energy dependence and dose response of Gafchromic EBT2 film over a wide range of photon, electron, and proton beam energies. *Med Phys* 2010;37:1942–1947.

19. Butson MJ, Mathur JN, Metcalfe PE. Radiochromic film as a radiotherapy surface-dose detector. *Phys Med Biol* 1996;41:1073–1078.

20. Nakano M, Hill RF, Whitaker M et al. A study of surface dosimetry for breast cancer radiotherapy treatments using Gafchromic EBT2 film. *J Appl Clin Med Phys* 2012;13:83–97.

21. Butson MJ, Cheung T, Yu PKN, Webb B. Variations in skin dose associated with linac bed material at 6 MV X-ray energy. *Phys Med Biol* 2002;47:N25–N30.

22. Paelinck L, De Wagter C, Van Esch A et al. Comparison of build-up dose between Elekta and Varian linear accelerators for high-energy photon beams using radiochromic film and clinical implications for IMRT head and neck treatments. *Phys Med Biol* 2005;50:413–428.

23. Butson MJ, Yu PKN, Metcalfe PE. Measurement of off-axis and peripheral skin dose using radiochromic film. *Phys Med Biol* 1998;43:2647–2650.

24. Butson MJ, Yu PKN, Metcalfe PE. Extrapolated surface dose measurements with radiochromic film. *Med Phys* 1999;26:485–488.

25. Quach KY, Morales J, Butson MJ et al. Measurement of radiotherapy X-ray skin dose on a chest wall phantom. *Med Phys* 2000;27:1676–1680.

26. Bilge H, Cakir A, Okutan M, Acar H. Surface dose measurements with GafChromic EBT film for 6 and 18 MV photon beams. *Med Phys* 2009;25:101–104.

27. Chiu-Tsao ST, Chan MF. Photon beam dosimetry in the superficial buildup region using radiochromic EBT film stack. *Med Phys* 2009;36:2074–2083.

28. Almberg SS, Lindmo T, Frengen J. Superficial doses in breast cancer radiotherapy using conventional and IMRT techniques: A film-based phantom study. *Radiother Oncol* 2011;100:259–264.

29. De Wagter C, De Maeseneire N, Goethals N et al. Exploiting the asymmetric layer geometry of EBT2 radiochromic film for assessment of skin dose and steep gradients. *Radiother Oncol* 2015;115:S429–S430.

30. Nobah A, Aldelaijan S, Devic S et al. Radiochromic film based dosimetry of image-guidance procedures on different radiotherapy modalities. *J Appl Clin Med Phys* 2014;15:229–239.

31. Butson MJ, Cheung T, Yu P, Metcalfe P. Effects on skin dose from unwanted air gaps under bolus in photon beam radiotherapy. *Radiat Meas* 2000;32:201–204.

32. Constantinou C, Attix FH, Paliwal BR. A solid water phantom material for radiotherapy X-ray and gamma-ray beam calibrations. *Med Phys* 1982;9:436–441.

33. De Puysseleyr A, Lechner W, De Neve W et al. Absorbed dose measurements in the build-up region of flattened versus unflattened megavoltage photon beams. *Z Med Phys* 2016;26:177–183.

34. Yu PK, Butson M, Cheung T. Does mechanical pressure on radiochromic film affect optical absorption and dosimetry? *Australas Phys Eng Sci Med* 2006;29:285–287.

35. Suchowerska N, Hoban P, Butson M et al. Directional dependence in film dosimetry: Radiographic and radiochromic film. *Phys Med Biol* 2001;46:1391–1397.

36. Bufacchi A, Carosi A, Adorante N et al. In vivo EBT radiochromic film dosimetry of electron beam for Total Skin Electron Therapy (TSET). *Med Phys* 2007;23:67–72.

37. Rudat V, Nour A, Alaradi AA et al. In vivo surface dose measurement using GafChromic film dosimetry in breast cancer radiotherapy: Comparison of 7-field IMRT, tangential IMRT and tangential 3D-CRT. *Radiat Oncol* 2014;9:156.

38. van Battum LJ, Hoffmans D, Piersma H, Heukelom S. Accurate dosimetry with GafChromic (TM) EBT film of a 6 MV photon beam in water: What level is achievable? *Med Phys* 2008;35:704–716.

39. Qin SB, Chen T, Wang LL et al. Angular dependence of the MOSFET dosimeter and its impact on in vivo surface dose measurement in breast cancer treatment. *Technol Cancer Res Treat* 2014;13:345–352.

40. Velkley DE, Manson DJ, Purdy JA, Oliver GDJ. Build-up region of megavoltage photon radiation sources. *Med Phys* 1975;2:14–19.

41. Gerbi BJ, Khan FM. Measurement of dose in the buildup region using fixed-separation plane-parallel ionization chambers. *Med Phys* 1990;17:17–26.

42. Nilsson B, Montelius A, Andreo P. Wall effects in plane-parallel ionization chambers. *Phys Med Biol* 1996;41:609–623.

43. Nilsson B, Montelius A. Fluence perturbation in photon beams under nonequilibrium conditions. *Med Phys* 1986;13:191–195.

44. De Puysseleyr A, De Neve W, De Wagter C. A patient immobilization device for prone breast radiotherapy: Dosimetric effects and inclusion in the treatment planning system. *Med Phys* 2016;32:758–766.

45. Rawlinson JA, Arlen D, Newcombe D. Design of parallel plate ion chambers for buildup measurements in megavoltage photon beams. *Med Phys* 1992;19:641–648.

46. Jong WL, Wong JHD, Ung NM et al. Characterization of MOSkin detector for in vivo skin dose measurement during megavoltage radiotherapy. *J Appl Clin Med Phys* 2014;15:120–132.

47. Moylan R, Aland T, Kairn T. Dosimetric accuracy of Gafchromic EBT2 and EBT3 film for in vivo dosimetry. *Australas Phys Eng Sci Med* 2013;36:331–337.

48. Avanzo M, Rink A, Dassie A et al. In vivo dosimetry with radiochromic films in low-voltage intraoperative radiotherapy of the breast. *Med Phys* 2012;39:2359–2368.

49. Corradini N, Presilla S, Sterpin E. Study on the effective depth of measurement for Gafchromic EBT2 and EBT3 films. *Med Phys* 2013;40:221.

50. Reinhardt S, Hillbrand M, Wilkens JJ, Assmann W. Comparison of Gafchromic EBT2 and EBT3 films for clinical photon and proton beams. *Med Phys* 2012;39:5257–5262.

51. Rakowski JT, Laha SS, Snyder MG et al. Measurement of gold nanofilm dose enhancement using unlaminated radiochromic film. *Med Phys* 2015;42:5937–5944.

52. Marieb EN, Hoehn K. *Human Anatomy and Physiology*. London UK: Pearson Education/ Benjamin Cummings; 2008.

In vivo dosimetry

MARTIN BUTSON, GWI CHO, DAVID ODGERS, AND JOEL PODER

17.1 INTRODUCTION

In vivo dosimetry is an essential tool in quality-assurance programs. It is important to check the dose delivered to patients undergoing radiation treatment to avoid over or underdosing in areas in which assurance of planning accuracy may be needed. Radiochromic films are uniquely able to measure not only point doses but also two-dimensional maps of applied doses in vivo to a patient. In this section, we will briefly discuss the qualities of an ideal dosimeter and some of the processes and typical applications of radiochromic films to in vivo dosimetry techniques.

17.2 THE IDEAL DOSIMETER—PROPERTIES

Except for calorimetry, no other dosimeter directly detects absorbed dose delivered by ionizing radiation [1]. The quantity measured by the dosimeter, for example, ionization, thermal, or chemical changes, and others has to be converted into absorbed dose. The relation of these quantities to absorbed dose can be defined by a number of characteristics by which dosimeters can be compared. Relative to other dosimeters, the ideal dosimeter compares favorably for all of these features.

17.2.1 ACCURACY

Accuracy is the most important feature of any dosimeter used for in vivo dosimetry, that is, the ability to correctly indicate the true value of the quantity being measured. It is a measure of the collective effect of errors in all of the parameters that influence the measurement, including both systematic and random errors. Systematic errors will not be reduced by repeated measurement and systematically shift the measured result in one direction. Stochastic errors however can be reduced by multiple measurements as they result from random variations that may go in all directions [2].

17.2.2 PRECISION

The precision of a measurement specifies the reproducibility of the result under identical conditions and can be estimated from the data obtained in repeated measurements [2]. The dosimeters precision is usually stated in terms of a standard deviation. The ideal dosimeter for in vivo dosimetry has a small standard deviation.

17.2.3 DETECTION LIMIT (SENSITIVITY)

The detection limit is defined as the lowest dose detectable for a certain dosimeter type. Some dosimeters are more sensitive to low-dose compared with others, and these detectors can be useful for in vivo dosimetry, especially in low-dose measurements such as under shields or out of the primary field.

17.2.4 MEASUREMENT RANGE

The dosimeters' ability to measure very low to very high doses with a high precision and accuracy makes it extremely useful for in vivo dosimetry. The range of doses that may need to be quantified is extremely large (10–10,000 cGy), especially in radiotherapy applications, and the ability of one dosimeter to measure doses over this entire range is extremely valuable. The low-dose limit is usually defined as double the background reading (i.e., the reading obtained without radiation present) [3]. The upper limit of the range is usually imposed by a decrease in sensitivity to an unacceptable value, for example, through exhaustion of the supply of atoms being acted upon by the radiation to produce the reading.

17.2.5 LINEARITY WITH DOSE

Ideally, the response of the dosimeter should be linearly proportional to the radiation dose, or in other words, constant dose sensitivity throughout its measurable range. This characteristic is not essential, however, as long as the dose response is reproducible. A linear dose response does however enable faster processing of results and may reduce the likelihood of errors [1].

17.2.6 DOSE-RATE DEPENDENCE

The ideal dosimeter, if being used to measure integrated dose, should be independent of the delivered dose rate within the range of dose rates to be encountered. The upper limit usually occurs when charged particle tracks are created close enough together in space and time to allow the ions to interact before they are detected by the dosimeters [3].

17.2.7 ENERGY DEPENDENCE

The energy dependence of the dosimeter is defined as the change in response of the dosimeter due to changes in radiation beam quality. The ideal in vivo dosimeter has minimal change in dose response with radiation energy. This is important for in vivo dosimeters in which the dosimeter may be measuring the dose from a specific radiation quality, which may be different than the beam quality used during calibration in a reference condition.

17.2.8 SPATIAL RESOLUTION

As delivered dose can be expressed as a point quantity, the ideal dosimeter should be able to measure the dose to an extremely small volume (approaching a point) [2]. The size of this volume is usually limited by the ability of the dosimeter to create a measurable signal of a significant level above background (detection limit). Spatial resolution becomes particularly important in regions of high dose gradients and small radiation fields.

17.3 IN VIVO DOSIMETRY REQUIREMENTS/RADIOCHROMIC PROPERTIES

Radiochromic film is a practical media for in vivo dosimetry as it performs well when compared with the list of ideal characteristics. In particular, for in vivo dosimetry, the energy dependence and spatial resolution qualities of radiochromic film are standouts amongst its competitors such as thermoluminescent dosimeters (TLD's), optically stimulated luminescence dosimeters (OSLDs), or in vivo diodes.

Part of any in vivo dosimetry process is the preparation of a protocol to formally develop a routine procedure to follow for film preparation, irradiation technique, and readout analysis. The development and routine use of this protocol is essential for accurate and reproducible in vivo dosimetry with radiochromic films.

- Film choice: The first requirement for in vivo dosimetry is knowing the dose-level requirements of the measurement and selecting the appropriate film type for analysis. The radiochromic film range provides many film types that can measure doses of the order of a few cGy up to 50 Gy. The choice of a right film is the first step in this process.
- In vivo mark up: The films should be marked for appropriate positioned or orientation when on the patient to allow for accurate alignment with predicted dose evaluation from your planning

computer. This can be performed before packaging the film, or sometimes it is more appropriate to mark the film (packaging) after placement on the patient with reference to the patient anatomy.

- Packaging: Once the film is selected, the film must be packaged appropriately to protect the film from damage as well as from patient contamination as discussed in Chapter 3. All in vivo dosimeters should be packaged and covered with some kind of protective material such as plastic wrap or equivalent when used in vivo. Extra wrapping or protection may be needed for sites in which bodily fluids are present.

- Collection of standards: Following an in vivo dosimetry procedure, standard doses should be collected to produce a calibration curve to compare the results with. The collection of these data is essential due to the nonlinear nature of the films response and the time difference between the in vivo irradiation/readout process, and the calibration irradiation/readout process is critical to ensure the highest level of accuracy for dose assessment. Although these times can be different, corrections may need to be applied to results to improve dose-measurement accuracy.

- Cleanliness: The films must be cleaned appropriately before analysis to remove any dirt, oil, or contamination fluids to provide optimal dosimetry assessment. Of course, these types of procedures should be performed whilst wearing appropriate personal protection, for example, gloves. The scanner used to analyze the film must also be appropriately clean and free from dirt or hand oils, sticky tape marks, and others as discussed in Chapter 4.

- Scan resolution and area selection: As radiochromic film has a high intrinsic detection resolution, the film can be used for assessment of dose in high-dose gradients as well as areas in which uniform dose is delivered. The appropriate selection of scanner resolution is needed to match the requirements of the procedure being performed. For example, to assess an in vivo dose estimate to the center of field of a large electron field will only require a low scanning resolution and a high scanning resolution will not provide any extra data but only make the in vivo dose-assessment process harder, whereas the assessment of dose at a field edge near a critical structure, such as an eye, would benefit from a high-resolution scanning. Again, these types of decisions are dependent on the situation, and the medical physicist should use common sense to match the requirements to the job at hand. Similarly, if a point-dose assessment is required, a smaller piece of radiochromic film is all that is required, whereas if you are trying to assess the dose delivered over an entire field, a larger film will be appropriate.

As mentioned earlier, the essential process for in vivo dosimetry is the development of a protocol and sticking to it. These protocols may vary, depending on the type of in vivo dosimetry required. For example, dosimetry protocol using radiochromic external beam therapy (EBT3) film and an Epson 10000XL desktop scanner for total-body irradiation (TBI) point-dose assessment. Using such a protocol allows the user to streamline the in vivo dosimetry process as well as minimize errors that may be created by ad-hoc dosimetry processes.

17.3.1 TBI DOSIMETRY USING EBT FILM AND A TEMPLATE SYSTEM

A simple-to-do process as described in the following could facilitate optimum in vivo dosimetry that is accurate and time saving. There may be many ways to do in vivo dosimetry, but this is the author's choice as it has been tested and proven to be accurate and reliable.

1. First, tape the radiochromic film to the grid template using sticky tape on both sides (Figure 17.1).
2. Cut the film into strips with the grid template still attached.
3. Remove the grid template and sticky tape the naming template to the bottom of each strip with a small overlap of the tape onto the radiochromic film.

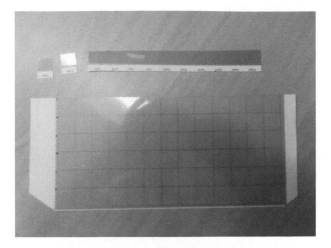

Figure 17.1 Example of film orientation and cutting template. (Courtesy of Martin Butson.)

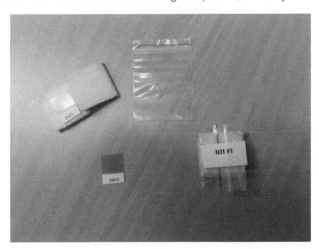

Figure 17.2 Packaging requirements for TBI dosimetry with radiochromic film. (Courtesy of Martin Butson.)

4. Cut the EBT film into squares for TBI dosimetry (Figure 17.2).
5. Place each TBI film (2 per packet) in between the TBI wax bolus and place inside a labeled TBI plastic bag.
6. Prepare dosimeters for all fractions required.
7. Irradiate the standard films to the given doses of 0 cGy (control), 150 cGy, 180 cGy, 200 cGy, 250 cGy, and 300 cGy for calibration.
8. Perform film calibration and patient dose analysis using the Epson scanning procedure described in the following section.

17.3.2 EPSON SCANNER PROCEDURE FOR READOUT OF TBI FILMS

The TBI films can be scanned and analyzed using the Epson scanner for direct input into the TBI in vivo dose-assessment calculation worksheet. The procedure for scanning is given as follows.

1. Place the TBI black cardboard template on top of the Epson Scanner with the star marking in the top left-hand corner of the scanning plate.

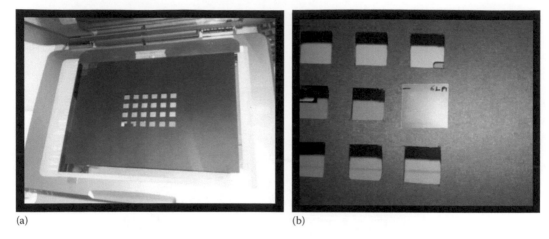

(a)   (b)

Figure 17.3 Scanning template for in vivo process. (Courtesy of Martin Butson.)

Figure 17.4 Irradiated film on template. (Courtesy of Martin Butson.)

2. Place the TBI films on the template windows in the positions marked for the appropriate film, for example, Right hand film 1 (RH 1) in the RH 1 window. Place film with appropriate orientation and face up (Figure 17.3).
3. Perform a scan of the whole scanner region using the standard settings on the Epson scanner, that is, film positive, 48-bit color, thumbnail off, 72 DPI, no color correction, professional mode, and tiff image. Copy the scan into the patient folder with appropriate naming convention, for example, Patientname_Fraction 1 (Figure 17.4)*.
4. Open ImageJ to perform film analysis.
5. Open the appropriate image file (Figure 17.5).
6. It is important to note that the image is showing the RED channel results, and this is indicated by the red writing and information in the top left-hand corner of the image. The green or blue channel will have green or blue writing and names, respectively (Figure 17.6).

---

* As the image is a transmission scan, it will appear "flipped" when viewed in Image J (mirror image).

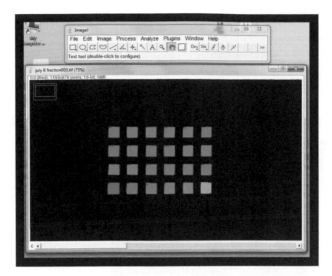

Figure 17.5 ImageJ analysis of film template. (Courtesy of Martin Butson.)

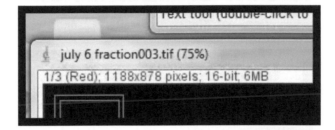

Figure 17.6 Selecting rechannel analysis. (Courtesy of Martin Butson.)

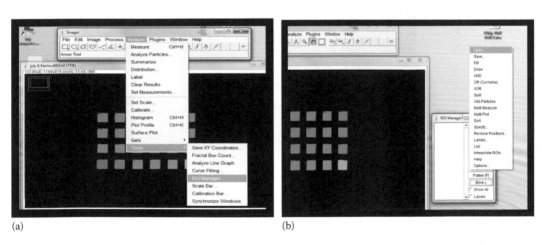

(a)   (b)

Figure 17.7 Region of interest manager in ImageJ analysis. (Courtesy of Martin Butson.)

7. Now that the appropriate image has been selected, you will need to apply a region of interest manager (ROI Manager) and template for the TBI film. Thus, follow the sequence as below, Select Analyze -> Tools -> ROI Manager.

8. The ROI Manager panel will appear. Select "More" -> "Open". Then select the TBI_Dosimetry.zip template (Figure 17.7).

9. Click "Show All" to reveal the ROIs to analyze. Make sure that ROIs lie within the windows and do not include a region with marker writing (Figure 17.8).

10. Select "Measure" and a Window will be created with the results for mean pixel density and standard deviation for each region. Select: Edit -> Select "All" then "Edit" -> Copy to select all data (Figure 17.9).

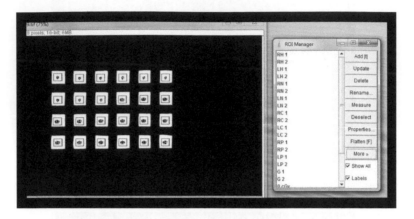

Figure 17.8 Created region of interests. (Courtesy of Martin Butson.)

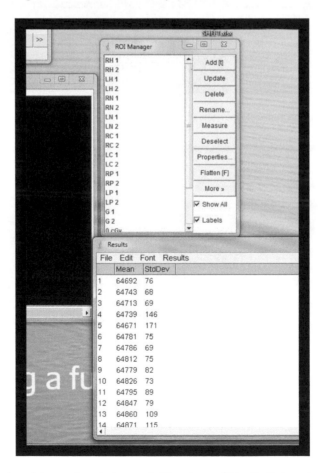

Figure 17.9 Pixel value data capture in ImageJ. (Courtesy of Martin Butson.)

Paste these data in the appropriate worksheet in the TBI patient Excel spreadsheet (e.g., FX 1 worksheet for fraction 1 film data or the film-calibration worksheet for calibration films). The excel spread sheet will automatically calculate optical densities and populate all other data including the patient film dose results page.

17.4 OTHER RADIOTHERAPY APPLICATIONS

17.4.1 LOW DOSE

The knowledge of the dose delivered to sensitive structures under shields in radiation therapy is critical in determining the risk of late complications. Treatment-planning systems have historically not been able to accurately predict this dose, and often an in vivo dose measurement is performed. Traditionally, TLDs have been used for this purpose due to their small size. However, in recent years, radiochromic film has been shown to provide an accurate and reproducible measure of the dose beneath these shields [4,5].

Butson et al. [4] showed that measurements made with radiochromic EBT films match predicted doses at the center of an eye shield irradiated with 50 and 150 kVp to within 2%. The results of the study showed that, compared with TLDs, the radiochromic films had the added advantage of being able to measure a two-dimensional dose map under the eye shield. This allows the radiochromic films to give a prediction of not only the dose to the lens but also the cornea and retina.

Chiu-Tsao and Chan [5] performed a feasibility study using EBT film and a prototype low-dose film, which has a higher inherent sensitivity to measure dose at peripheral points outside of primary 6-MV photon beams. The dose measured by the radiochromic film devices was then compared with the dose measured by an ionization chamber under identical conditions. General agreement (within 5%) between the radiochromic films and ionization chamber was found in the study at 15-cm off axis. The results of this study confirm the accuracy of radiochromic films in measuring low doses in megavoltage photon beams, demonstrating their suitability to be used for low-dose in vivo measurements.

17.4.2 CONVENTIONAL DOSE

Not only is the knowledge of the dose delivered to sensitive structures important in radiotherapy applications, but also dose to the target. There are instances in radiotherapy in which the target may be at or close to the surface of the patient, and in these instances the accuracy of the dose predicted by the planning system cannot be guaranteed. It is therefore useful to have a dosimeter that can be used in vivo to confirm that the target is being irradiated with sufficient dose to achieve the desired reaction. The properties of radiochromic film, such as its excellent spatial resolution, small effective point of measurement, and high sensitivity, make it ideal for this purpose. There have been a number of publications that have demonstrated the effectiveness of radiochromic film for the purpose of validating the prescribed dose close to the surface of a patient [6–14]. The majority of these publications center specifically on breast and head-and-neck applications of radiotherapy. Details of surface and buildup dose are presented in Chapter 14.

Alashrah et al. [15] compared radiochromic MD-55 film with TLD and Monte Carlo methods in predicting surface dose for a 6-MV photon beam. The results of the study showed that both MD-55 film, and TLD were able to predict the surface dose to within 1% when compared with Monte Carlo.

Radiotherapy treatments of the breast are often performed using parallel opposed tangent fields to avoid delivering too much dose to the lung and heart of the patient. Due to this beam configuration, a combination of entrance and exit dose may result in unnecessarily high dose to the skin. Several studies, including those by Almberg et al. [7], Price et al. [10], and Rudat et al. [13], successfully utilized radiochromic film to compare superficial skin doses in breast-cancer radiotherapy using conventional and intensity modulated radiation therapy (IMRT) techniques. Despite minor differences in techniques used in each study, the conclusions were the same, that is, the use of IMRT beams for tangents in breast-cancer therapy does not significantly change the dose to the surface of the breast. This demonstrates the applicability of radiochromic film for measuring in vivo surface dose in breast-cancer therapy.

Radiochromic film can also be utilized to estimate the dosimetric effect of immobilization devices in breast-cancer therapy. Thermoplastic casts are often employed in breast-cancer therapy to secure the breasts, maintain setup reproducibility, and limit the presence of skin folds. However, these thermoplastic casts may also effectively act as bolus, increasing the dose to the skin of the breast by bringing the maximum depth dose closer to the surface. Kelly et al. [7] successfully demonstrated that EBT films can be used to determine the change in surface dose due to the addition of a thermoplastic cast. The study showed that one particular thermoplastic cast increased the dose at the surface of the breast by up to 45%.

Head-and-neck radiotherapy-treatment plans also consist of high-dose regions close to the surface of the patient, and despite many recent advances in treatment-planning system dose-calculation algorithms, there remains considerable dosimetric uncertainty in the surface and buildup region in head-and-neck-treatment plans. Therefore, as for breast-cancer radiotherapy treatments, there also exists a need for measurement to confirm the prescribed dose close to the surface of the patient. Several studies have been published in recent years that have utilized radiochromic film for this purpose. Chung et al. [8] used GAFchromic™ HS film to estimate the dose to the surface of an anthropomorphic phantom for head-and-neck-treatment plans for two different planning systems. The study concluded that both treatment-planning systems overestimated the dose to the surface of the phantom by 7.4%–18.5% within the first 2-mm depth. Qi et al. [11] performed a similar study on an anthropomorphic phantom comparing EBT films with metal–oxide–semiconductor field-effect transistor (MOSFET) and an ionization chamber. The study found general agreement within measurement uncertainty for all three detector types. When comparing with the treatment-planning system predicted dose, each detector showed an overestimation of up to 8.5% within the first 5 mm by the treatment-planning system.

17.4.3 HIGH DOSE

Stereotactic body radiotherapy has been introduced in recent years for the treatment of medically inoperable patients with early stage non small-cell lung cancer, liver, pancreas, and spinal cord. Treatments are delivered using a hypofractionated schedule with fraction doses as high as 7–20 Gy. At this dose level, skin toxicity is a critical tolerance point, and attempts are made to spread this skin dose using noncoplanar beam setups. It is common practice, however, to perform in vivo measurements to quantify the skin dose, due to the planning system limitations mentioned earlier. Holt et al. [16] performed skin-dose measurements using EBT2 films, mounted on the surface of an anthropomorphic phantom to determine the skin dose when changing the treatment technique from IMRT to volumetric-modulated arc therapy (VMAT). The group found that the measured skin dose for both IMRT and VMAT was the same on both the dorsal and ventral regions of the phantom, within experimental uncertainty. No correlation between the measured skin dose and dose predicted by the treatment-planning system was found, again demonstrating the limitations of treatment-planning systems in predicting skin dose and underscoring the need for a dose measurement with a suitable dosimeter such as radiochromic film.

17.4.4 NICHE APPLICATIONS OF GAFCHROMIC™ FILM FOR IN VIVO DOSE MEASUREMENTS

The properties of GAFchromic film, its sensitivity and dose response range, energy independence, good spatial resolution, and flexibility, make it an excellent choice for in vivo measurement. In particular, its ability to give dose information over a broad, two-dimensional area is valuable and unique amongst most common dosimeters.

On account of this, radiochromic film has found use in an extensive range of in vivo applications. Not only is it being used across the range of routine clinical measurements already discussed, it has also found use in a number of more niche applications.

As outlined already, radiochromic film is commonly used for measuring skin dose in vivo, in particular, in radiotherapy of the breast [17]. Film has also found a very specific in vivo application here in measuring directly the impact of tissue expanders on the radiotherapy treatment of patients who have previously had such a device implanted. These tissue expanders typically contain a high-density rare-earth magnet, which leads to a reduction in transmission of the beam that passes through the magnet, and create a localized dose reduction at skin surface. Srivastava et al. [18] provided extensive data on dosimetry; however, surface dose was not evaluated. Gee et al. have used EBT3 film to measure in vivo the effect of these magnets on the skin dose and consistently measured these localized dose reductions in 15 out of 16 patients on whom the measurements were performed [19]. Figure 17.10 [19] shows how a two-dimensional dose map measured using radiochromic film can

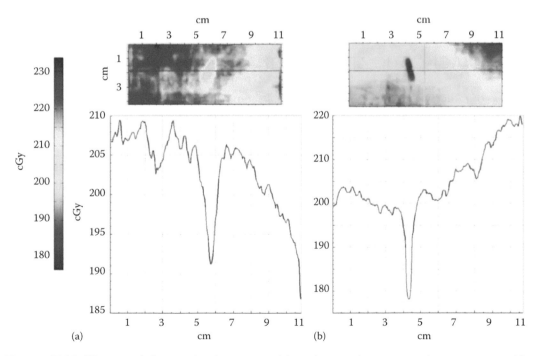

Figure 17.10 2D map of dose reductions caused by a breast tissue expander as measured by radiochromic film. (Adapted from Gee, H.E. et al., *J. Med. Imag. Radiat. Oncol.*, 60, 138–145, 2016.)

ascertain such variations in dose and provides the confidence in dose delivery. The significance of film as an in vivo dosimeter is seen by comparing these measurements with similar work performed by Damast et al. [20] using TLDs to quantify the dose reduction to skin caused by an expander. In measurements across six patients, only one set of readings clearly indicated the presence of a local dose reduction, even though the effect had previously been verified ex vivo in a solid water phantom. This illustrates the significant advantage conveyed by film's capability for two-dimensional measurement and also describes here the findings for Figure 17.10.

In another application to breast radiotherapy, two different types of radiochromic films have been used to verify intraoperative radiotherapy breast treatments both with megavoltage electrons as well as kilovoltage photons [21–22]. A more extensive discussion on electron beam can be found in Chapter 14. In the study by Avanzo et al. [22], this included placing films wrapped in sterile envelopes into the excision cavity, and on the shielded pectoralis fascia, as well as on the skin. Again, radiochromic film is uniquely suited to complex treatment situations such as these in which 2D dosimetric map is needed.

The versatility of radiochromic film has also led to its use in in vivo measurements for both TBI treatments and total skin electron therapy, in which achieving dose homogeneity can be a real challenge [23–25]. Monitoring the delivered dose each fraction allows quantification of underdosed regions to be detected so that they can be accounted for in subsequent fractions. Other niche areas using radiochromic films within radiotherapy include, but are not limited to, brachytherapy [26–29], stereotactic body radiotherapy [30], and TomoTherapy® [31] that has been discussed in various chapters in the current book.

Although radiochromic films have not found any significant use within diagnostic or therapeutic nuclear medicine applications, they have been used to assess dose from radioactive sources such as phosphorus-32 for hemangiomas [32] and liposome transport of Rhenium-186 [33]. In summary, radiochromic film has been used for many beneficial dosimetric comparisons and analysis requirements in niche areas of medical dosimetry.

17.5 DIAGNOSTIC APPLICATIONS

The use of radiochromic film in diagnostic applications has increased significantly over the last decade [34–41], and a discussion can be found in Chapter 5. This is largely due to the fact that radiochromic film is now being recognized as one of the premier dosimeters for measurement of skin dose, of which there is substantial interest in diagnostic applications such as cardiac catheterization. The quantification of entrance surface dose is vital in diagnostic applications to be able to quantify the subsequent lifetime-attributed risk of cancer incidence due to the patient being exposed to radiation from a particular procedure.

17.5.1 IMAGING AND SKIN DOSE

Several studies published recently have made use of radiochromic film for in vivo measurements to determine the dose to the skin during diagnostic studies. Loader et al. [34] used radiochromic XRCT film to measure the radiation distribution and magnitude of the skin dose from a CT coronary angiography. They found that the skin dose remained broadly constant (7–9 mGy) when averaged over the circumference of a phantom. Gotanda et al. [36] used a similar approach with

EBT film to measure the skin-dose distribution during a pediatric CT examination of the head. The study concluded that the GAFchromic EBT film was able to reliably predict the maximum (skin) dose during pediatric CT head examinations.

Multislice CT scanners allow the estimation of the effective dose to the patient from the dose-length product parameter, the value of which is displayed on the CT console during acquisition. This method only represents an approximation based on standard circular phantoms and ignores the actual size of the patient. De Denaro et al. [38] however made use of radiochromic XR-QA films to derive the effective dose to the patient during CT acquisitions and found that the effective dose predictions as made by the film were extremely reliable.

Due to the increased use of radiochromic film in diagnostic applications, efforts have been made to optimize film for kilovoltage energies typical of these diagnostic procedures. One example of this is the development of cHXR film, which has a sensitivity one order of magnitude larger than the XR film typically used in diagnostic applications. Gorny et al. [42] have published a study that shows the HXR film could be calibrated with a precision of 1% over the exposure range and with an X-ray beam spectrum relevant to diagnostic CT examinations.

17.5.2 FLUOROSCOPY

Monitoring of skin entrance dose radiation exposure during lengthy interventional procedures is common due to the potential for skin injury [43]. Chu et al. [40] performed a study in which GAFchromic XR films were placed on the skin of 20 patients and attempted to relate the measured skin dose to the fluoroscopy duration and dose-area product, both of which are readily available real-time measurements. The study found that there was a weak correlation between skin dose and dose area product (DAP)/fluoroscopy duration and concluded that determination of skin dose during fluoroscopic procedures should be made from direct measurement [40]. They published a further study in which skin-dose measurements made with the XR films were used to develop dose indices that may be used as a predictor for skin injury due to cardiac catheterization procedures [41].

Giordano et al. [37] also found a good correlation between maximum skin dose and DAP/fluoroscopy duration using GAFchromic XR films. The authors of the study concluded that the AP values as derived from skin-dose measurements using radiochromic films were suitable for online skin dosimetry and may therefore be used to avoid radiation-induced skin dose injuries. These derived DAP values may also be used to compare local values against published reference levels.

17.6 SUMMARY

Radiochromic films are available in various types and suitable dose range that can be used in radiation oncology and diagnostic radiology applications for monitoring in vivo dosimetry. Some applications are skin dose in breast, head and neck, groin, exit dose, hypofractionated treatment such as stereotactic body radiation therapy (SBRT), extended distance treatment such as total body irradiation (TBI), total skin electron irradiation (TSEI), and intraoperative and in diagnostic applications. The level of accuracy in doismetry can be improved by selecting the appropriate film type for the application required and following guidelines for accurate dosimetry procedures as outlined in this book.

REFERENCES

1. Metcalfe P, Kron T, Hoban P. *The Physics of Radiotherapy X-Rays and Electrons*. Madison, WI: Medical Physics Publishing; 2007.
2. Podgorsak EB. *Radiation Oncology Physics: A Handbook for Teachers and Students*. Vienna, Austria: International Atomic Energy Agency; 2005.
3. Attix FH. *Introduction to Radiological Physics and Radiation Dosimetry*. Weinheim, Germany: Wiley-VCH; 1991.
4. Butson MJ, Cheung T, Yu PKN et al. Measurement of radiotherapy superficial X-ray dose under eye shields with radiochromic film. *Med Phys*. 2007;24:29–33.
5. Chiu-Tsao ST, Chan MF. Use of new radiochromic devices for peripheral dose measurement: Potential in vivo dosimetry application. *Biom Imag Interv J*. 2009;5:1–12.
6. Almberg SS, Lindmo T, Frengen J. Superficial doses in breast cancer radiotherapy using conventional and IMRT techniques: A film-based phantom study. *Radiother Oncol*. 2011;100:259–264.
7. Andrew K, Nicholas H, Peter M et al. Surface dosimetry for breast radiotherapy in the presence of immobilization cast material. *Phys Med Biol* 2011;56:1001.
8. Chung H, Jin H, Dempsey JF et al. Evaluation of surface and build-up region dose for intensity-modulated radiation therapy in head and neck cancer. *Med Phys* 2005;32:2682–2689.
9. Devic S, Seuntjens J, Abdel-Rahman W et al. Accurate skin dose measurements using radiochromic film in clinical applications. *Med Phys* 2006;33:1116–1124.
10. Price S, Williams M, Butson M, Metcalfe P. Comparison of skin dose between conventional radiotherapy and IMRT. *Australas Phys Eng Sci Med* 2006;29:272–277.
11. Qi ZY, Deng XW, Huang SM et al. In vivo verification of superficial dose for head and neck treatments using intensity-modulated techniques. *Med Phys*. 2009;36:59–70.
12. Quach KY, Morales J, Butson MJ, Rosenfeld AB, Metcalfe PE. Measurement of radiotherapy X-ray skin dose on a chest wall phantom. *Med Phys*. 2000;27:1676–1680.
13. Rudat V, Nour A, Alaradi AA et al. In vivo surface dose measurement using GafChromic film dosimetry in breast cancer radiotherapy: Comparison of 7-field IMRT, tangential IMRT and tangential 3D-CRT. *Radiat Oncolo*. 2014;9:1–9.
14. Srivastava RP, De Puysseleyr A, De Wagter C. Skin dose assessment in unmodulated and intensity-modulated radiation fields with film dosimetry. *Radiat Meas*. 2012;47:504–511.
15. Alashrah S, Kandaiya S, Maalej N, El-Taher A. Skin dose measurements using radiochromic films, TLDS and ionisation chamber and comparison with Monte Carlo simulation. *Radiat Prot Dosim*. 2014;162:338–344.
16. Holt A, van Vliet-Vroegindeweij C, Mans A et al. Volumetric-modulated arc therapy for stereotactic body radiotherapy of lung tumors: A comparison with intensity-modulated radiotherapy techniques. *Int J Radiat Oncol Biol Phys*. 2011;81:1560–1567.
17. Cheung, T, Butson MJ, Perter KN. Multilayer Gafchromic film detectors for breast skin dose determination in vivo. *Phys Med Biol*. 2002;47:N31.
18. Srivastava SP, Cheng CW, Andrews J, Das IJ. Dose perturbation due to metallic breast expander in electron and photon beam treatment of breast cancer. *J Radiat Oncol*. 2014;3:65–72.
19. Gee HE, Bignell F, Odgers D et al. In vivo dosimetric impact of breast tissue expanders on post-mastectomy radiotherapy. *J Med Imaging Radiat Oncol*. 2016;60(1):138–145.
20. Damast S, Beal K, Ballangrud Å et al. Do metallic ports in tissue expanders affect post mastectomy radiation delivery? *Int J Radiat Oncol Biol Phys*. 2006;66(1):305–310.
21. Ciocca M et al., In vivo dosimetry using radiochromic films during intraoperative electron beam radiation therapy in early-stage breast cancer. *Radiother Oncol*. 2003;69:285–289.

22. Avanzo M, Rink A, Dassie A et al. In vivo dosimetry with radiochromic films in low-voltage intraoperative radiotherapy of the breast. *Med Phys.* 2012;39:2359–2368.

23. Su FC, Shi C, Papanikolaou N. Clinical Application of Gafchromic EBT film for in vivo dose measurements of total body irradiation radiotherapy. *Appl Radiat Isot.* 2008;66:389–394.

24. Bufacchi A, Carosi A, Adorante N et al. In vivo EBT radiochromic film dosimetry of electron beam for Total Skin Electron Therapy (TSET). *Med Phys.* 2007;23(2):67–72.

25. Gamble LM, Farrell, TJ, Jones GW, Hayward, JE. Two-dimensional mapping of under dosed areas using radiochromic film for patients undergoing total skin electron beam radiotherapy. *Int J Radiat Oncol Biol Phys* 2005;62(3):920–924.

26. Pai S, Reinstein LE, Gluckman G et al. The use of improved radiochromic film for in vivo quality assurance of high dose rate brachytherapy. *Med Phys* 1998;25(7):1217–1221.

27. Seo H, Haque M, Hill R, Baldock C. In vivo dosimetric verification of a HDR brachytherapy surface mould. *Australas Phys Eng Sci Med* 2008;31(4):519.

28. Lin R, Haque M, Odgers D et al. In vivo dosimetric verification of skin dose using GAFCHROMIC EBT2 film. *Journal of Medical Imaging and Radiation Oncology; Proceedings of the 1st Joint Meeting of RANZCR, AIR, FRO and ACPSEM*, Brisbane, Australia October 22–25, 2009.

29. Poder J, Corde S. I-125 ROPES eye plaque dosimetry: Validation of a commercial 3D ophthalmic brachytherapy treatment planning system and independent dose calculation software with GafChromic® EBT3 films. *Med Phys.* 2013;40:121709.

30. Cho GA, Ralston A, Tin MM et al. In vivo and phantom measurements versus Eclipse TPS prediction of near surface dose for SBRT treatments. *J Phys Conf Ser.* 2014;489:012008.

31. Avanzo M, Drigo A, Kaiser SR et al. Dose to the skin in helical tomo therapy: Results of in vivo measurement with radiochromic films. *Med Phys.* 2013;29(3):304–311.

32. Shi CB, Yuan B, Lu JR et al. Continuous low-dose-rate radiation of radionuclide phosphorus-32 for hemangiomas. *Cancer Biother Radiopharm.* 2012;27:198–203.

33. Medina LA, Goins B, Rodríguez-Villafuerte M et al. Spatial dose distributions in solid tumors from 186Re transported by liposomes using HS radiochromic media. *Eur J Nucl Med Mol Imaging.* 2007;34(7):1039–1049.

34. Loader RJ, Gosling O, Roobottom C et al. Practical dosimetry methods for the determination of effective skin and breast dose for a modern CT system, incorporating partial irradiation and prospective cardiac gating. *Br J Radiol* 2012;85:237–248.

35. Jones AK, Ensor JE, Pasciak AS. How accurately can the peak skin dose in fluoroscopy be determined using indirect dose metrics? *Med Phys* 2014;41:071913.

36. Gotanda R, Katsuda T, Gotanda T et al. Dose distribution in pediatric CT head examination using a new phantom with radiochromic film. *Australas Phys Eng Sci Med.* 2012;31:339–344.

37. Giordano C, D'Ercole L, Gobbi R et al. Coronary angiography and percutaneous transluminal coronary angioplasty procedures: Evaluation of patients' maximum skin dose using Gafchromic films and a comparison of local levels with reference levels proposed in the literature. *Med Phys.* 2010;26:224–232.

38. de Denaro M, Bregant P, Severgnini M, de Guarrini F. In vivo dosimetry for estimation of effective doses in multislice CT coronary angiography. *Med Phys.* 2007;34:3705–3710.

39. D'Alessio D, Giliberti C, Soriani A et al. Dose evaluation for skin and organ in hepatocellular carcinoma during angiographic procedure. *J Exp Clin Cancer Res.* 2013;32:1–22.

40. Chu RYL, Thomas G, Maqbool F. Skin entrance radiation dose in an interventional radiology procedure. *Health Phys* 2006;91:41–46.

41. Chu RYL, Schechtor E, Chu N. Dose indices of radiation to skin in fluoroscopically guided invasive cardiology procedures. *Health Phys.* 2007;93:S124–S127.

42. Gorny KR, Leitzen SL, Bruesewitz MR et al. The calibration of experimental self-developing Gafchromic® HXR film for the measurement of radiation dose in computed tomography. *Med Phys* 2005;32:1010–1016.

43. delle Canne S, Carosi A, Bufacchi A et al. Use of GAFCHROMIC XR type R films for skin-dose measurements in interventional radiology: Validation of a dosimetric procedure on a sample of patients undergone interventional cardiology. *Med Phys* 2006;22:105–110.

18

Postal and clinical trial dosimetry

TANYA KAIRN, JOERG LEHMANN, SCOTT CROWE, JESSICA E. LYE,
PAOLA ALVAREZ, DAVID FOLLOWILL, AND TOMAS KRON

18.1 INTRODUCTION

The accurate planning and treatment of radiotherapy patients in accordance with a treatment protocol has been shown to impact treatment outcome and patient's survival [1]. External audits of treatment plan quality and dosimetric accuracy are therefore very important tools in radiotherapy. Multicenter audits are used by radiotherapy centers and auditing institutions to provide valuable information on the accuracy and reliability of radiotherapy treatment-planning and delivery systems and processes. Audit results can be used by audited centers to validate or improve their systems and processes. This is especially important in centers where new treatment techniques are being introduced, and staff are unfamiliar with new processes [2]. The value of audits in the radiotherapy treatment quality-improvement process is indicated by the improvements in treatment-delivery accuracy that have been observed when centers are audited more than once [3–6].

With the power of their collected data, audits can uncover larger trends and problems relating to radiation therapy planning system algorithms or delivery systems [7,8]. When results are aggregated and published by auditing bodies, the benefits of the audit are extended to unaudited radiotherapy centers, who are able to modify and improve their local procedures based on audit recommendations.

Audits are also frequently used by clinical trials administrators as part of the credential of radiotherapy centers for participation in specific trials [9]. Audits are used in this context, because it is especially important that the treatment outcomes arising from the trial interventions are associated with the doses that were actually delivered by those interventions [10]. For clinical trials to produce valid results, the centers contributing to them need to be demonstrably capable of planning treatments that follow the trial protocol and delivering treatments that are dosimetrically and geometrically matched to their treatment plans [3,10].

As radiotherapy-treatment planning and delivery technologies have become more sophisticated, the deliverable dose distributions and the treatment plans that produce them have become more complex and modulated, to effectively treat targeted tumors while sparing surrounding healthy tissues [11]. At the same time, there has been a widespread increase in the adoption of stereotactic techniques, in which hypofractionated doses are used in combination with reduced margins, and where the geometric accuracy of the treatment delivery carries increased importance [12]. When implementing these techniques, or running a trial investigating their effects, it is especially important to use a radiation dosimeter that allows multidimensional dose measurements to be obtained with a high spatial resolution while simultaneously providing an indication of geometric accuracy. Radiochromic film (RCF) is ideal for this purpose (see Chapters 8 and 9).

18.2 FILM CHARACTERISTICS AND LIMITATIONS

The choice of detector in an audit setting depends on the required accuracy, spatial requirements, reading uncertainty, costs, and whether 1D, 2D, or 3D information is required. RCF is well suited to use in multicenter audits of radiotherapy treatment-planning and delivery, especially when measurements of dose distributions or evaluations of geometric accuracy are required. This is because RCF is well known to produce reliable measurements of two-dimensional dose distributions, which can be compared with planned dose distributions and checked against alignment marks drawn onto the film, to confirm treatment-delivery accuracy.

Additional advantages of using RCF for multicenter audits include the following:

- Pieces of RCF are lightweight, easy to pack, and inexpensive to send through the mail, making them easy to use for remote audits (such as optically stimulated luminescence dosimeters (OSLDs), thermoluminescent dosimeters (TLDs) and alanine, but unlike ionization chambers and diodes).
- RCF can be used to make multidimensional measurements of the geometric distribution of dose (unlike OSLDs, TLDs, and alanine).
- RCF can be directly marked to indicate intended alignment and cut into large and small pieces, or cut to fit specific phantoms (unlike radiographic film).
- Measurements made using RCF do not need to be corrected for influence quantities such as temperature, pressure, or humidity (unlike ionization chambers) and do not need to be corrected using temperature-dependent factors (unlike diodes).
- RCF can be used to obtain measurements that are accurate enough to detect meaningful dose variations (see Section 18.4).
- RCF is very thin and is approximately tissue equivalent, which means that several pieces of film can be placed within one phantom and used to make measurements in different planes simultaneously [13,14], because the presence of each piece of film has a negligible effect on the dose measured by each of the other pieces of film (unlike OSLDs and TLDs).

- With appropriate selection of scanning resolution and handling of noise, RCF can be used to measure small-field dose without volume averaging (unlike most other types of dosimeter).
- With appropriate calibration, RCF can be used to make accurate measurements of the doses used in standard radiotherapy fractions [15,16] as well as the much higher doses used in hypofractionated (stereotactic) radiotherapy [17,18], without saturation.
- RCF displays no detectable angular dependence (unlike OSLDs, radiographic film, and many electronic point dosimeters) making it suitable for use in measuring doses delivered using rotational or arc-based modalities, as well as for use in making measurements at different orientations relative to the beam axis (e.g., measuring depth–dose profiles).
- The dose response of RCF shows minimal energy-dependence in the megavoltage range (unlike radiographic film, TLDs, and some types of diode dosimeters), making it straightforward to use at different depths in different types of phantoms.
- When planning or completing a measurement using RCF, it is not necessary to carefully avoid regions of high-dose gradient (unlike point dosimeters). Rather, film can be specifically used for the purpose of measuring high-dose gradients [16].

The only other dosimetry system that offers advantages similar to these is gel dosimetry, which has seen limited clinical adoption [3].

There are, however, challenges associated with the use of RCF for dosimetry audits which are as follows:

- RCF is a non-reusable and comparatively expensive dosimeter.
- Except in cases in which film is used to obtain a qualitative indication of the location or shape of an irradiated region, RCF does not provide an immediate result.
- RCF must be calibrated using a range of doses that extend substantially above and below the maximum and minimum doses that are expected to be delivered in the audit.
- Point-dose measurements made using RCF may be subject to higher uncertainties than dose measurements made using carefully positioned ionization chambers or diodes.
- Although relatively insensitive to spectral variations within the megavoltage range, RCF has a noticeably different response when irradiated by beams in the kilovoltage range [19–21]. This slight energy dependence may affect measurement accuracy in audits in which the nominal beam energy is allowed to vary between centers. (Methods for managing this effect are discussed in Section 18.3.4.)
- Unlike TLDs and OSLDs (which are both used routinely in personal dose-monitoring programs), RCF is not well suited to measuring low doses (such as imaging doses or doses more than 20 cm outside the radiotherapy beam), due to dramatically decreased signal-to-noise. When planning the use of RCF for an audit where low doses are expected, repeated irradiations or scaling up of the planned irradiation dose should be considered.
- As RCF continues to darken with time (albeit at a decreasing rate), days and weeks after irradiation, some uncertainty may be introduced into film-audit results by the long periods of time over which audits are conducted and the unpredictability of the timing at which irradiated films may be returned to the auditor for analysis. (Methods for managing this effect are discussed in Section 18.3.4.)
- Different types of RCFs display different levels of sensitivity to optical and ultraviolet light. For example, when exposed to bright light (1500 lx) for a period of 2 h, GAFchromic™ EBT2 film was found to darken by an amount equivalent to a 6.1-cGy exposure, whereas older EBT film was found to darken by an amount equivalent to a 11.9-cGy exposure [22]. However, all RCFs should be kept in light-tight packaging as much as possible.

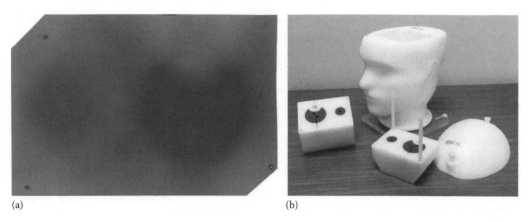

(a)   (b)

Figure 18.1 Example of RCF: (a) used in an IMRT audit setting inside a head-and-neck phantom and (b) by IROC Houston.

RCF has been successfully used in many postal audits by the Imaging and Radiation Oncology Core (IROC) Houston Quality Assurance Center (formerly the Radiological Physics Center) in the United States [23,24] (see Figure 18.1). It has been used for onsite auditing in the United Kingdom [18,25] and introduced in several ad-hoc audits in Australia and New Zealand [14,26,27].

Successful application of the formidable advantages of RCF, so that accurate and valuable results can be obtained from radiotherapy audits, requires that care is taken to ensure that the film is prepared, handled, irradiated, and analyzed appropriately. Considerations beyond those for general use of RCF as discussed elsewhere in this book need to be made when film is used in an audit. These pertain to the circumstances of an audit that can involve extended and unpredictable transportation including air travel, variable time lines, a range of staff handling the film, and to the implications of the results of the film-based measurements in an audit. The methods by which the use of RCF can be managed, within the audit context, are discussed throughout Section 18.3.

18.3 FILM PREPARATION, HANDLING, AND ANALYSIS

18.3.1 DEVISING METHODS AND WRITING INSTRUCTIONS

Multicenter audits are, by definition, major projects. Audits can involve large numbers of people with different equipment, local procedures and levels of experience, and who may be separated by great distances.

The completion of an audit therefore requires detailed planning and active project management. All audit procedures should be well documented. A quality-management system according to international standards is highly recommended.

The purpose of the audit must be clearly defined and should be documented as part of a comprehensive project plan. Resources (equipment, staff, travel, etc.) required for the audit should be evaluated and documented, and a timeline for completing all the steps in the audit process should be drafted. The timeline should nominate dates for the completion of specific milestones, while making generous allowance for potential delays. A risk analysis may also help one to identify critical tasks and improve the efficiency and deliverability of the audit.

Human research ethics approval is also required from all participating centers, if the audit involves the use of patient information (including planning CT images).

Expressions of interest from potential participating centers should be sought early in the planning process, and formal commitments to participate should be obtained before the audit begins.

When undertaking a well-resourced audit, or an audit involving a small number of radiotherapy centers within one city or state, it may be possible for auditors to visit all centers in person, carrying all phantoms and dosimetry equipment, and either supervising or completing all measurements personally [28,29]. In these cases, the auditor "ensures the consistent and accurate execution of the audit procedure among the participating centers" [29].

In many cases, however, it is not possible or desirable for auditor to visit each center in person. In these situations, audit materials and instructions are mailed to the participating centers and mailed back to the auditor after measurements are completed.

Whether the audit is completed in person or remotely, auditors and participating centers must be supplied with clear and detailed instructions for completing the measurements required by the audit. The provision of clear and complete instructions helps one to ensure that all measurements are carried out correctly and consistently, and that the film is handled appropriately. Staff at audited centers should also be provided with honest and specific estimates of the time commitment, and especially the linear accelerator (Linac) time commitment, required by the audit.

The overall methods by which audits are undertaken depend on the goals of those audits. Kron et al. [30] proposed an audit-stratification system that has subsequently been adopted for use by the Australian Clinical Dosimetry Service and is increasingly used to categorize dosimetry audits. The following three levels defined in this system (I, II, and III [30]) are also useful for describing the importance of RCF measurements in dosimetric audits:

- *Level I dosimetry audit*: The level I audit involves Linac output measurements under reference conditions [3,30]. The use of film in a level I audit is unlikely, unless the audit is specifically focused on small field dosimetry.
- *Level II dosimetry audit*: This audit involves measurements of dose distributions in simple (planar) phantoms, with and without density heterogeneities [3,30]. RCF is not widely used in level II audits but has the potential to be adopted for use in measuring dose distributions downstream of density heterogeneities and in explaining anomalous point-dose measurements.
- *Level III dosimetry audit*: This audit involves full end-to-end testing of the radiotherapy chain, including measurements of complex dose distributions in humanoid phantoms [3,30]. Level III audits are often used in the credentialing process for radiotherapy centers entering clinical trials [3,14,31]. The use of RCF to provide detailed and accurate measurements in level III audits is well established [3,13,14,32,33].
- *Geometry audit*: RCF is very well suited to use in audits of the geometric reliability of radiotherapy treatment-delivery systems, as it allows two-dimensional dose planes from multiple static images to be summed together and recorded for remote evaluation. RCF can be used to record and intercompare the results of isocenter tests such as star-shots and Winston-Lutz tests, or to evaluate the consistency of collimation systems on Linacs at different centers [34].
- *Relative-dosimetry audit*: RCF may be used without calibration, to compare levels of fluence modulation produced by different treatment-planning (and especially inverse-planning optimization) systems or to evaluate the geometric accuracy of radiotherapy treatment delivery, with absolute dose measurements (traceable to primary standards) either made using other dosimeters or not made at all.

- *Evaluation of local quality assurance (QA) process*: Local film quality-assurance processes may also be evaluated via audits, where centers are either required to scan and analyze their own measurements of test fields, or to analyze samples of preirradiated film, and provide their results for intercomparison and evaluation.

In all cases except the last, all film preparation, scanning, and analysis are done by the auditor. However, auditors may also advise audit participants to attempt to replicate the audit measurement, to confirm the result, and as a means to evaluate their local film-dosimetry techniques.

Different audits require the use of very different film-irradiation setups and analysis techniques. Despite the differences between the goals and specific methods used by different audits, it is possible to make several suggestions that broadly apply to the use of RCF for radiotherapy auditing:

1. The film calibration and analysis method used by the audit should not be dependent on users irradiating their film on the expected measurement date or sending their film back to the auditor immediately after irradiation. More generally, the film calibration and analysis method used by the audit should not be dependent on all film being scanned the same amount of time after irradiation (see Section 18.3.4).

2. If possible, the audit procedure should be designed to be similar to routine clinical procedures, to increase the likelihood of compliance and also make the audit more indicative of clinical reality.

3. It should be understood that audit participants may not be familiar with RCF and the care that is required when handling it. Instructions to participants should provide clear guidance on the handling of the film (keep the film out of the light as much as possible, keep the film away from dust, touch the film only at the edges, keep the film out of the bunker and away from radiation sources when it is not being irradiated, etc.).

4. Understand that different centers in different geographical areas use different conventions when describing directions in relation to the Linac. When describing the setup of the phantom for irradiation, try to use standard medical terms (such as *toward the superior end of a head-first supine patient* rather than local conventions such as *in the gun direction* or *toward the south*) when describing directions. Alternatively, diagrams and photographs may be used as well as (or instead of) text-based directions (Figure 18.2). Video-based instructions, distributed electronically or made accessible online, should also be considered.

5. Be aware that different calibration conventions are used by different centers [35,36]. Some centers use isocentric calibrations (1 MU is 1 cGy at 10-cm depth in water at 90-cm source-to-surface distance (SSD), or similar) and some use SSD-based calibrations (1 MU is 1 cGy at the depth of maximum dose in water at 100-cm SSD, or similar). This will result in different doses being measured at different centers for the same number of MU. If the delivery of specific doses is required by the audit, then this difference should be taken into account when recommending the number of MU to be used (Figure 18.2).

6. When the measurement method has been finalized and the instructions for audited centers have been drafted, feasibility testing of the proposed method should be undertaken. This testing should include inviting uninvolved physicists (including physicists who have not previously performed film-dosimetry measurements) to attempt to follow the audit instructions. Auditors should respond to any problems or points of confusion by appropriately revising the method and/or clarifying the instructions.

7. A pilot study should be undertaken to provide an indication of expected results and to assist in the selection of tolerances on audited parameters (see Section 18.4 for further discussion of the selection of tolerances).

SMALL FIELD FILM AUDIT

Thanks for volunteering to help with our film audit of small field collimator reproducibility. This work will investigate the variation in collimator positions and resultant output factors for small fields across various systems.

The 6 requested irradiations should be able to be performed within 20 minutes. The experimental conditions are pictured on the next page. The following steps describe the set-up:

Summary:
- 5 cm backscatter material
- Film at isocentre
- 5 cm build-up material
- Deliver approx. 120 cGy each field
- Irradiations taking approx. 20 minutes per machine

1. Place the backscatter on the couch and use a calibrated front pointer to set this to 100 cm SSD.
2. Place the marked film on the phantom and align the centre of the film to the crosshairs. Ensure the film label (e.g. "A1") is facing ANT and is located in the target (couch)/B (right) corner. (i.e. nearest the left foot of a head first supine patient).
3. Carefully place the included 1 cm of RW3 build-up material on top of the film.
4. Place an additional 4 cm of water equivalent phantom material on top of the existing 1 cm build-up.
5. Ensure that the ODI reads 95 cm (or within your tolerance). If there is disagreement, check again with your front point before continuing.

The film should be kept in a cool dark location before and after irradiation. A temperature indicator is included in the package – please record the indicator colour on the measurement form. Remember to remove film from the bunker when not being irradiated.

Included is 1 cm of RW3 water equivalent material, to be placed directly upstream of the film. For the remaining 4 cm of build-up material and the 5 cm of backscatter material, please use whatever water-equivalent material is available within your department, with preference given to RW3 (Goettingen water 3).

6 pieces of film will be irradiated for each linear accelerator, listed on the attached form.

We want approximately 120 cGy delivered to each piece of film. We ask each centre to calculate the number of monitor units required using measured output factors (and TPR data, where necessary), if possible:

$$MU = 120 \text{ cGy}/OF \qquad OR \qquad MU = 120 \text{ cGy}/(OF * TPR)$$

Where the centre has not accurately measured the output factor and/or TPR data for the field sizes of interest, we recommend the use of the following MU:

Figure 18.2 Example of a simple set of instructions provided for use in a geometry audit that was conducted in Australia in 2015. Note that these instructions briefly describe the purpose of the study, provide text-based instructions on making the required measurements, and provide participants with general advice on film handling. These instructions were provided along with a form for recording results (and errors) and a page of diagrams clarifying the film and phantom setup. (Document drafted by Emma Inness and Paul H. Charles.)

8. Audit participants should be given the contact details of one or more of the auditors, who will take responsibility for providing advice and clarification regarding the audit process or the measurement method.
9. Audit participants must be encouraged to record and report problems or errors in the irradiation of the audit films. Space to record errors or deviations from the audit instructions should be provided on all audit data-recording forms.

10. Auditors should accept that audit participants will make mistakes and be vigilant for these during analysis. It is expected that out-of-tolerance audit results arise from problems with the treatment-planning or delivery systems, but out-of-tolerance results may also be caused by user error.

18.3.2 CUTTING AND LABELING OF FILM

When undertaking a dosimetry audit using RCF, every aspect of the measurement process must be carefully planned, starting from the selection, cutting, and labeling of the film.

The type of RCF to be used for the audit should be selected with careful consideration of the purpose of the measurement (as well as the expense, robustness, and other features of the film). Some audits require accurate dosimetry measurements, and therefore require the use of a film designed for dosimetry purposes [14,37], whereas other audits may permit the use of a film that is designed for qualitative evaluation [34]. When evaluating a specific type of RCF for suitability as an audit dosimeter, it is important to check the manufacturer's recommendations and review the recent literature for analyses of the film's properties or suggestions for its use. It is also advisable to specifically evaluate and commission the film and scanning system to be used for the audit [38] (see Chapter 4). Audit-specific demands to be investigated include temperature and light dependency of unexposed film and of film exposed to range of relevant radiation doses as well as post-irradiation darkening effects at different dose levels.

Given the importance and visibility of audits and the potentially large quantities of film required, it may be possible for an auditing organization to communicate directly with the manufacturer and ask for special consideration in terms of additional quality checks during manufacture or for larger batches of film with a similar sensitivity to be delivered to them.

The selection of the film readout system, or scanner, is also an important consideration, when planning an audit. When selecting a reader, not only the size of film to be scanned needs to be considered but also the uniformity of readout across the film area and its potential dependency on film orientation. It might be advantageous to obtain a reader with a larger film bed or wider scan width and only use the central part of the latter for increased uniformity in response. Use of a cutout for reproducible film placing is recommended [39] (Figure 18.3). Custom-build film readers

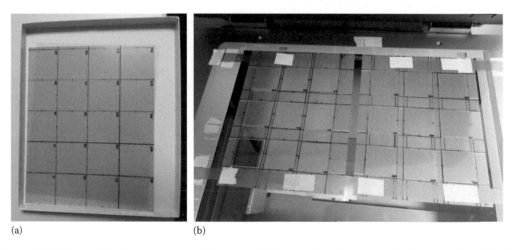

(a) (b)

Figure 18.3 Example of (a) cutting pattern and (b) scanning frame, used for preparing film for distribution as part of the geometry audit described in Figure 18.2.

are another option for audit operations with specific needs, such as a large audit volume. The IROC Houston Quality Assurance Center uses a reader based on a charge-coupled device camera and a light-emitting diode (LED) light box from Photoelectron Corporation (North Billerica, MA). The reader shows high reproducibility in measurements, features simple operation and produces images that need minimal corrections. The software used for the evaluation of the images was also developed by IROC Houston, to meet the specific needs of the IROC program.

Film can be costly, so when a large number of measurements are planned, it is especially important to optimize the film-cutting pattern, with consideration for the measurement method to be used for the audit. If the audit involves a two-dimensional measurement of a dose plane, then it is important to decide whether the film should be cut to encompass only the treated area or whether doses to out-of-field regions are also important to the audit. If a particular phantom is to be used for the audit, then the geometry of that phantom may dictate the cutting pattern [13]. Alternatively, if the audit involves the measurement of a large number of disconnected dose points, then it may be necessary to cut the film into a large number of small pieces.

In all cases, RCF should be cut with care to avoid or minimize delamination of the film layers [40] that may be exacerbated by rough handling during transit. Clean and sharp scissors are highly recommended. Laser cutting devices and services are available. If a commercial phantom is used, laser cut film sold by the phantom manufacturer might be used [41]. The nature of the audit and the need to acquire as much information as possible in a single exposure may make laser cutting a sensible proposition for an auditing organization. The areas of film lost through cutting and due to marking and labeling of the film need to be considered in the planning of the irradiation.

Reference marks, including an indication of orientation, should be written onto the film surface before cutting. Each piece of film should be given a different mark, so that the specific pieces of film sent to each center and irradiated for each measurement can be tracked. These marks should be as clear and unambiguous as possible, as they must be understood by all users. Avoid labeling with visually similar characters (e.g., avoid using both letter "O" and numeral "0"), and ensure that all labels are asymmetric (e.g., use "A." rather than just "A") to make orientation instructions easier to interpret. Alignment marks, such as dots or pinpricks indicating the intended position of Linac cross wires or lines indicating the intended placement of the film within a phantom, may also be needed. The accuracy of the position marking will directly impact the achievable positional accuracy of the audit and its uncertainty.

18.3.3 Packaging for a postal audit

One of the advantages of using RCF for dosimetry audits is that it can easily be packaged and mailed to distant locations, so that audits can be undertaken remotely. The film may be sent while contained within the phantom, or packaged with the phantom, or without any phantom at all.

Placing the film within the phantom and then sending the entire phantom to the user involves some financial cost (which may be borne by the auditing organization, a clinical trial administering institution, or the audited center), some risk of damage or loss of the phantom (which may be mitigated by paying for insurance and/or a courier service), and the possibility that measurement results may be slightly affected by film-compression effects [42]. However, this method allows the auditor to control the positioning of the film within the phantom and removes some opportunities for user error. This method is especially advantageous when the placement of the film within the phantom is complex or unfamiliar to users and when the geometric accuracy of the irradiation is a measured parameter in the audit.

For example, the IROC Houston Quality Assurance Center, which maintains one of the largest dosimetric audit programs in the world, ships RCF prepared for use and placed inside tissue equivalent *dosimetry inserts* that are designed to be placed inside accompanying phantoms [13,33]. These inserts generally contain film that is cut and positioned to allow measurements to be made in two intersecting planes and marked to allow the position of any measured high-dose region to be compared with the planned location of the treatment target [13].

When simpler measurement geometries are used, or when multiple measurements at the same position are required, the advantages of shipping the film inside the phantom are reduced. In these situations, it may still be advantageous to ship all or part of the measurement phantom with the film, if the audit requires control over the specific phantom material used for the measurement. For example, a recent Australian audit of small field-collimation reproducibility shipped a 1-cm thick sheet of water-equivalent plastic (RW3, PTW Freiburg GmbH, Freiburg, Germany), with each set of measurement films, and instructed users to place the RW3 directly upstream of the film, whereas the remaining water-equivalent plastics that were used to construct the 10-cm thick phantom needed for the measurements were supplied by the audited centers, according to their local resources (see Figure 18.2). Alternatively, it may be worth creating a cassette to load the film into. This allows staff at audited centers to handle film with ease and ensures protection of the film at all times.

Auditors may choose to ship the film without any phantom material at all. In these cases, careful instructions should be given to users, regarding the phantoms or materials and geometries to be used for the measurements, and audit participation should be limited to those centers that have access to the equipment required to set up the measurements as instructed.

Many radiotherapy centers own or have access to similar equipment, including water-equivalent plastic slabs, lung- and bone-equivalent materials, and more-or-less humanoid phantoms. The International Atomic Energy Agency's (IAEA) summary recommendations for commissioning and quality assurance of radiotherapy treatment-planning systems (IAEA TEC-DOC 1583 [43]) recommend the use of a particular thorax phantom (the CIRS IMRT thorax phantom, model 002LFC, Computerized Imaging Reference Systems, Norfolk, USA), and this has led to numerous centers purchasing that same phantom [44], or one very similar. The CIRS 002LFC thorax phantom is routinely used for the IAEA's own international radiotherapy treatment-planning audits [29].

Whether RCF is shipped within, with, or without the measurement phantom, it needs to be carefully packaged, for protection from damage including superficial damage (scratches, abrasions), structural damage (bending, folding), exposure to ultraviolet light, exposure to radiation (X-ray screening, flights), and overheating. Therefore, the following precautions should be taken when mailing or couriering RCF (Figure 18.4):

- Avoid scratches by using careful packaging, including the use of nonabrasive envelopes around the film.
- Avoid bending by securing the film within the package, using appropriate packing so that other objects do not move around within the parcel, and using sufficient padding and wrapping to ensure that the contents of the parcel are not damaged by objects outside the parcel.
- Minimize light exposure by using light-tight packaging and instructing users to keep the film out of the light as much as possible.
- Monitor and correct for unintended radiation exposure by sending a reference piece of film that should be kept with the measurement films at all times, but not irradiated during the measurement.

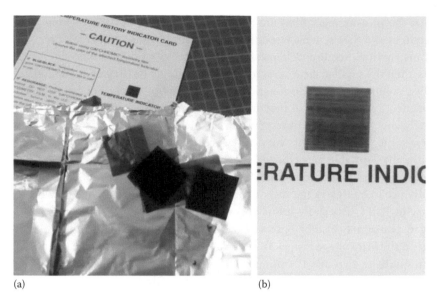

(a)   (b)

Figure 18.4 Photographs showing (a) film returned from pre-audit pilot study, with aluminum foil used as light-tight packaging and (b) temperature history indicator card showing questionable thermal history (temperature indicator should be blue or black but was instead striated with red lines). This result indicated further action needed to be taken to protect the film from heat; the audit was rescheduled for a cooler time of year.

- Monitor for exposure to high temperatures by including a vendor-supplied *temperature card* inside all packages.
- Carefully extract the film from the phantom and/or package and inspect it for damage, after the return from the audit. For possible further analysis, the film should continue to be stored in a dark environment.

All audit participants should be asked to supply a delivery address and phone number. It may be advisable to write to all participants at their nominated delivery addresses, before dispatching the film, to verify that correspondence is received within an appropriate time frame. Packages sent to individuals at large institutions may be subject to delays at internal mail rooms or loading docks.

In preparing for the audit, it may also be advisable to test the packaging of the film by sending a sample package containing spare film to and from a distant location. A subsequent visual inspection of the film and its packaging will allow issues with film packaging to be identified early. This test may also provide a valuable indication of the duration of delays that may be encountered during transit.

18.3.4 CONTROL OF CALIBRATION AND MEASUREMENT CONDITIONS

Reliable RCF measurements require the irradiation and scanning conditions of the measurement and calibration films to follow a consistent protocol. Ideally, the measurement and calibration films should be irradiated using the same beam and scanned the same amount of time after irradiation. However, control over beam quality and scanning time can be lost during a postal audit. Although the calibration films may be irradiated and scanned under controlled conditions at the central auditing site, the measurement films are likely to be irradiated at different times using different radiation sources and then sent back at different amounts of time after irradiation.

If audit instructions require all centers to use the same nominal beam energy for their film irradiations, then the small in-field variations in beam energy (arising from differences in primary beam quality, collimator scatter, or phantom heterogeneity) should not be expected to have an appreciable effect on the results of the film-dosimetry measurements, due to the film having an approximately energy-independent response in the megavoltage range [45].

However, some level III audits may allow treatments to be planned by the participating centers using any available megavoltage energy. For example, Gillis et al.'s multicenter IMRT audit involved centers that used nominal beam energies ranging from 6 to 20 MV [46]. Across an unusually broad energy range such as this, some energy dependence may be observable in the response of RCFs, especially if low doses are used for the measurements [21]. An energy-dependent response may also be observed if RCF is used in audits of superficial radiotherapy treatments (in the kilovoltage range). In such cases, it is important to perform film calibrations using beam qualities that are as similar as possible to all of the beam qualities used for the measurements. If this is not possible, then the use of a film calibration and analysis method that minimizes energy dependence (such as that proposed by Tamponi et al. [47] and applied by Peet et al. [21]), or the use of a correction for energy dependence [21], would be advisable (Figure 18.5).

For some audits, it may not be possible to avoid variations in development time, that is, the time between film exposure and digitization. The polymerization of the active component in RCF continues after irradiation, at a decreasing rate. RCF may stabilize sufficiently for routine clinical use within hours after irradiation [48], leading to variations of just 0.5% if film samples are scanned an hour apart after 24 h of development [49]. However, if development times differ by weeks or months, as may occur during an audit, then the resulting measurement uncertainties

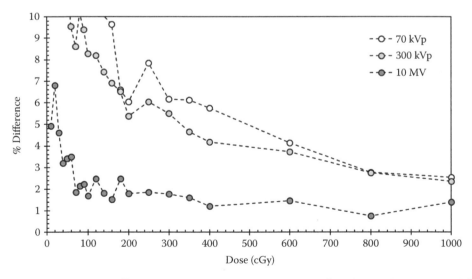

Figure 18.5 Percentage difference between net optical density for photon beams with different nominal energies (70 kVp, 300 kVp, and 10 MV) and net optical density for nominal 6-MV photon beam, in which net-optical densities are measured using GAFchromic™ EBT3 film. Note that differences between results for nominal 6- and 10-MV beams are generally within 2%, whereas differences increase substantially for kilovoltage beams. (Data courtesy of Samuel C. Peet; From Peet, S.C. et al., *Med. Phys.*, 43, 5647–5652, 2016.)

may become unacceptable. There are several methods that can be employed to minimize this uncertainty which are as follows:

1. Providing each audited center with additional film pieces and instructions on performing calibration irradiations, and requesting that each center performs a calibration at approximately the same time that they irradiate the measurement film [50]. This method has several obvious disadvantages, including the expense of the additional film, the additional Linac time needed for completing the calibration, the increased opportunities for user error, and the loss of an independent dose check.

2. The irradiation, to a known dose, of a piece of reference film, at the same time as the irradiation of the measurement film [37]. This practice follows the dosimetry protocol proposed by Lewis et al. [48], wherein the calibrated dose is scaled to bring the measured dose values of the reference films to their actual value.

3. Scanning the calibration film repeatedly, after multiple development times, and analyzing each measurement film using the calibration scan acquired at the appropriate time. Note that scanning the film multiple times can cause additional darkening of the film, possibly up to 1% [51]; so this uncertainty should be weighed against the uncertainties arising from different development times, before this method is adopted.

4. Limiting the doses used in the calibration and measurement to the comparatively low-dose range, in which the RCF response is approximately linear, and then using the film to make relative dose measurements only. This removes the need for calibration irradiations, but potentially limits the value of the film measurements, and leads to the requirement for a second dosimeter to be used to measure dose in Gy.

Film is often used as a relative dosimeter, with absolute dose measurements (traceable to primary standards) performed using other dosimeters. Dosimeters that can be cross calibrated with reference ionization chambers and are suitable for postal audits include thermoluminescence [52] and optically stimulated luminescence dosimeters [53] and alanine [54,55]. In addition to absolute dose determination, these other dosimeters may also be included in film audits to allow imaging dose to be measured and taken into account, for example, by attaching TLDs to the exterior of the phantom and removing them after imaging, prior to the audit irradiation [23,33].

18.3.5 FILM ANALYSIS

Analysis of films from an audit needs to be based on a well-established and tested protocol. Development of the protocol should begin by assessing the radiation modality under controlled conditions and evaluating the film-scanning technique to be used for the audit, with consideration to scanner warmup and other spatial and temporal variations in scanner output (see Chapter 4), the possible need to scan each piece of film several times, and the need to transmit the scanned film image to the analysis software using a standard, lossless format, such as tiff. The details of the protocol should be informed by comprehensively commissioning the film, the scanner, and the procedures used for the scanning and analysis, which should be treated as a combined system and kept constant [39]. Decisions on which color channel(s) and algorithms to use for the dose conversion should be based on local testing of published methods.

As discussed in the previous Section 18.3.4, postal audits often use a separate, absolute dosimeter placed adjacent to the film in the phantom. Conversion of the film signal to dose will be based on the batch-specific dose response of the film, as established in the commissioning, with normalization to the dose from the absolute dosimeter at the point of contact.

Generally, it will be necessary to perform smoothing of the film-based two-dimensional dose data. The specifics of the filter to be used and its parameters will depend on the film type and the dose distribution recorded in the audit. They should be determined and their impact should be tested as part of the audit design.

If the film-based dose data is to be compared with the corresponding dose plane from the treatment-planning system, then specific and detailed export instructions need to be given to the audited sites regarding the export of a dose plane. Alternatively, the entire three-dimensional dose distribution can be exported and send to the auditing organization (or reference center), where the auditors then extract the required dose plane(s).

In order to complete the required comparison, the two dose distributions need to be aligned. This should be done in the analysis software using the positional markers (pin pricks or others) used. Using automatic alignment can be considered as a trouble-shooting tool. It might be necessary to exclude some pixels from the analysis, for instance, because of markings or labeling on the film. Software tools should be available to do this efficiently and with documentation of the omitted pixels.

Various approaches can be used for the comparison of measured and calculated dose. Commonly, a gamma analysis is done [56]. The selection of the gamma criteria depends on a multitude of factors, including the clinical site and treatment/trial goals. Another key factor is the accuracy of the detector. Global 3%/3 mm gamma (threshold 10%–20%) are commonly used criteria for measurements performed with ion chamber and diode arrays [6]. These criteria may be too stringent when used for a postal audit relying on TLD/film measurements, in which the accuracy of the detectors adds a significant component to the total uncertainties (see Section 18.4).

18.4 MEASUREMENT UNCERTAINTIES AND TOLERANCES

Most dosimetric audits are undertaken with the aim of identifying whether specific types of radiotherapy treatments can be delivered accurately. Audits that contribute to the credentialing of radiotherapy centers for participation in clinical trials are particularly focused on identifying whether the centers can accurately plan and deliver treatments according to trial protocols. All these audits require a threshold for reporting failure; a tolerance value that defines the limit beyond which the difference between the planned and measured dose becomes unacceptable. The question to be answered by each dose measurement is, *Is the difference between the planned and measured dose less than the tolerance value allowed by this audit?* It is therefore very important that the tolerance value chosen for use in the audit takes account of the uncertainty in the measurement [37].

The selection of acceptance thresholds is one of the most critical aspects of an audit. Thresholds will depend on technical factors, as can be estimated with an uncertainty analysis, as well as on nontechnical factors, such as the circumstances of the audit and the implications of a false-negative or a false-positive result.

RCF measurements are affected by the same sources of uncertainty when they are used in dosimetry audits as when they are used in any other context. These include film and scanner non uniformity effects, scanning noise, and calibration uncertainty. As discussed in Section 18.3.4, the use of RCF for dosimetry audits may also be subject to additional uncertainties arising from differences between the irradiation conditions used for the calibration and for the measurements and differences in film development time. When film is used for remote audits, additional measurement setup uncertainties may also need to be considered.

Palmer et al.'s report on the planning and management of a multicenter audit of high-dose-rate brachytherapy treatment delivery provides an example of a detailed uncertainty budget for a film audit [37]. Relevant sources of uncertainty were listed by Palmer et al. as applicator reconstruction (2.25%), planned dose calculation (AAPM TG-43) (0.5%), high-dose-rate brachytherapy equipment (source strength 0.4% and dwell position 1%), and film dosimetry (calibration dose delivery on a Linac 0.5%, calibration fit 0.5%, film position 0.9%, energy dependence 1.5%, phantom scatter 0.5%, and film scanning 0.2%), for a total uncertainty of 3.2% (one standard deviation) or 6.4% (two standard deviations) [37]. Based on their uncertainty budget, Palmer et al. suggested that a clinically relevant deviation from the intended dose distribution could be assessed using a gamma evaluation with criteria of 5%/3 mm [37].

Uncertainty budgets for the use of RCF may produce different results depending on which particular type of RCF is used, how the film is used, which uncertainties are investigated, what methods are used to investigate those uncertainties, and how conservatively the results are interpreted. Examples of total uncertainties for megavoltage photon beam dosimetry using RCF that have been explicitly derived from uncertainty budgets cover a wide range, including 0.55% [57], 0.9% [39], 1.6% [58], 2.8% [49], and 4% [59].

In practice, the achievable accuracy of film-dosimetry measurements arising from an audit may be substantially different from the level of accuracy predicted by the uncertainty budget analysis [37]. Consequently, when planning a dosimetry audit using RCF (or any other dosimeter), it may be advisable to undertake a local measurement reproducibility study or a small pilot audit, to more reliably identify the magnitude of uncertainties affecting the specific measurements that will be used in the audit [13,20,33,37].

For example, after pilot audits at two centers showed that almost all measurements (for four different treatments delivered at each center) agreed with planned doses at more than 97% of points when a gamma evaluation with 3%/3 mm was used, Palmer et al. tightened their gamma evaluation criteria from 5%/3 to 3%/3 mm and set a threshold for investigation at 80% [37].

The pilot audit conducted and subsequently validated by Molineu et al. [13,60], in preparation for their audit of IMRT head- and neck-treatment planning and delivery [33], led to the adoption of two levels of criteria. A 7%-dose criterion for points located within the target and 7%/4 mm gamma evaluation criteria over the entire dose distribution.

Healy et al. similarly used a trial run of a method proposed for use in credentialing centers for participation in an IMRT-based clinical trial [14]. Healy et al. observed that when a gamma evaluation was performed using 7%/4 mm criteria, their RCF measurements were found to agree with the planned dose planes at 100% of points [14]. Healy et al. ultimately selected 5%/3 mm as the gamma evaluation criteria for use in their credentialing audit, as these produced very high agreement (greater than 98%) under the controlled conditions of the trial run, while also allowing regions of difference between the planned and measured doses to be identified [14].

It is advisable that the planning of a dosimetry audit using RCF (or any other dosimeter) should include a comprehensive evaluation of the magnitude of all foreseen uncertainties that may affect the measurement results, in combination with some form of pilot study, to identify useable and informative tolerance values for audited parameters. For longer running audits, tolerances should be evaluated over time and adjusted to identify true outliers.

Finally, the tolerances allowed by the audit must take into account the known uncertainties in radiotherapy treatment-dose calculations, which may include uncertainties of up to 3% in homogeneous media [61], as well as the overall goal of treating the patient using a dose that is accurate to within 5% [62].

18.5 REPORTING OF RESULTS

At the completion of a set of audit measurements for a particular radiotherapy center, that center should be supplied with a report describing the results of the audit. This requires the preparation and use of a standard report form for the audit.

The report form should incorporate all significant details of the audit. In addition to all relevant dates, these include information about the audited organization, technical details about the planning system and delivery system used, and delivery parameters. Specifics about the audit should include a description of the tests performed, type of detectors and locations, serial numbers of devices and batch number of the film, information on how the reported parameters were calculated or determined, and criteria for the evaluation of the results and the final condition of the test (pass or fail).

The use of images to illustrate results is recommended (see Figure 18.6). The images that are available or desirable for this purpose will vary and depend on the specifics of the audit. If a gamma analysis has been used, then a gamma map (a two-dimensional display of the calculated gamma distribution) should be supplied. One-dimensional dose and dose-difference profiles, orthogonally placed through a central point, can also be used to illustrate the level of agreement between planned and measured dose distributions.

Results from any other measurements, such as absolute dosimeters, should also be reported separately.

An example for a postal audit that uses film and TLD is the Stereotactic Ablative Body Radiotherapy lung-audit developed in Melbourne, Australia [63]. Figure 18.7 shows the phantom and film exposed to small fields. The objective of this part of the audit is to assess small field dosimetry and the ability of the planning system to predict penumbra and output of the small field in the presence of an inhomogeneity. The use of TLDs for point-dose assessment complements the film results.

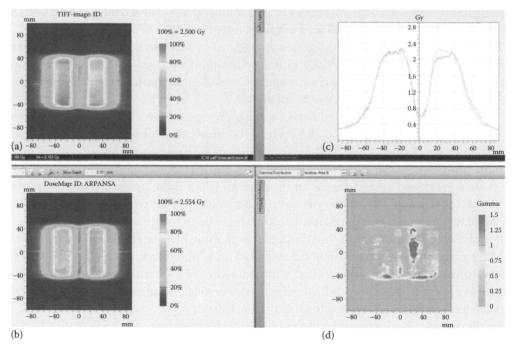

Figure 18.6 Example of images useful in reporting audit results. In this image, taken from the Verisoft software (PTW Freiburg, Germany), panels (a) and (b) show the dose plane measured with film and the calculated dose plane from the treatment plan, respectively. Panel (c) shows dose profiles, and panel (d) shows the gamma map.

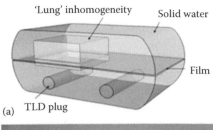

Figure 18.7 Mail-able phantom with RCF to test small-field dosimetry in the context of SABR [63]. Phantom design (a), phantom with RCF exposed to a 3 × 3 cm² field within a cork inhomogeneity (b), and travel case for the phantom (c).

If the audit is being completed on behalf of a third party, such as a clinical trials administering organization, then that organization should also be supplied with the result of the audit. The level of detail required in this report (what, if anything, more than a statement of *pass* or *fail*, is required) should be agreed between the auditor and the third party prior to the commencement of the audit. Willingness to agree to this sharing of information should be a condition of participation in the audit. The audit data must be anonymized (including the removal of the names and locations of all institutions and individuals, including staff and patients) before it can be shared more broadly, including via scientific publication.

18.6 SUMMARY AND OUTLOOK

RCF is well suited to use in radiotherapy audits because it is lightweight and easy to send through the mail, can be marked and cut to size, shows negligible angular dependence and a small spectral dependence, is approximately tissue equivalent, and provides accurate two-dimensional dose measurements with a high spatial resolution.

The successful use of RCF to produce accurate and valuable audit results requires that care is taken to ensure that the film is prepared, handled, irradiated, and analyzed appropriately. This requires careful management, planning, and implementation of all aspects of the audit.

Specifically, managers of a RCF-based audit should

1. Clearly define the purpose of the audit and plan all aspects of the audit with reference to that purpose.
2. Adopt a measurement or analysis procedure that accounts for variations in development time among film returned from different centers (or include an appropriate uncertainty in the audit tolerance).

3. Adopt an analysis procedure that accounts for any large differences in beam quality between the calibration and measurement irradiations (or include an appropriate uncertainty in the audit tolerance).

4. Draft a set of instructions to participants that uses standard medical terms or diagrams and photographs to give directions for measurement setup, specifies the dose or MU to be used, includes guidance on film handling for participants who lack prior experience with RCF, and encourages recording and reporting of measurement errors.

5. Undertake feasibility testing of the audit method and instructions, preferably with assistance from physicists who have not previously performed film-dosimetry measurements.

6. Consider including another type of dosimeter with the film, to verify the accuracy of film measurements of dose in Gy, to provide a measurement in Gy in audits in which film is only used for relative dose measurements, or to account for imaging dose delivered during phantom setup.

7. When preparing film for use in the audit, optimize the cutting pattern, cut with care to avoid delamination, and label the film thoughtfully (including alignment marks if necessary) to aid in film setup, tracking, and analysis.

8. For remote audits, consider sending the film inside the phantom if the film setup is especially complex or if the geometric accuracy of the irradiation is an audited parameter and consider sending the film with a specific part of the phantom if the audit requires control over a particular aspect of the measurement.

9. Package film carefully to avoid scratches, bending, and light exposure. Monitor for exposure to high temperatures.

10. Complete a thorough uncertainty analysis and/or use a pilot study to inform the selection of appropriate tolerances on audited parameters.

If the audit is well planned and thoughtfully documented, then recommendations for participants are simple and few:

1. Follow all audit instructions (including basic film handling instructions) closely, even when instructions differ from local film dosimetry practices.

2. Always raise questions of auditors when instructions seem ambiguous and seek advice from auditors when confused.

3. Always advise auditors of known errors made during phantom setup or film irradiation.

4. Try to irradiate and return film in a timely manner.

In the future, it is possible that three-dimensional dosimeters (such as dosimetry gels [3]) may take the place of two-dimensional dosimeters in some audit applications. Currently, the use of the Linac's built-in electronic portal imaging device (EPID) is emerging as a promising audit option. The use of EPIDs adds flexibility to the audit process by eliminating the need for any materials to be shipped or for staff to travel [64].

Increases in the use of competing audit techniques such as the EPID (or other dosimeters) might lead to decreases in the use of RCF. Alternatively, increases in the global participation in audits, arising from the availability of new audit techniques and technologies, might lead to increases in the overall use of RCF for auditing purposes. Thus far, RCF has proven to be a useful tool for postal and clinical trial dosimetry.

ACKNOWLEDGMENTS

The authors wish to thank Emma Inness and Paul H. Charles for providing the document shown as Figure 18.2 and Samuel C. Peet for providing the data that was used to produce Figure 18.5.

REFERENCES

1. Peters LJ, O'Sullivan B, Giralt J et al. Critical impact of radiotherapy protocol compliance and quality in the treatment of advanced head and neck cancer: Results from TROG 02.02. *J Clin Oncol* 2010;28(18):2996–3001.
2. Institute of Physics and Engineering in Medicine. Guidance for the clinical implementation of intensity modulated radiation therapy. IPEM Report No 96. York, UK: IPEM; 2008.
3. Kron T, Haworth A, Williams I. Dosimetry for audit and clinical trials: Challenges and requirements. *J Phys Conf Ser* 2013;444:012014.
4. Izewska JE, Bera P, Vatnitsky S. IAEA/WHO TLD postal dose audit service and high precision measurements for radiotherapy level dosimetry. *Radiat Prot Dosim* 2002;101:387–392.
5. Williams I, Kenny J, Lye J et al. The Australian clinical dosimetry service: A commentary on the first 18 months. *Australas Phys Eng Sci Med* 2012;35(4):407–411.
6. Clark CH, Hussein M, Tsang Y et al. A multi-institutional dosimetry audit of rotational intensity-modulated radiotherapy. *Radiother Oncol* 2014;113(2):272–278.
7. Kry SF, Alvarez P, Molineu A et al. Algorithms used in heterogeneous dose calculations show systematic differences as measured with the radiological physics center's anthropomorphic thorax phantom used for RTOG credentialing. *Int J Radiat Oncol Biol Phys* 2013;85(1):e95–e100.
8. Dunn L, Lehmann J, Lye J et al. National dosimetric audit network finds discrepancies in AAA lung inhomogeneity corrections. *Med Phys* 2015;31(5):435–441.
9. Ibbott GS, Haworth A, Followill DS. Quality assurance for clinical trials. *Front Oncol* 2013;3:311.
10. Dixon P, O'Sullivan B. Radiotherapy quality assurance: Time for everyone to take it seriously. *Eur J Cancer* 2003;39:423–429.
11. Nutting C, Dearnaley DP, Webb S. Intensity modulated radiation therapy: A clinical review. *Brit J Radiol* 2000;73:459–469.
12. Benedict SH, Yenice KM, Followill D et al. Stereotactic body radiation therapy: The report of AAPM Task Group 101. *Med Phys* 2010;37:4078–4101.
13. Molineu A, Followill DS, Balter PA et al. Design and implementation of an anthropomorphic quality assurance phantom for intensity-modulated radiation therapy for the radiation therapy oncology group. *Int J Radiat Oncol Biol Phys* 2005;63:577–583.
14. Healy B, Frantzis J, Murry R et al. Development of a dosimetry inter-comparison for IMRT as part of site credentialing for a TROG multi-centre clinical trial for prostate cancer. *Australas Phys Eng Sci Med* 2011;34:195–202.
15. Kairn T, Hardcastle N, Kenny J et al. EBT2 RCF for quality assurance of complex IMRT treatments of the prostate: Micro-collimated IMRT, RapidArc, and TomoTherapy. *Australas Phys Eng Sci Med* 2011;34:333–343.
16. Kairn T, Papworth D, Crowe SB et al. Dosimetric quality, accuracy and deliverability of modulated radiotherapy treatments for spinal metastases. *Med Dosim* 2016:41;258–266.
17. Wilcox EE, Daskalov GM. Evaluation of GAFCHROMIC® EBT film for CyberKnife® dosimetry. *Med Phys* 2007;34:1967–1974.

18. Palmer AL, Dimitriadis A, Nisbet A, Clark CH. Evaluation of Gafchromic EBT-XD film, with comparison to EBT3 film, and application in high dose radiotherapy verification. *Phys Med Biol* 2015;60:8741–8752.
19. Butson MJ, Cheung T, Yu PKN. Weak energy dependence of EBT gafchromic film dose response in the 50 kVp–10 MVp X-ray range. *Appl Radiat Isot* 2006;64:60–62.
20. Butson MJ, Yu PKN, Cheung T, Alnawaf H. Energy response of the new EBT2 RCF to X-ray radiation. *Radiat Meas* 2010;45:836–839.
21. Peet SC, Wilks R, Kairn T et al. Calibrating RCF in beams of uncertain quality. *Med Phys* 2016;43:5647–5652.
22. Andres C, Del Castillo A, Tortosa R et al. A comprehensive study of the Gafchromic EBT2 RCF. A comparison with EBT. *Med Phys* 2010;37:6271–6278.
23. Faught AM, Kry SF, Luo D et al. Development of a modified head and neck quality assurance phantom for use in stereotactic radiosurgery trials. *J Appl Clin Med Phys* 2013;14(4):4313.
24. Ibbott SI, Followill DS. The radiological physics center: 40 years of vigilance and quality assurance for NCI sponsored clinical trials, in *SSDL Newsletter* 2010. pp. 25–30.
25. Hussein M, Tsang Y, Thomas RA et al. A methodology for dosimetry audit of rotational radiotherapy using a commercial detector array *Radiother Oncol* 2013;108(1):78–85.
26. Healy B, Frantzis J, Murry R et al. Results from a multicenter prostate IMRT dosimetry intercomparison for an OCOG-TROG clinical trial. *Med Phys* 2013;40(7):071706.
27. Miller J, Ball D, Kron T. Technical aspects of the CHISEL randomised phase III trial of highly conformal hypofractionated image guided (stereotactic) radiotherapy for inoperable early stage lung cancer. *J Med Imaging Radiat Oncol* 2009;53:A50.
28. Middleton M, Frantzis J, Healy B et al. Successful implementation of image-guided radiation therapy quality assurance in the trans Tasman radiation oncology group 08.01 PROFIT Study. *Int J Radiat Oncol Biol Phys* 2011;81:1576–1581.
29. Gershkevitsh E, Pesznyak C, Petrovic B et al. Dosimetric inter-institutional comparison in European radiotherapy centres: Results of IAEA supported treatment planning system audit. *Acta Oncol* 2014;53:628–636.
30. Kron T, Hamilton C, Roff M, Denham J. Dosimetric intercomparison for two Australasian clinical trials using an anthropomorphic phantom. *Int J Radiat Oncol Biol Phys* 2002;52:566–579.
31. Followill DS, Evans DR, Cherry C et al. Design, development, and implementation of the radiological physics center's pelvis and thorax anthropomorphic quality assurance phantoms. *Med Phys* 2007;34:2070–2076.
32. Ibbott GS, Thwaites DI. Audits for advanced treatment dosimetry. *J Phys Conf Ser* 2015;573:012002.
33. Ibbott GS, Molineu A, Followill DS. Independent evaluations of IMRT through the use of an anthropomorphic phantom. *Technol Canc Res Treat* 2006;5:481–487.
34. Kairn T, Asena A, Charles PH et al. Field size consistency of nominally matched linacs. *Australas Phys Eng Sci Med* 2015;38:289–297.
35. Andreo P, Burns DT, Hohlfeld K et al. Absorbed dose determination in external beam radiotherapy: An international code of practice for dosimetry based on standards of absorbed dose to water. IAEA Technical Report Series No 398. Vienna, Austria: IAEA; 2000.
36. Almond PR, Biggs PJ, Coursey BM et al. AAPM's TG-51 protocol for clinical reference dosimetry of high-energy photon and electron beams. *Med Phys* 1999;26:1847–1870.
37. Palmer AL, Lee C, Ratcliffe AJ et al. Design and implementation of a film dosimetry audit tool for comparison of planned and delivered dose distributions in high dose rate (HDR) brachytherapy. *Phys Med Biol* 2013;58:6623–6640.
38. Kron T, Lehmann J, Greer P. Dosimetry of ionising radiation in modern radiation oncology. *Phys Med Biol* 2016;61(14):R167-R205. doi: 10.1088/0031-9155/61/14/R167.

39. Martisíková M, Ackermann B, Jäkel O. Analysis of uncertainties in Gafchromic EBT film dosimetry of photon beams. *Phys Med Biol* 2008;53(24):7013–7027.

40. Moylan R, Aland T, Kairn T. Dosimetric accuracy of Gafchromic EBT2 and EBT3 film for in vivo dosimetry. *Australas Phys Eng Sci Med* 2013;36:331–337.

41. Wang L, Kielar KN, Mok E et al. An end-to-end examination of geometric accuracy of IGRT using a new digital accelerator equipped with onboard imaging system. *Phys Med Biol* 2012;57(3):757–769.

42. Yu PK, Butson M, Cheung T. Does mechanical pressure on RCF affect optical absorption and dosimetry? *Australas Phys Eng Sci Med* 2006;29:385–387.

43. International Atomic Energy Agency. Commissioning of radiotherapy treatment planning systems: Testing for typical external beam treatment techniques. IAEA TEC-DOC-1583. Vienna, Austria: IAEA; 2008.

44. Kairn T, Aland T, Crowe SB, Trapp JV. Use of electronic portal imaging devices for electron treatment verification. *Australas Phys Eng Sci Med* 2016;39:199–209.

45. Arjomandy B, Tailor R, Anand A et al. Energy dependence and dose response of Gafchromic EBT2 film over a wide range of photon, electron, and proton beam energies. *Med Phys* 2010;37:1942–1947.

46. Gillis S, De Wagter C, Bohsung J et al. An inter-centre quality assurance network for IMRT verification: Results of the ESTRO QUASIMODO project. *Radiother Oncol* 2005;76:340–353.

47. Tamponi M, Bona R, Poggiu A, Marini P. A new form of the calibration curve in radiochromic dosimetry. Properties and results. *Med Phys* 2016;43:4435–4446.

48. Lewis D, Micke A, Yu X, Chan MF. An efficient protocol for RCF dosimetry combining calibration and measurement in a single scan. *Med Phys* 2012;39:6339–6350.

49. Aland T, Kairn T, Kenny J. Evaluation of a Gafchromic EBT2 film dosimetry system for radiotherapy quality assurance. *Australas Phys Eng Sci Med* 2011;34:241–260.

50. Lee J, Mayles HM, Baker CR et al. UK SABR Consortium lung dosimetry audit; Relative dosimetry results. *Radiother Oncol* 2015;115:S74–S75.

51. Richley L, John AC, Coomber H, Fletcher S. Evaluation and optimization of the new EBT2 RCF dosimetry system for patient dose verification in radiotherapy. *Phys Med Biol* 2008;55:2601–2617.

52. Jafari SM, Bradley DA, Gouldstone CA et al. Low-cost commercial glass beads as dosimeters in radiotherapy. *Radiat Phys Chem* 2014;97:95–101.

53. Lye J, Dunn L, Kenny J et al. Remote auditing of radiotherapy facilities using optically stimulated luminescence dosimeters. *Med Phys* 2014;41:032102.

54. Budgell G, Berresford J, Trainer M et al. A national dosimetric audit of IMRT. *Radiother Oncol* 2011;99:246–252.

55. Schaeken B, Cuypers R, Lelie S et al. Implementation of alanine/EPR as transfer dosimetry system in a radiotherapy audit programme in Belgium. *Radiother Oncol* 2011;99:94–96.

56. Low DA, Harms WB, Mutic S, Purdy JA. A technique for the quantitative evaluation of dose distributions. *Med Phys* 1998;25(5):656–661.

57. Sorriaux J, Kacperek A, Rossomme S et al. Evaluation of Gafchromic EBT3 films characteristics in therapy photon, electron and proton beams. *Med Phys* 2013;29:599–606.

58. Van Battum L J, Hoffmans D, Piersma H, Heukelom S. Accurate dosimetry with GafChromic EBT film of a 6MV photon beam in water: What level is achievable? *Med Phys* 2008;35:704–716.

59. Hill R. Reporting uncertainties in measurement: What approach should be followed? *Australas Phys Eng Sci Med* 2013;36:1–3.

60. Molineu A, Hernandez N, Nguyen T et al. Credentialing results from IMRT irradiations of an anthropomorphic head and neck phantom. *Med Phys* 2013;40:022101.

61. Ahnesjö A, Aspradakis MM. Dose calculations for external photon beams in radiotherapy. *Phys Med Biol* 1999;44:R99–R155.

62. International Commission on Radiation Units & Measurements. Determination of absorbed dose in a patient irradiated in a patient irradiated by beams of X or gamma rays in radiotherapy procedures. ICRU Report No 24. Bethesda, MD: ICRU; 1976.

63. Kron T, Lonski P, Crain M et al. Credentialing for stereotactic ablative body radiotherapy (SABR) trials using a small mailable phantom. *Australas Phys Eng Sci Med* 2016;39(1):278.

64. Miri N, Lehmann J, Legge K, Vial P, Greer PB. Virtual EPID standard phantom audit (VESPA) for remote IMRT and VMAT credentialing. *Phys Med Biol* 2017;62:4293–4299.

19

Use of radiochromic film with synchrotron radiation

TOMAS KRON, ELIZABETH KYRIAKOU,
AND JEFFREY C. CROSBIE

19.1 INTRODUCTION

Synchrotron radiation has unique properties but has not yet found wide use in medicine. This is perhaps not too surprising; the main reason is that synchrotrons are scientific research tools in a research facility setting, not a clinical environment. Furthermore, of the 50 or so synchrotrons worldwide, fewer than six have the capability to handle human patients. However, synchrotron radiation has shown promise for both diagnostic and therapeutic applications [1,2]. There are several features that make this type of radiation unique for many applications as varied as crystallography in bioscience, forensic analysis, and advance material research P340. Given the usefulness of X-rays in many parts of medicine, it is not surprising that synchrotron radiation is being explored for medical applications such as radiography and radiotherapy.

Synchrotron radiation or synchrotron-generated radiation (commonly abbreviated to SR) spans the electromagnetic spectrum, including X-rays. SR is produced by changing the direction of the

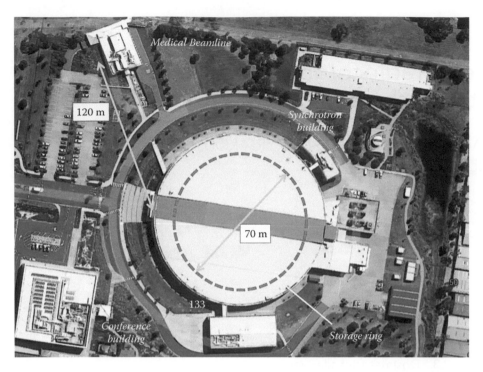

Figure 19.1 Aerial view of the Australian synchrotron in Clayton indicating its dimensions.

charged particles orbiting a storage ring at relativistic speeds as shown in Figure 19.1. Compared with conventional X-rays, synchrotron radiation is coherent, inherently collimated and extremely high in dose rate. Table 19.1 shows a comparison of synchrotron and conventional X-rays.

The relativistic electron beam is maintained in circular orbit, in a storage ring via focusing and steering magnets. In practice, generation of X-rays from electron beams is performed using bending magnets or insertion devices such as undulators and wigglers [3]. So-called third-generation synchrotrons are electron accelerators with storage rings in which the electron beam can be maintained for several hours. Figure 19.1 shows an aerial photograph of the Australian Synchrotron in Clayton, near Melbourne. The energy of the SR depends on the energy of the electron beam and the strength of the fluctuating magnetic field that forces the directional change on the beam, which in turn generates the SR beam down a beamline and into an experimental hutch [2]. Insertion devices known as wigglers and undulators with magnetic field strengths of several Tesla are used to generate X-rays useful for medical applications (>10 keV). Figure 19.2 schematically illustrates the generation of the X-ray beam. It is worth noting that most electron synchrotrons have many beamlines with different photon energies for different applications ranging from infrared to visible, UV, and X-rays.

The very high dose rate of the synchrotron beam allows one to select monoenergetic beams using crystal lattice reflection with still acceptable dose rates. Such monoenergetic beams are used in radiographic imaging applications of SR. However, a substantial portion of synchrotron radiotherapy and radiobiology studies use a polychromatic X-ray beam in which the *raw* or *white* wiggler beam is filtered by a set of absorbers (e.g., graphite, aluminum, copper) to remove the very low energy X-rays, typically less than 30 keV. The filtered or *pink* beams can have very high dose rates, in the region of 10 kGy/s or 600 kGy/min compared with 10 Gy/min in standard linear accelerator.

Table 19.1 Comparison of the physical properties of synchrotron and conventional X-radiation

	Conventional X-rays	Synchrotron radiation	Comments
Generation	X-ray tube with cathode and anode	Undulator or wiggler in very high energy electron storage ring (GeV)	
Typical energy	30–300 kVp	5–200 keV	Conventional X-rays described by maximum acceleration potential
Typical dose rate	2 Gy/ min (therapy) Variable (diagnostic)— fluoroscopy: 1 mGy/ min	10^6 Gy/min	Beam current and focal spot adjustable in diagnostics— exposures typically very short except fluoroscopy and CT
Coherence	No	Yes	
Field size	>20 × 20 cm^2 @ 50 cm	<1 × 5 cm^2 @ 20 m	
Homogeneity/ flatness	Reasonable	Substantial roll-off in vertical direction c.f. horizontal	
Divergence	Large	Very small	

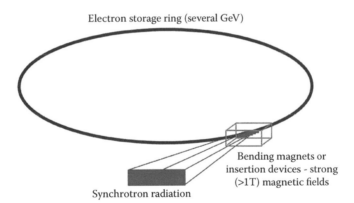

Electron storage ring (several GeV)

Bending magnets or insertion devices - strong (>1T) magnetic fields

Synchrotron radiation

Figure 19.2 Schematic illustration of a synchrotron storage ring and the X-ray extraction location. A combination of focusing and steering magnets (quadropoles and sextupoles) maintains the electron beam orbit. Bending magnets or insertion devices (wigglers and undulators) are used to generate synchrotron radiation which is channeled into an experimental hutch, thus constituting a beamline.

Due to such high dose rate, traditional radiation detectors are not suitable for dosimetry. The objective of the present chapter is to explore why and how radiochromic film can be used for dosimetry in medical applications of synchrotron radiation. A lot of research on dosimetry has been performed. The chapter is part of a book on radiochromic film to provide innovative and up-to-date information on such dosimetry. As such features of the film are only described as they are specific and relevant to synchrotron radiation. The main focus of the chapter is a description of synchrotron radiation and the challenges it poses for radiation dosimetry. It will be discussed where the special features of radiochromic film can address some of the challenges of intense dose-rate dosimetry. Reference to some preclinical and clinical studies will be also provided.

19.2 SYNCHROTRON RADIATION

19.2.1 SYNCHROTRONS IN MEDICINE

The modern third-generation synchrotron facilities are spread around the globe and include the Advanced Photon Source (Chicago, USA), the European Synchrotron Radiation Facility or ESRF (Grenoble, France), and the Australian Synchrotron Facility (Melbourne, Australia) shown in Figure 19.1. These facilities maintain relativistic electrons in orbit around a storage ring that has a diameter comparable with a football stadium (110 × 50 m). Much of the research conducted at synchrotrons relates to X-ray diffraction and scattering mainly used for protein crystallography, small angle X-ray scattering, X-ray absorption, or X-ray fluorescence of material samples including biological samples.

Only a minority (less than 10) of synchrotron facilities have dedicated beamlines for medical applications; however, this number is growing quickly as many applications have been identified ranging from diagnostic imaging to radiotherapy applications. For diagnostic applications, k-edge subtraction imaging has been one of the first applications used on human subjects [4]. High contrast images are acquired using a contrast medium with high atomic number and delivering radiation just above and just below the k-edge. A subtraction of the two images demonstrates the location of the contrast agent with exquisite spatial and contrast resolution. Of particular, interest for diagnostic imaging is also phase contrast imaging that utilizes the special properties of coherence and monochromatism to generate images that would not be possible with other modalities [5]. Figure 19.3 shows an example for a phase contrast image of the lung of a rabbit pup acquired at the SPRING 8 synchrotron in Japan.

Phase contrast is particularly powerful in demonstrating density contrast on a microscopic scale and in images in which other contrast options are limited. Not surprisingly, it has been of particular interest for lung [5,6] and breast [7,8] imaging. A research group led by Prof. Stuart Hooper from Monash University in Australia has been particularly active in reporting advances in lung imaging

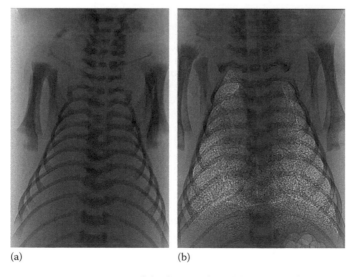

(a)   (b)

Figure 19.3 Phase contrast X-ray image of the lungs of a rabbit pup delivered at term and imaged (a) before the onset of lung aeration, when the airways are liquid-filled and (b) 1 h after the onset of air breathing. The trachea and the margins of the lung lobes can also be observed. (https://user. spring8.or.jp/sp8info/?p=3051—reproduction courtesy by the authors R Lewis and S Hooper.)

using synchrotron phase contrast imaging with applications in neonatal care [9,10]. The developments of phase contrast imaging using synchrotron radiation also show an interesting effect: because phase contrast was considered useful commercial vendors developed methods to generate images with the similar features without the cost of a synchrotron [11].

For radiation therapy, several approaches have been taken to use the special features of synchrotron radiation. One can maximize the local energy deposition by the tuning the X-rays to deliver radiation just above the k-edge of the absorption spectrum for a high-Z element or compound (e.g., iodine). This approach has been used by using iodinated contrast solution as well as gold nanoparticles [12,13]. A French team led by the Grenoble University Hospital (CHU) in partnership with the University of Grenoble and the ESRF are in the middle of a Phase 1 clinical trial of stereotactic synchrotron radiation therapy (SSRT) for metastatic brain tumors [14]. To date, a total of twelve cancer patients have received a single 5-Gy fraction of SSRT to their tumors. For the SSRT technique, a monochromatic and uniform X-ray beam (80 keV) is used for the irradiation, and patients also receive iodine contrast to assist with localized dose enhancement around the tumor as has been described by many investigators for a variety of conventional sources [15,16,17]. Given the more tightly controlled X-ray energy with large dose rates, this should be a very promising technique [13,18,19]. The patients at the ESRF receive conventional radiotherapy fractions in the local CHU clinic as well. There are strict inclusion criteria for the patients on the trial. The first publications that specifically describe the trial are expected by 2017.

19.2.2 MICROBEAMS

Microbeam radiotherapy (MRT) using synchrotron-generated X-rays is a novel, preclinical form of radiotherapy that promises significant advantages over conventional methods if successfully translated to clinical practice. A useful review of the preclinical MRT literature was published recently (2016) by Smyth et al. [20]. An earlier review (2010) was reported by Brauer-Krisch et al. [21,20]. For the synchrotron MRT technique, the high flux, polychromatic X-ray beam from the synchrotron is segmented into a lattice of narrow, microplanar beams, 25–50-µm wide and separated by 200–400 µm. Figure 19.4 shows the image of a piece of radiochromic film (HD-810) that had been exposed to a microbeam. For comparison, the dose distribution of a broad beam is shown as well. Typical in-beam radiation doses are 300–800 Gy, with doses in the valley between the microbeams of the order of 5–20 Gy. This MRT technique is only practical on a synchrotron.

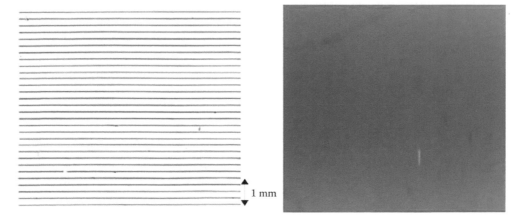

1 mm

Figure 19.4 Comparison of microbeam (left) and broad beam (right) exposure as assessed with HD-810 radiochromic film.

Conventional X-ray sources do not possess the necessary physical properties of very high flux rate with minimally divergent beams to produce these microbeam *wafers* of radiation.

At the ESRF in France and the imaging and medical beam line (IMBL) in Australia, MRT is delivered by scanning the sample (e.g., rodent, cells, dosimeter) vertically through the microplanar beam [22,23]. The absorbed dose to the sample is then an function of the motor scan speed; a slow scan delivers a higher dose than a fast scan.

The heterogenous dose distribution and high dose rate of MRT results in distinctly different biological responses within targeted tissues when compared with conventional radiotherapy. MRT challenges a number of fundamental radiobiology paradigms. Evidence from preclinical models achieve tumor control despite the fact that only a small proportion of the tumor receives peak microbeam doses. Normal tissues however, even serial organs, appear to tolerate MRT far better than conventional radiotherapy (CRT), presumably due to repair of tissue damaged by peak doses and by undamaged adjacent tissue receiving valley doses.

19.2.3 DOSIMETRIC CHALLENGES

The unique imaging and potentially revolutionizing treatment possibilities with beams created by a synchrotron have been accompanied by the development of first principles dosimetry systems for this unique irradiation system [24,25]. In addition, the use of the synchrotron has enabled new insight into spatial response of ionization chambers on a microscopic level leading to explanations of their performance as tools for reference dosimetry on a macroscopic level [26].

MRT is the most prominent example for the dosimetric challenges posed by the radiation that can be produced by synchrotrons [14,27,28]. When several hundred Gy are delivered using kilovoltage photons with minimal divergence in less than a second, dose can be meaningfully specified within several micrometers.

As in complex intensity modulated radiation therapy (IMRT) dose distributions, the dosimetry associated with MRT is determined by dose and dose distribution (in the case of MRT on a micrometer scale). These components include the in-beam or *peak* dose, the valley dose (between adjacent microbeams), and the integrated dose (the dose averaged over the entire irradiation area) [29]. The so-called peak-to-valley dose ratio (PVDR) in MRT is related to biological effect and determined by the following: (1) the collimator geometry (i.e., the microbeam width and the center-to-center spacing), (2) the overall size of the MRT array, (3) the energy of the incident X-ray beam, and (4) the density and the depth of the materials in question.

Dosimetry for imaging applications is also challenging as imaging using monochromatic X-rays has the potential to perform high quality imaging at comparatively low doses. When performing dose measurements, it is also important to consider any element's k-edge of which for X-ray absorption is in the range of interest. As such it is prudent to avoid radiochromic film doped with high-Z materials to enhance sensitivity [30].

19.2.4 DOSIMETRIC OPPORTUNITIES

When considering synchrotron radiation, one should also appreciate the benefits it can bring to the scientific field of dosimetry. This also applies in the context of radiochromic film dosimetry. The very high dose rate in synchrotron radiation allows researchers to test variations in dose response with dose rate. However, the high brilliance also allows the use of monochromators to extract only a narrow bandwidth of X-ray energy, typically with a full width half maximum of less than 1% of the selected energy. These monoenergetic X-rays are well suited to test radiation detectors for variations of response with X-ray energy. Kron et al. [31] used this method to investigate a large number of dosimeters, including MD-55 and MD-870 films. They found an under-response of the film compared with water between 20 and 50 keV. This is similar to fat and compatible with the elemental composition of radiochromic film consisting mostly of hydrogen and carbon. These results were later confirmed for MD-55 film by Nariyama et al. [32]

EBT film was studied by Brown et al. [33] using 25, 30, and 35-keV X-rays from a synchrotron. They found that EBT3 demonstrated a remarkable improvement in variation of dose response with energy compared with the older versions of film.

19.3 RADIOCHROMIC FILM FOR SYNCHROTRON DOSIMETRY

For synchrotron irradiation, radiochromic film has several potential applications with high spatial resolution and the immediate visibility of the dose distribution is of particular relevance [34]. Film is used for assessment of complex dose distribution, for example, in microbeams and often exposed at the same time as other targets to verify the dose delivery. It is significant to note that it is not just film that can verify synchrotron radiation but that synchrotron radiation can also be used to verify the properties of radiochromic film. Testing for dose and dose-rate dependence is something in which synchrotron radiation can assist as well as in assessing variations of response with radiation quality using monoenergetic X-rays [31]. The high spatial resolution of radiochromic film makes it a useful choice for a precise measurement of the irradiated area. Such measurements are important when conducting reference dosimetry in which accurate knowledge of the volume (and ultimately mass) of media irradiated is critical to absolute dose measurements [35,22].

19.3.1 FILM TYPES AND PRINCIPLES OF DOSIMETRY

As discussed in Chapter 2, radiochromic film has undergone a rapid transition from one type of film to the next, and many improvements have been made since the publication of an American Association of Physicists in Medicine (AAPM) task group report [36] that provided information of radiochromic film used in radiotherapy. Different films are available for different applications ranging from single emulsion low sensitivity films with high spatial resolution to highly sensitive films with high-Z material doping to enhance sensitivity and dual emulsion films.

Radiochromic film dosimetry is attractive because the film is self-developing, and readout can be performed using desktop scanners designed for photographic and other common applications. This makes film dosimetry despite the relatively high cost of the film itself overall cheap and easy to set-up. The radiochromic film dosimeter needs to be seen as the combination of film, readout device (scanner), and analysis software, and procedures need to consider the scanner and its limitations as much as the film material [37,38].

The accuracy of film dosimetry is limited by the consistency of the thickness of the active film layer and of the thickness of the support material around the active layer. Although the optical density (OD) of the active layer changes as a function of dose, the OD of the supportive layer remains constant, independent of dose. As only the combined OD of all layers can be measured, several methods have been used to manage the small spatial nonuniformities of the active and the supportive layers in the calculation of the dose received by the film. Double irradiation is the time-honored method [39] and a process that can even be refined by using multiple exposures [40]. Multichannel analysis on the other hand is based on the differences in change of OD with dose for different colors of light [41,42]. Several commercial software packages are available to support radiochromic film dosimetry and its dose response characteristics.

The HD, MD, and EBT variations of radiochromic film have been used to measure the characteristic of microbeam dose distribution. The EBT3 variety of radiochromic film is popular in the radiotherapy clinic and has a working dose range of about 0.2–5 Gy [43,44]. However, it has also been shown to be useful for the higher dose levels beyond 20 Gy that are characteristic for novel stereotactic radiotherapy procedures [45]. This EBT film has been used to measure the valley dose in MRT; however, the uncertainties reported in 2008 by Crosbie et al. [27] were on the order of ±25%, which is too high for a clinical purpose. The other common type of radiochromic film is MD-55 and a newer variation MD-V3. This film type has a nominal working dose range of about 2–80 Gy and is particularly good at relative dose measurements such as assessing area profiles of the synchrotron X-ray beam.

The HD varieties of radiochromic film (HD-810 and more recently HD-V2) have a practical dose range from about 20–1000 Gy. This makes them the most commonly used types of radiochromic film for synchrotron MRT dosimetry [46]. The HD films are often used to measure the microbeam *peak* dose in which the peak dose is on the order of hundreds of Gy in the peak. The HD film is not so useful at simultaneously measuring the valley dose, except if one exposes the film to very high *peak* doses (>2000 Gy), thus saturating the peak region but delivering measureable *valley* doses of tens of Gy. The newer HD-V2 films have an improved, flatter spectral response to X-ray photons with energies less than 100 keV compared with the older HD-810 films, which exhibited also a marked variation in response with energy. The HD-V2 films possess a more homogenous/uniform active layer with less variation in response for the orientation of the film (i.e., portrait vs. landscape).

A disadvantage of HD-V2 film compared with the older HD-810 film is increased statistical noise. The images of the HD-V2 films appear *dirty* to the eye when scanning with the microscope at high resolutions (Figure 19.5). The origin of the *dirtiness* appears to be in the silicone grains that have been added to the film by the manufacturer to counter the effects of the Newton Rings interference patterns. These dispersing grains work well when high spatial resolutions are not required. However, for MRT dosimetry, the addition of these grains and the resulting increase in statistical noise is counterproductive because its effects are on a similar scale as typical microbeam features as seen in Figure 19.5.

Figure 19.6 shows a comparison of the dose response of three film types for microbeam dosimetry. The differences in dose response curve can be clearly seen not only in the dose range on the vertical axis but also in the shape of the dose response curve. The dose response for EBT3 and MD films can be fitted using a single exponential ($y = a \exp(bx)$), whereas HD film was better approximated using a Rodbard curve ($y = a + (b - a)/(1 + (x/c)^d)$).

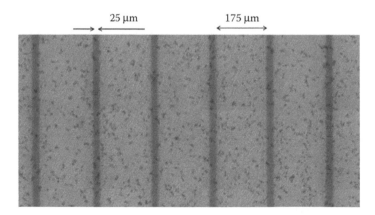

25 µm   175 µm

Figure 19.5 Microscope image of HD-V2 radiochromic film showing the microparticles embedded on the surface of the film to avoid the formation of Newton rings.

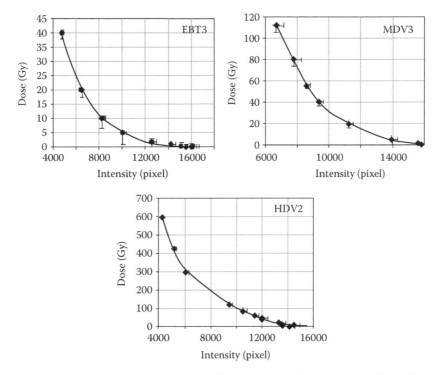

Figure 19.6 Comparison of the dose response of three types of radiochromic film with synchrotron radiation. The horizontal axis shows the intensity of transmitted light on a 14-bit scale. The curves show the best fit using the equations given in the text.

19.3.2 VARIATION OF RESPONSE WITH X-RAY ENERGY

As with most kilovoltage X-ray dosimetry methods, variations of detector response with radiation energy is important as the effective atomic number will affect the detector response. This variation in response is because of varying degrees of photoelectric absorptions and Compton scattering effects. Here synchrotron radiation provides a unique tool to probe the variation of response with energy because synchrotron radiation is of sufficient flux to filter out a wide range of mono energetic beams through crystal refraction. This has been used to study several detectors and develop a mathematical formula to describe dosimeter response with energy [31,47].

Also studied radiochromic film that was found to be closer to fat than water equivalence. This was confirmed by Nariyama et al. in 2002 for MD-55 [32].

19.3.3 HIGH- AND LOW-DOSE MEASUREMENT

Synchrotron radiation dosimetry is challenging because the dose range of interest is rather large. On the low dose side, one of the main advantages of mammography using synchrotron phase contrast imaging, is the potential for dose reduction [48]. The hope is to obtain high-quality diagnostic images with mean glandular doses below 1 mGy.

However, microbeam radiation is probably the most important challenge as high and low doses need to be assessed at the same time and in close proximity as can be seen in Figure 19.5. As the peak valley dose ratio is considered one of the key parameters to determine the efficacy of microbeam treatment [49,50], it is important to asses both peak and valley dose at the same time. An elegant method has been used by Crosbie et al. [27] who used two types of radiochromic film in combination, albeit with too high an uncertainty in the EBT film. A modified *double exposure* technique could also be used to deliver high and low doses to just one film type (HD-V2) as a function of sample scan speed; a fast scan speed for measurable peak doses (e.g., 50 Gy or 150 Gy) and a slow scan speed (100 times slower) for high but measureable valley doses (e.g., 50 Gy), with both fields delivered to HD film.

19.3.4 SPATIAL RESOLUTION REQUIREMENTS

Synchrotron radiation is characterized by X-rays with energies between 5 and 200 keV. The typical secondary electron range after these photons interact with matter by photoelectric or Compton interactions, is of the order of several tens of μm [51]. This is essential to ensure sharp edges of radiation beams and the spatial dose patterns that are characteristic for microbeams. The corollary of this is that any dosimeter used for determination of dose distributions must have a spatial resolution of the order of micrometers, which is hard to achieve.

The increased sensitivity afforded by the dual emulsion is not generally required for synchrotron experiments using microbeams; single emulsion films such as HD 810 are preferable. Also the evaluation technique becomes critical as discussed in the next section. The flatbed film scanners, which are commonly used for radiochromic film evaluation in radiotherapy, do not have the required spatial resolution. Therefore, microscopes are often the preferred option for evaluation.

Given the extreme dosimetric requirements in terms of dose range and spatial resolution, not many alternative detectors are available. Of mention are gels [52] and Presage [53,54] dosimeters that allow for three-dimensional dosimetry. A shortcoming of radiochromic film is that despite the fact that the film is self-developing, dosimetry using radiochromic film is not real time. As such semiconductor devices such as the dose magnifying glass developed by the University of Wollongong [55] are likely to also play a role in the future.

19.4 EVALUATION METHODS

Conventional densitometers and most film scanners do not have a spatial resolution sufficient to determine dose distributions created by microbeams. The Joyce-Loebl 3CS microdensitometer has served as an unofficial gold standard of film densitometry for MRT. However, the optomechanical components in these microdensitometers are highly sensitive, making reproducibility between different operators almost impossible. This is where optical microscopes as discussed earlier have a clear advantage [46]. Figure 19.7 shows the experimental setup of a confocal microscope for film evaluation. The film is attached to a microscope slide as can be seen in the figure and an automated scanning of the film across the field of view of the microscope can be achieved relatively quickly.

Crosbie et al. first reported on using an upright microscope with a charge coupled device (CCD) camera to photograph the MRT-irradiated films [27]. The spatial resolution of a typical life sciences microscope with an x10-objective is greater than 1 μm. When coupled with a high-quality, high bit-depth camera, an LED light source, and a motorized stage, the entire irradiated region of the film can be photographed as a true 2D digital image with high spatial resolution.

Nariyama et al. [56] reported in 2009 on the use of a microscope with various band pass filters to measure the OD distribution of radiochromic films exposed to synchrotron-generated microbeams. The authors reported their measured peak-to-valley dose ratios and obtained good agreement with Monte Carlo simulations. More recently, Bartzsch et al. [46] provided a comprehensive account of how they use a Zeiss AxioVert microscope to perform microdensitometry of

Figure 19.7 Inverted optical microscope (Olympus IX83) setup for evaluation of radiochromic film at RMIT University, Melbourne. The camera on the bottom left of the top figure is a 16-bit black and white camera (Hamamatsu Orca Flash 4.0) with very high sensitivity.

MRT-irradiated films. It is important to note that microscope-based densitometry is nontrivial and also requires careful setup, corrections, and an understanding of the interplay of light source (spectral response, uniformity) and camera response.

Figure 19.8 shows a summary of characteristics of different microscopes for evaluation of film in comparison with the more conventional flatbed scanner (Epson 700) that is commonly used in radiotherapy departments.

Microscope-based methods of densitometry offer an advantage over the more traditional methods of microdensitometry because one obtains a full 2D digital image that can be subject to further image-processing techniques. The microscope method lends itself nicely to automation and batch processing of multiple images using (typically) in-house scripts or macros running in computer programs such as MATLAB®, Python, ImageJ, and so on. The scripts can perform all the necessary corrections to the images and produce plots of peak and valley doses within a few minutes.

The spatial resolution offered by the J–L microdensitometer is on the order of 10 μm based on modulation transfer function evaluation [46,27]. This resolution is adequate for sampling of microbeams measuring 25 μm. The microscope method of densitometry, however, provides an order of magnitude greater resolution than the microdensitometer (and comes at a cost of large file sizes on a computer, typically hundreds of MB). With the microscope method, the *grain* size or rather the dimensions of the monomer crystals in the film itself becomes the limiting factor in determining spatial resolution. Bartzsch et al. [46] estimated the grain size of HD-V2 radiochromic film to be about 2 μm based on measurement and subsequent analysis.

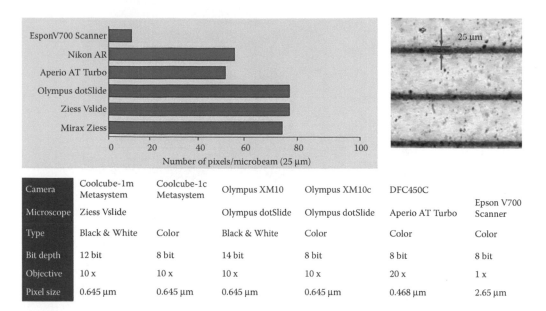

Camera	Coolcube-1m Metasystem	Coolcube-1c Metasystem	Olympus XM10	Olympus XM10c	DFC450C	
						Epson V700 Scanner
Microscope	Ziess Vslide		Olympus dotSlide	Olympus dotSlide	Aperio AT Turbo	
Type	Black & White	Color	Black & White	Color	Color	Color
Bit depth	12 bit	8 bit	14 bit	8 bit	8 bit	8 bit
Objective	10 x	10 x	10 x	10 x	20 x	1 x
Pixel size	0.645 μm	0.645 μm	0.645 μm	0.645 μm	0.468 μm	2.65 μm

Figure 19.8 Unpublished comparison of different microscopes with a flatbed scanner for evaluation of radiochromic film (microscopes courtesy of Monash Medical Imaging). Shown here are results for HD-V2 film.

It is now a common practice in clinical radiotherapy to use color scanners for evaluation of radiochromic film in order to extend the dose response curve and provide a potential measure of emulsion thickness. Color can also improve the evaluation of microbeam exposed films as can be seen in Figure 19.9. However, it is important to weigh up the additional information afforded by the three colors against the better bit depth often given by monochromatic scanners as illustrated in Figure 19.10.

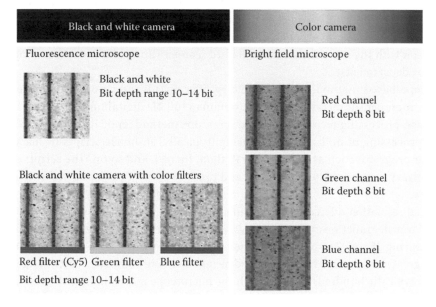

Figure 19.9 Evaluation of film using different color channels: visual comparison of images acquired from radiochromic film using different microscopic evaluation techniques.

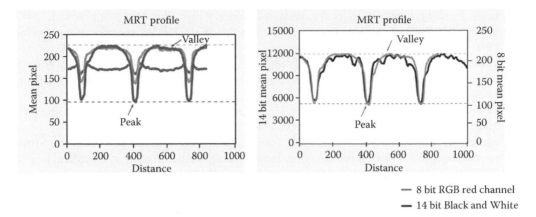

Figure 19.10 Profiles acquired of a radiochromic film exposed to a microbeam irradiation. Shown on the left are profiles using three different color channels. The figure on the right shows the difference in scan using 8- and 14-bit depth. The potential for *dose resolution* can be seen at the two vertical axes.

19.5 SUMMARY AND OUTLOOK

In summary, there is no existing type of radiochromic film which could be considered *perfect* for synchrotron MRT dosimetry studies. If a new version of radiochromic film were to be created for synchrotron microbeam applications, then it would have the following properties:

- A flat energy response for X-rays with a mean energy above 50 keV
- A working dose range of 0.5–500 Gy
- Low statistical noise (removal of the silicone grains)
- Homogenous coating of the active layer so that film orientation is not critical
- Symmetric active layer with respect to the front and rear faces of the film (EBT3)
- The option of small cuts of film that can be attached to a glass microscope slide

Some of them are already realised in HD-V2 films, but a new dose range and the silicone grains would need to be implemented. However, it is fair to say that the excellent spatial resolution and overall tissue equivalence afforded by radiochromic film is a good match to the challenges posed by synchrotron dosimetry, particularly in the context of Micro Beam Therapy. It will be interesting to see how film and its evaluation techniques develop in the years ahead; no doubt all of radiation dosimetry stands to benefit.

ACKNOWLEDGMENTS

We would like to thank the Peter Mac Foundation and the National Health & Medical Research Council of Australia for financial support of our synchrotron research program as well as being granted beamtime from The Australian Synchrotron.

REFERENCES

1. Kron T. Applications of synchrotron radiation x-rays in medicine. *Phys Med Biol* 1998; 43:215–216.

2. Lewis R. Medical applications of synchrotron radiation x-rays. *Phys Med Biol* 1997;42:1213–1243.

3. Winick H, Doniach S. *Synchrotron Radiation Research*. Heidelberg, Germany: Springer, 1980.

4. Takeda T, Itai Y, Wu J, Ohtsuka S, Hyodo K, Ando M et al. Two-dimensional intravenous coronary arteriography using above-k-edge monochromatic synchrotron x-ray. *Acad Radiol* 1995;2:602–608.

5. Lewis RA, Yagi N, Kitchen MJ, Morgan MJ, Paganin D, Siu KK et al. Dynamic imaging of the lungs using X-ray phase contrast. *Phys Med Biol* 2005;50:5031–5040.

6. Murrie RP, Stevenson AW, Morgan KS, Fouras A, Paganin DM, Siu KK. Feasibility study of propagation-based phase-contrast X-ray lung imaging on the imaging and medical beamline at the Australian synchrotron. *J Synchrotron Radiat* 2014;21:430–445.

7. Arfelli F, Bonvicini V, Bravin A, Cantatore G, Castelli E, Palma LD et al. Mammography with synchrotron radiation: Phase-detection techniques. *Radiology* 2000;215:286–293.

8. Longo R, Arfelli F, Bellazzini R, Bottigli U, Brez A, Brun F et al. Towards breast tomography with synchrotron radiation at elettra: First images. *Phys Med Biol* 2016;61:1634–1649.

9. Siew ML, Wallace MJ, Allison BJ, Kitchen MJ, te Pas AB, Islam MS et al. The role of lung inflation and sodium transport in airway liquid clearance during lung aeration in newborn rabbits. *Pediatr Res* 2013;73:443–449.

10. Te Pas AB1, Kitchen MJ, Lee K, Wallace MJ, Fouras A, Lewis RA et al. Optimizing lung aeration at birth using a sustained inflation and positive pressure ventilation in preterm rabbits. *Pediatr Res* 2016;80:85–91.

11. Olivo A, Gkoumas S, Endrizzi M, Hagen CK, Szafraniec MB, Diemoz PC et al. Low-dose phase contrast mammography with conventional X-ray sources. *Med Phys* 2013;40:090701.

12. Edouard M, Broggio D, Prezado Y, Estève F, Elleaume H, Adam JF. Treatment plans optimization for contrast-enhanced synchrotron stereotactic radiotherapy. *Med Phys* 2010;37:2445–2456.

13. Rahman WN, Corde S, Yagi N, Abdul Aziz SA, Annabell N, Geso M. Optimal energy for cell radiosensitivity enhancement by gold nanoparticles using synchrotron-based monoenergetic photon beams. *Int J Nanomedicine* 2014; 9:2459–2467.

14. Bräuer-Krisch E, Adam JF, Alagoz E, Bartzsch S, Crosbie J, DeWagter C et al. Medical physics aspects of the synchrotron radiation therapies: Microbeam radiation therapy (mrt) and synchrotron stereotactic radiotherapy (ssrt). *Phys Med* 2015;31:568–583.

15. Das IJ. Forward dose perturbation at high atomic number interfaces in kilovoltage x-ray beams. *Med Phys* 1997;24:1781–1787.

16. Herold DM, Das IJ, Stobbe CC, Iyer RV, Chapman JD. Gold microspheres: A selective technique for producing biologically effective dose enhancement. *Int J Radiat Biol* 2000;76:1357–1364.

17. Solberg TD, Iwamoto KS, Norman A. Calculation of radiation dose enhancement factors for dose enhancement therapy of brain tumours. *Phys Med Biol* 1992;37:439–443.

18. Hugtenburg RP. Microdosimetry in X-ray synchrotron based binary radiation therapy. *Eur J Radiol* 2008;68:S126–128.

19. Rahman WN, Wong CJ, Ackerly T, Yagi N, Geso M. Polymer gels impregnated with gold nanoparticles implemented for measurements of radiation dose enhancement in synchrotron and conventional radiotherapy type beams. *Australas Phys Eng Sci Med* 2012;35:301–309.

20. Smyth LM, Senthi S, Crosbie JC, Rogers PA. The normal tissue effects of microbeam radiotherapy: What do we know, and what do we need to know to plan a human clinical trial? *Int J Radiat Biol* 2016;92:302–311.

21. Bräuer-Krisch E, Serduc R, Siegbahn EA, Le Duc G, Prezado Y, Bravin A et al. Effects of pulsed, spatially fractionated, microscopic synchrotron X-ray beams on normal and tumoral brain tissue. *Mutat Res* 2010;704:160–166.

22. Lye JE, Harty PD, Butler DJ, Crosbie JC, Livingstone J, Poole CM et al. Absolute dosimetry on a dynamically scanned sample for synchrotron radiotherapy using graphite calorimetry and ionization chambers. *Phys Med Biol* 2016;61:4201–4222.

23. Pelliccia D, Poole CM, Livingstone J, Stevenson AW, Smyth LM, Rogers PA et al. Image guidance protocol for synchrotron microbeam radiation therapy. *J Synchrotron Radiat* 2016;23:566–573.

24. Duane S, Lee ND, Graber F, Brauër-Krisch E, Crosbie JC. Absolute dosimetry for synchrotron microbeam radiation therapy using a graphite calorimeter. *Radiother Oncol* 2014;111:S201.

25. Harty PD, Lye JE, Ramanathan G, Butler DJ, Hall CJ, Stevenson AW4 et al. Absolute X-ray dosimetry on a synchrotron medical beam line with a graphite calorimeter. *Med Phys* 2014;41:052101.

26. Butler DJ, Stevenson AW, Wright TE, Harty PD, Lehmann J, Livingstone J et al. High spatial resolution dosimetric response maps for radiotherapy ionization chambers measured using kilovoltage synchrotron radiation. *Phys Med Biol* 2015;60:8625–8641.

27. Crosbie JC, Svalbe I, Midgley SM, Yagi N, Rogers PA, Lewis RA. A method of dosimetry for synchrotron microbeam radiation therapy using radiochromic films of different sensitivity. *Phys Med Biol* 2008;53:6861–6877.

28. Doran SJ, Abdul Rahman AT, Bräuer-Krisch E, Brochard T, Adamovics J, Nisbet A et al. Establishing the suitability of quantitative optical CT microscopy of presage(r) radiochromic dosimeters for the verification of synchrotron microbeam therapy. *Phys Med Biol* 2013;58:6279–6297.

29. Ibahim MJ, Crosbie JC, Yang Y, Zaitseva M, Stevenson AW, Rogers PA et al. An evaluation of dose equivalence between synchrotron microbeam radiation therapy and conventional broad beam radiation using clonogenic and cell impedance assays. *PLoS One* 2014;9:e100547.

30. Williams M, Metcalfe P. Radiochromic film dosimetry and its applications in radiotherapy. In: Rozenfeld A, Kron T, D'Erico F and Moscovitch M, (Eds.) *4th Solid State Dosimetry Summerschool*. Wollongong, Australia: American Institute of Physics, 2011;1345.

31. Kron T, Duggan L, Smith T, Rosenfeld A, Butson M, Kaplan G et al. Dose response of various radiation detectors to synchrotron radiation. *Phys Med Biol* 1998;43:3235–3259.

32. Nariyama N, Namito Y, Ban S, Hirayama H. Response of gafchromic md-55 radiochromic film to synchrotron radiation. *Radiat Prot Dosim* 2002;100:349–352.

33. Brown TA, Hogstrom KR, Alvarez D, Matthews KL 2nd, Ham K, Dugas JP. Dose-response curve of ebt, ebt2, and ebt3 radiochromic films to synchrotron-produced monochromatic X-ray beams. *Med Phys* 2012;39:7412–7417.

34. Soares CG. New developments in radiochromic film dosimetry. *Radiat Prot Dosim* 2006;120:100–106.

35. Crosbie JC, Rogers PA, Stevenson AW, Hall CJ, Lye JE, Nordström T et al. Reference dosimetry at the Australian synchrotron's imaging and medical beamline using free-air ionization chamber measurements and theoretical predictions of air kerma rate and half value layer. *Med Phys* 2013;40:062103.

36. Niroomand-Rad A, Blackwell CR, Coursey BM, Gall KP, Galvin JM, McLaughlin WL et al. Radiochromic film dosimetry: Recommendations of AAPM radiation therapy committee task group 55. American association of physicists in medicine. *Med Phys* 1998;25:2093–2115.

37. Ferreira BC, Lopes MC, Capela M. Evaluation of an Epson flatbed scanner to read gafchromic ebt films for radiation dosimetry. *Phys Med Biol* 2009;54:1073–1085.

38. Paelinck L, De Neve W, De Wagter C. Precautions and strategies in using a commercial flatbed scanner for radiochromic film dosimetry. *Phys Med Biol* 2007;52:231–242.

39. Klassen NV, van der Zwan L, Cygler J. Gafchromic MD-55: Investigated as a precision dosimeter. *Med Phys* 1997;24:1924–1934.

40. Roozen K, Kron T, Haworth A, Franich R. Evaluation of ebt radiochromic film using a multiple exposure technique. *Australas Phys Eng Sci Med* 2011;34:281–289.
41. Micke A, Lewis DF, Yu X. Multichannel film dosimetry with nonuniformity correction. *Med Phys* 2011;38:2523–2534.
42. Perez Azorin JF, Ramos Garcia LI, Marti-Climent JM. A method for multichannel dosimetry with EBT3 radiochromic films. *Med Phys* 2014;41:062101.
43. Chan E, Lydon J, Kron T. On the use of gafchromic ebt3 films for validating a commercial electron Monte Carlo dose calculation algorithm. *Phys Med Biol* 2015;60:2091–2102.
44. Sorriaux J, Kacperek A, Rossomme S, Lee JA, Bertrand D, Vynckier S et al. Evaluation of Gafchromic® EBT3 films characteristics in therapy photon, electron and proton beams. *Phys Med* 2013;29:599–606.
45. Palmer AL, Dimitriadis A, Nisbet A, Clark CH. Evaluation of gafchromic EBT-XD film, with comparison to EBT3 film, and application in high dose radiotherapy verification. *Phys Med Biol* 2015;60:8741–8752.
46. Bartzsch S, Lott J, Welsch K, Bräuer-Krisch E, Oelfke U. Micrometer-resolved film dosimetry using a microscope in microbeam radiation therapy. *Med Phys* 2015;42:4069–4079.
47. Kron T, Smith A, Hyodo K. Synchrotron radiation in the study of the variation of dose response in thermoluminescence dosimeters with radiation energy. *Australas Phys Eng Sci Med* 1996;19:225–236.
48. Keyriläinen J, Bravin A, Fernández M, Tenhunen M, Virkkunen P, Suortti P. Phase-contrast X-ray imaging of breast. *Acta Radiol* 2010;51:866–884.
49. Crosbie JC, Anderson RL, Rothkamm K, Restall CM, Cann L, Ruwanpura S et al. Tumor cell response to synchrotron microbeam radiation therapy differs markedly from cells in normal tissues. *Int J Radiat Oncol Biol Phys* 2010;77:886–894.
50. Siegbahn EA1, Stepanek J, Bräuer-Krisch E, Bravin A. Determination of dosimetrical quantities used in microbeam radiation therapy (MRT) with Monte Carlo simulations. *Med Phys* 2006;33:3248–3259.
51. Reich H. *Dosimetrie ionisierender strahlung.* Stuttgart, Germany: Teubner, 1990.
52. Dilmanian FA1, Romanelli P, Zhong Z, Wang R, Wagshul ME, Kalef-Ezra J et al. Microbeam radiation therapy: Tissue dose penetration and BANG-gel dosimetry of thick-beams' array interlacing. *Eur J Radiol* 2008;68:S129–S136.
53. Annabell N, Yagi N, Umetani K, Wong C, Geso M. Evaluating the peak-to-valley dose ratio of synchrotron microbeams using PRESAGE fluorescence. *J Synchrotron Radiat* 2012;19:332–339.
54. Gagliardi FM, Cornelius I, Blencowe A, Franich RD, Geso M. High resolution 3d imaging of synchrotron generated microbeams. *Med Phys* 2015;42:6973–6986.
55. Wong JH, Knittel T, Downes S, Carolan M, Lerch ML, Petasecca M et al. The use of a silicon strip detector dose magnifying glass in stereotactic radiotherapy QA and dosimetry. *Med Phys* 2011;38:1226–1238.
56. Nariyama N, Ohigashi T, Umetani K, Shinohara K, Tanaka H, Maruhashi A et al. Spectromicroscopic film dosimetry for high-energy microbeam from synchrotron radiation. *Appl Radiat Isot* 2009;67:155–159.

20

Ultraviolet radiation dosimetry

MARTIN BUTSON, SAMARA ALZAIDI, TINA GORJIARA,
AND MAMOON HAQUE

20.1 INTRODUCTION

Ultraviolet (UV) radiation is the portion of the electromagnetic spectrum in between x-rays and visible light, that is, between 40 and 400 nm. The UV spectrum is broadly divided into vacuum UV (40–190 nm), far UV (190–220 nm), UVC (220–290 nm), UVB (290–320 nm), and UVA (320–400 nm). Solar radiation is our primary natural source of UV; however, artificial sources can include, but are not limited to, tanning booths, black lights, curing lamps, germicidal lamps, mercury vapor lamps, halogen lights, high-intensity discharge lamps, fluorescent and incandescent sources, and some lasers and welding apparatus.

Vacuum UV, far UV, and UVC are almost never observed in nature because they are absorbed completely in the atmosphere. Germicidal lamps are designed to emit UVC radiation because of their ability to kill bacteria. Overexposure to UVC can cause corneal burns, which is commonly referred to as welders' flash or snow blindness.

UVB is a highly important form of UV for biological applications as it has enough energy to cause photochemical damage to cellular DNA, yet not enough to be completely absorbed by the atmosphere. UVB is needed by humans for synthesis of vitamin D but also causes harmful effects such as erythema (sunburn), cataracts, and development of skin cancer. Individuals working outdoors are at the greatest risk of UVB effects. Most solar UVB is blocked by ozone in the atmosphere; however, reductions in atmospheric ozone cause increase in UVB intensities in various parts of the world.

UVA is the most commonly encountered type of UV light. UVA exposure has an initial pigment-darkening effect (tanning) followed by erythema if the exposure is excessive. Atmospheric ozone absorbs very little of this part of the UV spectrum. UVA exposure has been associated with toughening of the skin, suppression of the immune system, and cataract formation. UVA light is often called black light. Most phototherapy and tanning booths use UVA lamps.

Protection from UV is provided by clothing, polycarbonate, glass, acrylics, and plastic diffusers used in office lighting. Alternatively, sun-blocking lotions can offer protection against UV exposure. Dosimetry is essential to quantify the amount of radiation and to control deleterious effect of UV light. This chapter provides the use of radiochromic films for UV dosimetry.

20.2 USE OF RADIOCHROMIC FILMS

Exposure guidelines for UV radiation have been established by the American Conference of Governmental Industrial Hygienists [1] and by the International Commission on Nonionizing Radiation Protection [2]. Handheld meters to measure UV radiation are commercially available; however, they only give a point exposure measurement and do not necessarily measure a personal UV exposure to a certain part of the body or biological system. An in vivo radiochromic film detector could, however, be worn close to the site of interest and integrate the exposure level over a given time period to fully and accurately assess UV-exposure levels.

People are constantly exposed to natural or artificial UV radiation in different ways, unintentionally or intentionally, at their workplace or in their spare time. Both quantity and quality of exposure are crucial for the induction of beneficial or detrimental photobiological UV effects. Quantification of the individual level of UV exposure requires personal UV dosimetry due to a permanent change of the position of people to UV sources. There is a strong wavelength dependence of action spectra of the different UV-induced photobiological effects on human beings or biological systems. A suitable personal-dosimeter system has to respect this fact by supplying a spectral response identical to the action spectrum in question; in this case, measurements are independent of the spectral irradiance of the UV source or changes in the spectrum (e.g., spectral irradiance of the sum). Actually, this means that a special sensor is necessary for each investigated photobiological effect. However, it is difficult to develop physical, photochemical, or photobiological sensors with spectral responses identical or nearly identical to the respective action spectra. In addition such sensors have to fulfill basic requirements to UV sensors in general and to personal dosimeters in particular. To this extent, various radiochromic film detectors have been developed to match specific types of action spectra for assessment of UV exposure in specific measurement conditions. A few of these will be briefly discussed in the following.

20.3 COMMERCIALLY AVAILABLE FILMS

20.3.1 POLYSULPHONE FILMS

Polysulphone films have been used in UV assessment for more than 40 years [3]. Polysulphone films were first introduced to monitor UV radiation in 1976 by Diffey and Davis [4] and are the most commonly used films in human-related exposure assessment to solar UV exposure and its health impacts [5]. The polysulphone polymer undergoes a UV-induced photodegradation when exposed to UV. Polysulphone's action spectra match quite well with that of erythemal responses of human skin when it is made approximately 30 to 40 micrometers thick [6]. Polysulphone polymer structures contain aromatic rings and are the cause for the high absorption of UV radiation [7].

The degradation causes changes to the absorption properties of the polysulphone film, and specifically, measurement can be made at the wavelength of 330 nm [3] to assess exposure levels matched to human erythema response. The highest sensitivity is recorded in the UVB wavelengths and falls quickly for wavelengths greater than 300 nm. There is no response to wavelengths greater than approximately 330 to 340 nm [8]. Polysulphone films have been used extensively for the quantification of erythemal UV exposure. Some examples include, but are not limited, to the following: exposure of infants to UV [9], investigations of outdoor worker's solar exposures [10], and exposure of outdoor cyclists [11].

As polysulphone films are quite thin, they are normally attached or mounted onto a frame that can be worn as a badge on any area of concern for personal UV-exposure measurement.

20.3.2 PHENOTHIAZINE/MYLAR RADIOCHROMIC FILMS

To perform UVA dosimetry, one option is the use of radiochromic phenothiazine with a Mylar film covering. Phenothiazine [12] is known to have a response to both UVA and UVB; however, by the incorporation of a Mylar film covering, the UVB component can be removed [13]. To produce the film, manufacturers cast the phenothiazine directly onto the Mylar film to create a radiochromic film sensitive to UVA radiation only if the UV source passes directly through the Mylar but can also measure both UVA and UVB if the film turned around to the Mylar is directly behind the UV source. Absorbance changes are measured in the wavelength region of approximately 370 nm to assess UV-exposure levels.

20.3.3 GAFCHROMIC™ FILM PRODUCTS

Although GAFchromic™ film products have been specifically developed for the measurement of ionizing radiation such as photons, electrons, and protons, they also exhibit sensitivity to UV radiation due to polymerization of the chemicals. These properties were discussed by Niroomand-Rad et al. [14]; however, Butson et al. [15] provided quantitative data on the UV exposure versus absorbance color change relationship for MD-55-2. The GAFchromic film product range has changed considerably over the past 15 years and each film type has exhibited different UV sensitivities. Normally, a reduced sensitivity to UV has been provided with each new interaction of the GAFchromic family as the company attempts to minimize the effects of UV radiation or ambient light on ionizing radiation dosimetry. Various studies have been performed to evaluate these responses. In 2010, Butson et al. [16] investigated the UV exposure response of GAFchromic external beam therapy (EBT) film as well as EBT2 film in 2013 [17]. In 2014, Chun et al. [18] performed similar studies for EBT3 film, and Katsuda et al. [19–20] investigated UV response of GAFchromic XR-RV3 and XR-SP2 films as well as utilizing UV wavelengths for uniform preexposure of EBT films for dosimetry purposes.

20.4 ULTRAVIOLET ACTION SPECTRA AND RADIOCHROMIC FILMS

An action spectrum is defined as the rate of change/reaction of some form of physiological activity compared with the wavelength of the incident light [21]. In relation to UV radiation action spectra, for biological reactions, it defines a parameter function that describes the relative effect of energy at different wavelengths in producing a certain biological response. These effects may be at a molecular level, such as DNA damage or erythema response of skin through to effects on a whole organism, such as plant growth. An action spectrum is used as a weighting function for the UV spectrum in an integration of the monochromatic UV irradiance.

Some specific medically useful action spectrum includes, but is not limited to, the following:

1. The Commission Internationale de l'\'Eclairage (CIE) action spectrum for erythema (reddening of the skin due to sunburn), proposed by McKinlay & Diffey in 1987 [22] and adopted as a standard by the Commission Internationale de l'\'Eclairage (International Commission on Illumination). The erythemal UV index is integration between 280 and 400 nm of the UV irradiance at ground level, weighted with the erythemal action spectrum.
2. Erythemal action spectrum, which has been determined by Parrish et al. 1981 [23], is shown in Figure 20.1 and was parameterized by Bernard and Seckmeyer in 1997 [24].
3. The mammalian nonmelanoma skin cancer action spectrum [25], which lies between the erythemal and DNA-damage action spectrum, shows a sine-like wavelength dependency between 340 and 400 nm.

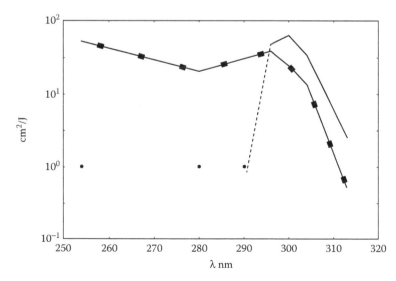

Figure 20.1 Action spectrum for Erythemal response. (Adopted from Parrish J.A. and Jaenicke K.F., *J. Invest. Dermatol.*, 76, 359–362, 1981.)

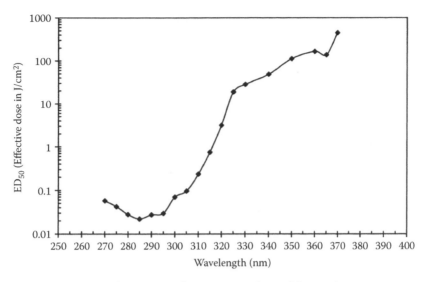

Figure 20.2 Action spectrum for cataract formation. (Adopted from Olanrewaju, M. et al., *Invest. Ophthalmol. Vis. Sci.*, 42, 2596–2602, 2001.)

4. The melanoma induction in platyfish-swordtail hybrids action spectrum [26], which shows a more significant UVA component than other action spectra.
5. The cataract formation action spectrum that is shown in Figure 20.2, used to evaluate damage to eyes from UV radiation [27].

Madronich et al. [28] show some other useful action spectra. It is important to note that the action spectra only give an indication of the relative wavelength dependency of biological effects. The actual biological response is determined by the actual dose, that is, the UV irradiance weighted with the action spectrum integrated over the wavelength range and the exposure time. This principle is similar to photon radiation response of certain organs or tissues during radiotherapy. It should also be noted that the dose-response relation may not be linear [29].

As these action spectra can vary considerably; the type and sensitivity of the radiochromic films used to measure UV radiation becomes very important. The dose or exposure response of the film compared with the action spectra of the biological or medical measurement being made will determine its accuracy, or limitation as such. A few examples of films that have been used for UV radiation exposure include, but are not limited to, polysulphone film, polyphenylene oxide (PPO) film, and the GAFchromic EBT film range. Some of these films will be discussed and reviewed for their UV radiation-exposure measurement characteristics in this chapter.

20.5 FILM UV SENSITIVITY RESPONSES

Film sensitivity is defined as the measured net optical density (OD) per the radiation exposure that caused the OD change [30–31]. A number of studies have been undertaken to investigate the sensitivity of radiochromic films for measurement of UV radiation.

Butson et al. [32] reported the feasibility of using EBT2 film for measurement of solar UV. It was shown that when exposed to UV from the thick laminate side (underside), the EBT2 film exhibits a measurable sensitivity to solar UV. The absorption peaks were observed at 585 and 636 nm. The results were analyzed with band pass wavelengths of 30 nm with centers at 535, 585, and 635 nm. The exposure response curves of these films are presented in Figure 20.3. The results indicate that

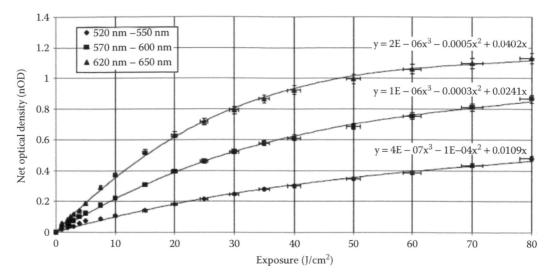

Figure 20.3 Sensitivity response of EBT2 film to various *band passes* of analysis shows the ability to vary the response of the film to UV exposure. (Adopted from Butson, E.T. et al., *Phys. Med. Biol.*, 58, N287–N294, 2013.)

different sensitivities can be produced, depending on the analysis technique. Higher sensitivities can be achieved as wavelengths get closer to the absorption peaks. It was also shown that the EBT 2 film begins to saturate when the exposure level reaches to 60 J cm^{-2} [32].

Aydarous et al. [31] characterized the EBT3 films for UV dosimetry. It was demonstrated that the sensitivity decreases as the UV exposure increases. Figure 20.4 shows sensitivity curves of EBT3 film as a function of UV radiation dose for wavelengths 254, 302, and 365 nm. It was shown that the EBT3 film is more sensitive to wavelengths of 302 and 365 nm. Nonetheless, the highest sensitivity for UVB radiation occurs at low levels of exposure (<10 J/cm^2). Their results show that

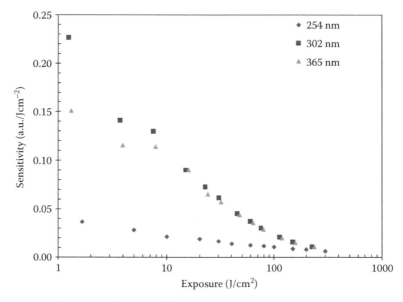

Figure 20.4 Sensitivity curves for EBT3 films exposed to UVA, UVB, and UVC radiation. (From Aydarous, A. et al., *Radiat. Eff. Defect. S.*, 169, 249–255, 2014.)

the EBT3 film is approximately 10 times more sensitive to UVA and UVB compared with UVC and the sensitivity response of the film may vary depending on the densitometer used to readout the film OD [31].

20.6 FILM DOSE/EXPOSURE RESPONSE

To have a personal UV dosimeter, the wavelength response of the dosimeter should be similar to the erythemal action spectrum of human skin, and also the dosimeter should have a monotonic response as UV dose increases [4]. Over the years, different types of radiochromic films and their responses to UV radiation have been investigated, including polysulphone film, PPO, GAFchromic MD-55-2, EBT, EBT2, and EBT3 films [31,33–40].

Davis et al. [6] investigated the dose and wavelength response of polysulphone film for UV dosimetry. The film was irradiated at five levels of radiation, ranging from 0.056 to 0.337 mW/cm², which is equivalent to 250, 500, 750, and 1000 mJ/cm² dose at a wavelength of 305 ± 5 nm. The resultant ΔA_{330} (the changes in absorbance measured at 330 nm) at each dose was within the experimental error of 5% of each other, indicating that the dose–response of the film is independent of dose rate [33].

Kolari et al. [34] evaluated the dose–response and accuracy of polysulphone film badges for monitoring UVB radiation using spectroradiometric measurement of the radiation source [34]. Parisi and Kimlin [36] used filtered and unfiltered polysulphone dosimeter for measurement of solar UV for a period of 3–6 days. They showed that a filtered polysulphone allowed measurements of approximately 100 minimum erythmal dose for a change of 0.35 in optical absorbance at 330 nm [36]. In the CIE report, polysulphone film has been considered as a practical personal dosimeter for UV radiation, which after exposure to the spectral range of 250–330 nm shows an increase in absorbance at 330 nm [41].

Kollias et al. [7] investigated spectral sensitivity of the polysulphone films using a series of monochromatic lights (±2 nm). They also compared the polysulphone-effective solar radiation (using solar UVB data obtained from polysulphone films) with the erythemally effective solar radiation (using data obtained from spectroradiometer). It was shown that the polysulphone films can be used to predict the erythema risk of solar UVB. It was also presented that measurements of solar UVB with polysulphone films were strongly correlated with the measurements with a spectroradiometer ($R^2 > 0.95$) [35]. The dose–response curves of polysulphone film measured as ΔA_{330} to different bands of UVB radiation is presented in Figure 20.5.

Butson et al. [40] used MD-55-2 film for the measurement of UVA and UVB, in addition to visible and infrared radiation. It was demonstrated that MD-55-2 radiochromic film responds mainly to broad-band UVA radiation. The coloration from UVB, visible and low-level infrared radiation, was almost negligible as shown in Figure 20.6 [40].

GAFchromic EBT film was evaluated by Butson et al. ([64] for the measurement solar UV radiation). It was demonstrated that color change of the film (changes in OD) is reproducible within ±10% at 5 kJ/m² UV exposure, under various solar radiation conditions [10].

Evaluation of EBT3 GAFchromic film for UV radiation dosimetry by Aydarous et al. [75] showed that EBT3 film has higher level of sensitivity for absorption peak of 633 nm compared with 582 nm. It was also demonstrated that the exposure response curves to UVA and UVB were nonlinear. These curves showed saturation after approximately 8 h of exposure, which is equivalent to an intensity of approximately 60 J/cm². The response curve of the EBT3 film to UVC radiation, however, remains unsaturated up to 300 J/cm² [31].

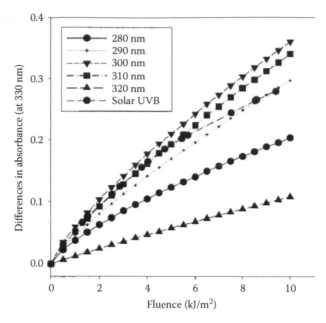

Figure 20.5 Dose–response of polysulphone film measured as ΔA_{330} (the changes in absorbance measured at 330 nm) to different bands of UVB radiation ranging from 280 to 320 nm. (From Kollias, N., *Photochem. Photobiol.*, 78, 220–224, 2003.)

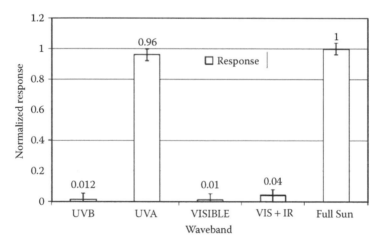

Figure 20.6 The response of MD-55-2 GAFchromic™ film to broadband UVA, UVB, visible, and infrared radiation. (From Butson, E.T., *Phys. Med. Biol.*, 55, N487–N493, 2010.)

20.7 SOLAR/PERSONAL UV EXPOSURE AND RADIOCHROMIC FILMS

UV radiation produced by the Sun or solar UV is the single most significant source of UV radiation and can reach a person on the ground from three sources, directly from the sun, scattered from the open sky, and reflected from the environment [42]. The level of UV is strongly dependent on many factors that include place on earth, time of day, humidity, pollution level, and season [43]. An example of the average incident levels around the world is shown in Table 20.1 that reports average UV

Table 20.1 Comparison of UV index around the world

Country/City	January	February	March	April	May	June	July	August	September	October	November	December
Argentina (Buenos Aires)	9	9	7	4	3	2	2	4	5	7	9	10
Australia (Darwin)	12	13	12	10	8	8	8	10	11	13	12	12
Australia (Sydney)	9	9	7	5	3	2	3	4	6	7	9	10
Canada (Vancouver)	1	1	3	4	6	7	7	6	4	2	1	1
France (Paris)	1	1	3	4	6	7	7	6	4	4	1	0
Greece (Iraklion)	3	4	5	8	9	9	10	9	7	4	3	2
Russia (St. Petersburg)	0	0	1	3	4	5	5	4	2	1	0	0
Singapore (Singapore)	11	12	13	13	11	11	11	11	12	12	11	10
USA (Los Angeles)	3	4	6	8	9	10	10	9	7	5	3	2
USA (New York)	2	3	4	6	7	8	9	8	6	3	2	1

Source: Courtesy of Martin Butson.

index values per month. The UV index is an international standard measurement of the strength of sunburn-producing UV radiation at a particular place and time. The calculations are weighted in favor of the UV wavelengths to which human skin is most sensitive, according to the CIE-standard McKinlay-Diffey erythemal action spectrum [44].

The level of UV exposure can also vary because of ground and building surfaces including type of paint, concrete or metallic surfaces. Clothing and eye wear also vary the level of UV exposure to a human thus causing significant variations in the amount and place of UV exposure we receive.

As such, personal UV dosimetry is an essential tool to increase knowledge based on the above-mentioned factors that influence UV exposure rates. Although UV detectors such as radiometers [45] are accurate devices for exposure measurement, they are often bulky and not suited for in vivo measurement of UV radiation exposure.

This type of measurement has been the field of radiochromic films for many years [46–49], and they have provided the medical and scientific world with significant data to establish known exposure levels for humans as well as other biological organisms.

Some commonly used photoactive chemical dosimeters that are used to create UV radiochromic films are PPO [50] and polysulphone [51]. PPO has seen many uses in UV-exposure measurement from underwater UVB assessment [52] through measurement of UV exposure of school teachers on playground duty [53].

Polysulphone radiochromic films have been extensively used and tested and have found usefulness in measurement of UV exposure in topics as wide as exposure during tropical beach holidays [54] through shade levels provided by certain tree types [55].

Other less commonly used photoactive chemicals are phenothiazine [56] that has been used as a solar UVA dosimeter [57]. Other chemicals that are used for UV detections are 8-methoxypsoralen [58] for DNA damage of human skin [59] and nalidixic acid [60], used to assess UV-exposure levels associated to melanoma induction [61].

In recent years, GAFchromic films have also been used for the measurement of UV exposure mostly for UVA wavelengths [62–66]. By combining products together, other films have been created to match specific action spectra for the required measurement site. As an example, Turnbull et al. [66] conducted measurements utilizing a dual-layer dosimeter made from a sheet of polysulphone and a sheet of nalidixic acid (in a polyvinyl chloride matrix) in an attempt to match the action spectra for melanoma induction. This work was enhanced by Wainwright et al. [67] and used to simultaneously measure both erythemal and vitamin D effective UV exposure.

20.8 AMBIENT LIGHT EFFECTS IN THE CLINIC

Different light sources have various levels of UV light, which may result in undesirable effects in the clinic when using radiochromic film for dosimetry. In many radiation therapy clinics around the world, radiochromic film dosimetry is frequently used for evaluation of radiotherapy dose received by the patient for a particular treatment. As mentioned in previous sections, different dosimetry films are available to use in the clinic. The commercially available GAFchromic MD-55-2, XR type T and type R, EBT, EBT2, and now EBT3 films are widely used for this purpose.

There are many factors that influence the radiochromic film dose–response, which may introduce dose readout errors [68]. One of these factors is susceptibility to ambient light and scanner light. This is because the active layers of the radiochromic films are sensitive to UV light. This sensitivity varies amongst the different types of radiochromic films based on the manufacturing of the film.

Many studies [68–75] presented the response characteristics of radiochromic dosimetry film by UV light exposure. Results from these studies vary because the measured UV sensitivity is dependent upon the type of film under investigation and the UV source. Although the studies indicate that a short exposure of radiochromic films to room light during the measurements will not have a significant impact on the radiation dosimetry, proper care and handling is still recommended to eliminate UV contamination.

20.8.1 ROOM LIGHTS

There are many types of indoor lighting. The majority can be categorized into tungsten incandescent bulbs, tungsten halogen incandescent bulbs, fluorescent tubes, compact fluorescent lamps (CFL), and light emitting diode (LED) lights. Although LED light might emit UV, it is considered low level compared with fluorescent and CFL.

There are characteristic differences between spectra emitted by incandescent lamps and fluorescent lamps because of the different principles of operation. The UV emission of incandescent lamps with tungsten filament is limited by the temperature of the filament and the absorption of the glass. Incandescent tungsten filament bulbs may emit radiation as short as 280 nm [76]. Tungsten halogen incandescent bulbs can emit UVA and UVB wavelengths. Some older models or fixtures with damaged or missing filters may also emit UVC as short as 200 nm [76].

For florescent lights, the UV content of the emitted spectrum depends on both the phosphor and the glass envelope of the fluorescent lamp. Most of the photons that are released from the mercury atoms have wavelengths in the UV regions at wavelengths of 253.7 and 185 nm. These are not visible to the human eye, so they are converted into visible light through fluorescence. The UV photons are absorbed by electrons in the atoms of lamp's interior coating, causing an energy jump then drop with emission of a further photon that has a lower energy at wavelengths visible to the human eye. Both daylight and cool-white fluorescent lamps may emit UVA and UVB radiation as short as 280 nm. Occasionally, the 254-nm mercury line is detected in the emission of a fluorescent source [76].

Some single-envelope CFLs emit UVB and traces of UVC radiation at wavelength of 254 nm, which is not the case for incandescent lamps as tested by Khazova et al. [77], in which 9 out of 53 single-envelope CFLs and none of the double-envelope lamps emitted UVC at 254 nm. The UV spectral irradiance for three CFLs is shown in Figure 20.7 for a double-envelope lamp and for single-envelope lamps. Experimental data show that CFLs produce more UVA irradiance than an incandescent tungsten lamp. Furthermore, the amount of UVB irradiance produced from single-envelope CFLs, from the same distance of 20 cm, was about 10 times higher than that irradiated by an incandescent tungsten lamp [78].

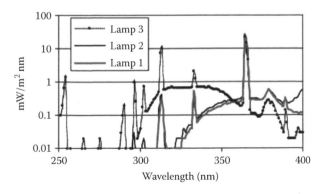

Figure 20.7 UV emission spectra of three CFLs (lamp one is double envelope, whereas lamps two and three are single envelope). (Reproduced from Khazova, M. and O' Hagan J.B., *Radiat. Prot. Dosim.*, 1–5, 2008. With permission.)

20.8.2 SCANNER LIGHTS

The most popular commercially available scanner for analyzing the radiochromic films is the following EPSON scanners. There are several models that come with different types of light sources. Different light sources mean that they will have different levels of UV light; however, due to the very short time of exposure to light while scanning, the effect of exposure is very minimal.

- V330: White LED
- V700: White cold cathode fluorescent lamp. IR LED for Digital ICE
- 10000XL: Xenon gas cold cathode fluorescent lamp

20.8.3 SUNLIGHT

Sunlight is attenuated as it travels through the earth's atmosphere. This means that all radiations with wavelength below 290 nm are filtered before it reaches the earth's surface [79]. Of the UV radiation that reaches the Earth's surface, more than 95% is the longer wavelengths of UVA, with the small remainder UVB. There is essentially no UVC. The fraction of UVB that remains in UV light after passing through the atmosphere is heavily dependent on cloud cover and atmospheric conditions. Figure 20.8 shows the UV spectra of different ambient light sources.

20.8.3.1 SENSITIVITY OF RADIOCHROMIC FILMS TO UV LIGHT

As mentioned earlier, the sensitivity of radiochromic dosimetry films depends on the type of film. Butson et al. [72] studied the response of MD-55-2 film to UVA and UVB from fluorescent and solar light, in which results showed that MD-55-2 responds almost exclusively to UVA with negligible coloration from UVB [72]. Figure 20.9 shows the OD as a function of UV exposure from fluorescent and solar UVA.

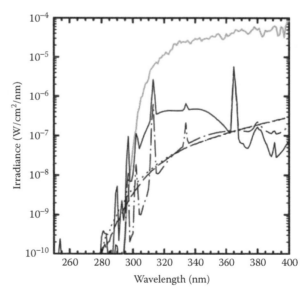

Figure 20.8 UV spectra comparison between different ambient light sources. Solar spectral irradiance (thick gray line), other lamps measured at 20 cm: tungsten incandescent (……..), tungsten halogen incandescent (--- ---), fluorescent light (-----), and CFL (---. ---). (Reproduced from Sayre, R.M. et al., *Photoch. Photobiol.*, 80, 47–51, 2004. With permission.)

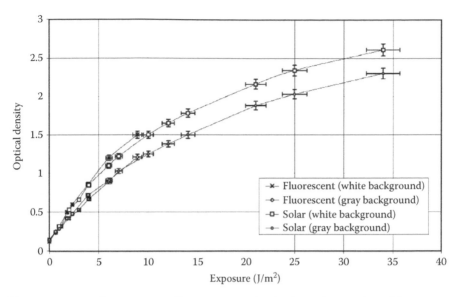

Figure 20.9 Response of MD-55-2 to fluorescent and solar UVA. (Reproduced from Butson, M.J. et al., *Phys. Med. Biol.*, 45, 1863–1868, 2000. With permission.)

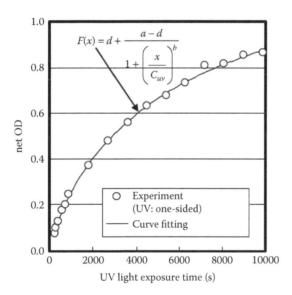

Figure 20.10 Curve fitting of net OD to UV light exposure time. (Reproduced from Tokura, S. et al., *IFMBE Proceedings*, 25, 273–276, 2009. With permission.)

Tokura et al. [73] tested the sensitivity of the EBT film using fluorescent light with peak wavelength of 352 nm. Figure 20.10 shows the curve fitting of net OD to UV light exposure time.

Butson et al. [71] also tested the sensitivity of EBT film for solar exposure under different climatic conditions. Figure 20.11 shows the results for the change in the net OD when exposed to solar UV under different climate conditions.

EBT2 has decreased sensitivity to ambient fluorescent light compared with EBT as investigated by William et al. [70]. This is due to the inclusion of the yellow dye in the manufacturing of the EBT2. The testing by William et al. was done for unexposed films as well as films that had already started polymerization as per Figure 20.12.

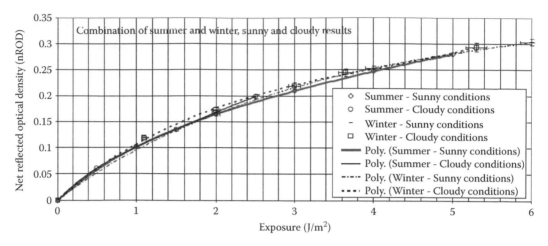

Figure 20.11 Comparison of the UV exposure versus darkening of EBT film under various climate conditions. (Reproduced from Butson, E. et al., *Phys. Med. Biol.*, 55, N487–N493, 2010. With permission.)

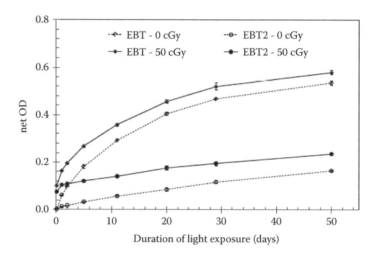

Figure 20.12 Sensitivity of EBT and EBT2 films to fluorescent light. (Reproduced from Williams, M. and Metcalfe, P., *4th SSD Summer School: Concepts and Trends in Medical Radiation Dosimetry*, AIP, Wollongong, Australia, 2011.)

20.9 SUMMARY

As radiochromic films are sensitive to UV radiation, they can also act as a UV dosimeter. The film properties including the active layer and materials used in the films construction all play roles in the sensitivity and wavelength response of the radiochromic film to UV radiation.

REFERENCES

1. http://www.acgih.org/. Accessed January 14th, 2016
2. http://www.icnirp.org/. Accessed January 18th, 2016.
3. Parisi AV, Turnbull DT. Solar UV dosimetry, UV Radiation and its effects– an update conference 2006, New Zealand.

4. Diffey BL, Davis A, Johnson M, Harrington TR. A dosimeter for long wave ultraviolet radiation. *Br J Dermatol* 1977;97(2):127–130.

5. Green AC, Whiteman DC. Solar Radiation. In D Schottenfeld (Ed.) *Cancer Epidemiol Prev.* pp. 294–305. Stanford CA: Oxford Pub.

6. Davis A, Deane GH, Diffey BL. Possible dosimeter for ultraviolet radiation. *Nature* 261;169–170.

7. Kollias K, Baqer A, Sadiq R et al. Measurement of solar UVB variations by polysulphone film. *Photochem Photobiol* 2003;78:220.

8. Parisi AV, Sabburg J, Kimlin MG. *Scattered and Filtered Solar UV Measurements.* 2004. Dordrecht the Netherlands: Kluwer Academic Publishers.

9. Moise AF, Gies HP, Harrison SL. Estimation of the annual solar UVR exposure dose of infants and small children in tropical Queensland, Australia. *Photochem PhotoBiol* 1999;69:457–463.

10. Kimlin GM, Parisis AV, Wong JC. Quantificiation of personal solar UV exposure of outdoor workers, indoor workers and adolescents at two locations in Southeast, Queensland. *Photodermatol Photoimmunol Photomed* 1998;14:7–11.

11. Kimlin MG, Martinez N, Green CC, Whiteman CD. Anatomical distribution of solar ultraviolet exposures among cyclists. *J Photochem Photobiol B* 2006;85:23–27.

12. Diffey BL, Jansen CT, Urbacn F, Wulf HC. The standard erythemal dose: A new photobiological concept. *Photodermatol photoimmunol Photomed* 1997;13:64–66.

13. Parisi AV, Kimlin GM, Turnbull DJ. Potential of phenothiazine as a thin film dosimeter for UVA exposures. *Photochem Photobiol Sci* 2005;4:907–910.

14. Niroomand-Rad A, Blackwell CR, Coursey BM et al. Soares CGRadiochromic film dosimetry: Recommendations of AAPM radiation therapy committee task group 55. American Association of Physicists in Medicine. *Med Phys* 1998;25(11):2093–2115.

15. Butson MJ, Cheung T, Yu PK et al. Ultraviolet radiation dosimetry with radiochromic film. *Phys Med Biol* 2000;45(7):1863–1868.

16. Butson ET, Cheung T, Yu PK, Butson MJ. Measuring solar UV radiation with EBT radiochromic film. *Phys Med Biol* 2010;55(20):N487–N493.

17. Butson ET, Yu PK, Butson MJ. Solar ultraviolet radiation response of EBT2 Gafchromic, radiochromic film. *Phys Med Biol* 2013;58(21):N287–N294.

18. Chun SL, Yu PK Note: Calibration of EBT3 radiochromic film for measuring solar ultraviolet radiation. *Rev Sci Instrum* 2014;85:10.

19. Katsuda T, Gotanda R, Gotanda T et al. Ultraviolet exposure of Gafchromic XR-RV3 and XR-SP2 films. *J Appl Clin Med Phys* 2015;16(5):5664.

20. Katsuda T, Gotanda R, Gotanda T et al. Comparing three UV wavelengths for pre-exposing Gafchromic EBT2 and EBT3 films. *J Appl Clin Med Phys* 2015;16(6):5663.

21. (http://www.biology-online.org/dictionary/Action_Spectrum): Last accessed 1st March 2016.

22. McKinley A, Diffey BL. A reference action spectrum for ultraviolet induced erythema in human skin. In Passchier WF and Bosnajakovic BFM. (Eds.) *Human Exposure to Ultraviolet Radiation: Risks and Regulations*, Amsterdam, the Netherlands: International Congress Series, Elsevier, 1987:83–87.

23. Parrish JA, Jaenicke KF. Action spectrum for phototherapy of psoriasis. *J Invest Dermatol* 1981;76(5):359–362.

24. Bernhard, G, Seckmeyer, G. Measurements of spectral solar UV irradiance in tropical Australia. *J Geoph Res* 1997;102(D7):8719–8730.

25. De Gruijl, FR, Van der Leun JC. Estimate of the wavelength dependency of ultraviolet cacinogenesis in humans and its relevance to the risk assessment of stratospheric ozone depletion. *Health Phys* 1994;67:319–325.

26. Setlow RB, Grist E, Thompson K, Woodhead, AD. Wavelengths effective in induction of malignant melanoma. *Proc Nat Acad Sci USA* 1993;90:6666–6670.

27. Oriowo OM, Cullen AP, Chou BR, Sivak JG. Action spectrum and recovery for in vitro UV-induced cataract using whole lenses. *Invest Ophthalmol Vis Sci* 2001;42:2596–2602.

28. Madronich S, McKenzie RL, Björn LO, Caldwell MM. Changes in biologically active ultraviolet radiation reaching the Earth's surface. *Photochem Photobiol* 1998;46:5–19.
29. http://www.temis.nl/uvradiation/info/uvaction.html). Accessed March 1, 2016.
30. Niroomand-Rad A. Radiochromic film dosimetry: Recommendations of AAPM radiation therapy committe task group 55. *Med Phys* 1998;25(11):2093–2115.
31. Aydarous A, Al-Omary EA, El Ghasaly M. Characterization of Gafchromic EBT3 films for ultraviolet radiation dosimetry. *Radiat Eff Defect S* 2014;169(3):249–255.
32. Butson ET, Yu PKN, Butson MJ. Solar ultraviolet radiation response of EBT2 Gafchromic, radiochromic film. *Phys Med Biol* 2013;58:N287–N294.
33. Davis A, Deane GHW, Diffey BL. Possible dosimeter for ultraviolet radiation. *Nature* 1976;261:169–170.
34. Kolari PJ, Hoikkala M, Lauharanta J. Assessment of the dose response of polysulphone film badges for the measurement of UV-radiation. *Photo-dermatology* 1986;3(4):228–232.
35. Kollias N. Measurement of solar UVB variation by polysulphone film. *Photochem Photobiol* 2003;78(3):220–224.
36. Parisi AV, Kimlin MG. Personal solar UV exposure measurements employing modified poly-sulphone with an extended dynamic range. *Photochem Photobiol* 2004;79(5):411–415.
37. Schouten PW, Parisi AV, Turnbull DJ. Evaluation of a high exposure solar UV dosimeter for underwater use. *Photochem Photobiol* 2007;83:931–937.
38. Schouten P., Parisi AV, Turnbull DJ. Applicability of the polyphenylene oxide film dosimeter to high UV exposures in aquatic environments. *J Photochem Photobiol B* 2009;96:184–192.
39. Butson ET. Measuring solar UV radiation with EBT radiochromic film. *Phys Med Biol* 2010; 55:N487–N493.
40. Butson MJ. Ultraviolet radiation dosimetry with radiochromic film. *Phys Med Biol* 2000;45:1863–1868.
41. CIE (International Commission on Illumination), Personal dosimetry for UV radiation, in CIE Publication No. 98. 1992. p. 611.
42. http://www.arpansa.gov.au/radiationprotection/basics/uvr.cfm Last Accessed February 15th, 2016.
43. (http://www.who.int/uv/intersunprogramme/activities/uv_index/en/index3.html). Last Accessed February 15th, 2016.
44. McKinlay AF, Diffey BL. A reference action spectrum for ultraviolet induced erythema in human skin. *CIE J.* 1987;6(1):17–22.
45. (http://www.uvprocess.com/manuals/manual_radiometer.pdf) Last Accessed February 15th, 2016.
46. Kolari PJ, Hoikkala M, Lauharanta J. Assessment of the dose response of polysulphone film badges for the measurement of UV-radiation. *Photodermatology* 1986;3(4):228–232.
47. Herlihy E, Gies PH, Roy CR, Jones M. Personal dosimetry of solar UV radiation for different outdoor activities. *Photochem Photobiol* 1994;60(3):288–294.
48. Sisto R, Lega D, Militello A. The calibration of personal dosemeters used for evaluating exposure to solar UV in the workplace. *Radiat Prot Dosimetry* 2001;97(4):419–422.
49. Siani AM, Casale GR, Sisto R et al. Short-term UV exposure of sunbathers at a Mediterranean Sea site. *Photochem Photobiol* 2009;85(1):171–177.
50. Davis A, Deane GHW, Diffey BL. Possible dosimeter for ultraviolet radiation. *Nature* 1976a;261:169–170.
51. Davis A, Deane GWH, Gordon D et al. A world-wide program for the continuous monitoring of solar UV radiation using poly(phenylene oxide) film, and consideration of the results. *J Appl Polym Sci* 1976b;20:1165–1174.
52. Schouten PW, Parisi AV. Underwater deployment of the polyphenylene oxide dosimeter combined with a neutral density filter to measure long-term solar UVB exposures. *J Photochem Photobiol B* 2012;112:31–36.

53. Downs NJ, Parisi AV, Igoe D. Measurements of occupational ultraviolet exposure and the implications of timetabled yard duty for school teachers in Queensland, Australia: Preliminary results. *J Photochem Photobiol B* 2014;131:84–89.
54. O'Riordan DL, Steffen AD, Lunde KB, Gies P. A day at the beach while on tropical vacation: Sun protection practices in a high-risk setting for UV radiation exposure. *Arch Dermatol* 2008;144(11):1449–1455.
55. Parisi AV, Willey A, Kimlin MG, Wong JC. Penetration of solar erythemal UV radiation in the shade of two common Australian trees. *Health Phys.* 1999;76(6):682–686.
56. Diffey BL, Davis A, Johnson M, Harrington TR. A dosimeter for long wave ultraviolet radiation. *Br J Dermatol* 1977;97:127–130.
57. Wong JC, Parisi AV. Measurement of UVA exposure to solar radiation. *Photochem Photobiol* 1996;63(6):807–810.
58. Diffey BL, Davis A. A new dosimeter for the measurement of natural ultraviolet radiation in the study of photodermatoses and drug photosensitivity *Phys Med Biol* 1978; 23:318–323.
59. Potten CS, Chadwick CA, Cohen AJ. et al. DNA damage in UV-irradiated human skin in vivo: automated direct measurement by image analysis (thymine dimers) compared with indirect measurement (unscheduled DNA synthesis) and protection by 5-methoxypsoralen. *Int J Radiat Biol* 1993;63(3):313–324.
60. Tate TJ, Diffey BL, Davis A. An ultraviolet radiation dosimeter based on the photosensitising drug nalidixic acid. *Photochem Photobiol* 1980;31:27–30.
61. Katsuda T, Gotanda R, Gotanda T et al. Ultraviolet exposure of Gafchromic XR-RV3 and XR-SP2 films. *J Appl Clin Med Phys* 2015;16(5):5664.
62. Butson MJ, Cheung T, Yu PK et al. Ultraviolet radiation dosimetry with radiochromic film. *Phys Med Biol* 2000;45(7):1863–1868.
63. Butson ET, Cheung T, Yu PK, Butson MJ. Measuring solar UV radiation with EBT radiochromic film. *Phys Med Biol* 2010;55(20):N487–N493.
64. Butson ET, Yu PK, Butson MJ. Solar ultraviolet radiation response of EBT2 Gafchromic, radiochromic film. *Phys Med Biol* 2013;58(21):N287–N294.
65. Chun SL, Yu PK. Calibration of EBT3 radiochromic film for measuring solar ultraviolet radiation. *Rev Sci Instrum* 2014;85(10):106103.
66. Turnbull DJ, Parisi AV. Dosimeter for the measurement of UV exposures related to melanoma induction. *Phys Med Biol* 2010;55(13):3767–3776.
67. Wainwright L, Parisi AV, Downs N. Dual calibrated dosimeter for simultaneous measurements of erythemal and vitamin D effective solar ultraviolet radiation. *J Photochem Photobiol B* 2016;157:15–21.
68. Girard F, Bouchard H, Lacroix F. Reference dosimetry using radiochromic film. *J Appl Clin Med Phys* 2012;13(6):3994.
69. Butson M, Yu P, Metcalfe P. Effects of readout light sources and ambient light on radiochromic film. *Phys Med Biol* 1998;43:2407–2412.
70. Williams M, Metcalfe, P. Radiochromic film dosimetry and its applications in radiotherapy. *4th SSD Summer School: Concepts and Trends in Medical Radiation Dosimetry* pp. 75–99. Wollongong, Australia: AIP, 2011.
71. Butson E, Cheung T, Yu PKN and Butson MJ. Measuring solar UV radiation with EBT radiochromic film. *Phys Med Biol* 2010;55:N487–N493.
72. Butson MJ, Cheung T, Yu PKN et al. Ultraviolet radiation dosimetry with radiochromic film. *Phys Med Biol* 2000;45:1863–1868.
73. Tokura S, Azuma Y, Aoyam H, Goto S. Density response characteristics of GafChromic EBT dosimetry film using ultraviolet light exposure. *IFMBE Proceedings* 2009;25/I:273–276.
74. Dini S A, Koona RA, Ashburn JR, Meigooni AS. Dosimetric evaluation of GAFCHROMIC® XR type T and XR type R films. *J Appl Clin Med Phy* 2005;6:114–134.

75. Aydarous A, Al-Omary EA, ElGhazaly M. Characterization of Gafchromic EBT3 films for ultraviolet radiation dosimetry. *Radiat Eff Defect S* 2014;169:249–255.
76. Sayre RM, Dowdy JC, Poh-Fitzpatrick M. Dermatological risk of indoor ultraviolet exposure from contemporary lighting sources. *Photoch Photobiol* 2004;80:47–51.
77. Khazova M, O' Hagan JB. Optical radiation emissions from compact fluorescent lamps. *Radiat Prot Dosim* 2008:1–5.
78. Chingwell CF, Sik RH, Bilski PJ. The photosensitizing potential of compact fluorescent vs incandescent light bulbs. *Photochem Photobiol* 2008.
79. Light Sensitivity, Scientific Committee on Emerging and Newly Identified Health Risks. Director-General for Health and Consumers, European Commission. 2008.

Index

Note: Page numbers followed by f and t refer to figures and tables, respectively.